Characterization of Nanocomposites

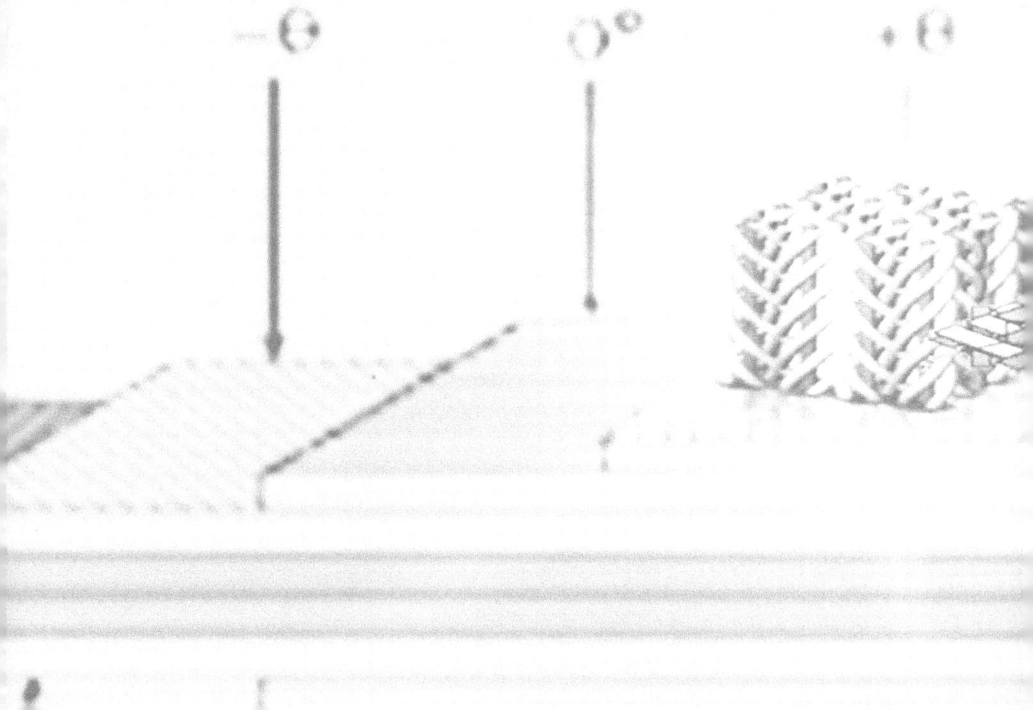

Characterization of Nanocomposites

Technology and Industrial Applications

edited by

Frank Abdi
Mohit Garg

PAN STANFORD PUBLISHING

Published by

Pan Stanford Publishing Pte. Ltd.
Penthouse Level, Suntec Tower 3
8 Temasek Boulevard
Singapore 038988

Email: editorial@panstanford.com
Web: www.panstanford.com

British Library Cataloguing-in-Publication Data
A catalogue record for this book is available from the British Library.

Characterization of Nanocomposites: Technology and Industrial Applications

ISBN 978-981-4669-02-3 (Hardcover)
ISBN 978-1-315-36489-6 (eBook)

Printed in the USA

To Dr. Tom Gates (1959–2008) of NASA Langley,
pioneer in nanocomposites

Contents

Acknowledgments

The technical contributions of the following individuals are gratefully acknowledged:

(1) The Alpha Star team, Dr. Jerry Housner, G. Abumeri, M. Garg, J. Surdenas, R. Dutton, Dr. Dade Huang, Dr. A. Najafi, and Dr. Rose Ragalini for their constructive advice, support, and valuable technical information, which was used throughout the book
(2) Dr. Bob Farahmand, Technical Horizon Inc., for providing information on the entire process of nanocomputation
(3) Prof. Levon Minnetyan, Clarkson University, for his chapter on the effect of interface and interphase in nanomodeling, and improvement in conductivity by nanocomposite insertion
(4) Dr. Christos C. Chamis, NASA/Glenn Research Center, for his chapter on composite nanomechanics, stress analysis and fracture in nanolaminates composites, and probabilistic simulation
(5) Dr. Stuart McHugh, Lockheed Martin, for his chapter on interlaminar shear prediction/validation
(6) Prof. Parviz Yavari and his students at California State University, Long Beach, for their chapter on simulation of material properties' characterization and validation of glass/epoxy/silica nanoparticles
(7) Mr. Mohit Garg, Alpha Star, for his valuable chapters on the influence of nanoparticles on mode I and II fracture toughness and impact resistance, and on material characterization of nanocomposites

(8) Prof. A. Mosallam, and Dr. Rashid Miraj, University of California (Irvine), for their chapters on the effect of carbon nanotube coating on fire resistance
(9) Prof. S. Panin and Dr. L. Kornienko, Institute of Strength Physics and Materials Sciences, SB RAS, Tomsk, Russia, for their chapter on the structure and the mechanical and tribotechnical properties of chemically modified UHMWPE-based nanocomposites
(10) Dr. M. Rouhi and Prof. M. Rohani on for their chapter on 3D modeling of nanocomposites
(11) Prof. G. M. Odegard, Michigan Technological University, for his chapter on the modeling of the mechanical properties of nanoparticle/polymer composites and on predicting the elastic properties of CNF/thermoset polymer composites considering the effect of interphase and fiber waviness
(12) Profs. Anthony J. Kinloch and Kamran Nikbin, Imperial College London, for their chapter on predicting the elastic properties of CNF/thermoset polymer composites considering the effect of interphase and fiber waviness
(13) Prof. Ambrose C. Taylor, Imperial College London, and David J. Bray, University of Warwick, for their chapter on dispersion of nanoparticles in polymers
(14) Prof. Samit Roy, University of Alabama, for his chapter on progressive failure of nanomaterials

—F.A.

Preface

Emerging nanocomposite materials in the form of polymer-based nanocomposites or metallic ultrafine-grained (UFG) alloys have shown tremendous increase in strength and ductility and have gained popularity in academia because of improved mechanical, thermal, and electrical properties. In the last few years, vast amounts of nanomaterials have been manufactured and used for commercial and military applications as improved advanced products with superior properties. The applications include (i) thermoset composites with an improved nano-inserted matrix or improved interphase, (ii) thermoplastic composites with long and short chopped fiber insertions, (iii) novel propellant formulations with nanoparticulate insertion, (iv) adhesive coating of individual graphite microfibers with a continuous, highly conductive metal layer, and (v) ultrafine-grained (UFG) alloys for potential armor and structural applications. Complex parts and superior joint structural performance can be obtained by improvements in interlaminar strength properties and matrix-dominated properties such as delamination, transverse tensile (TT), and in-plane shear.

Problems that adversely reduce the properties of nanocomposites are nonuniform distributions of particles inside the matrix, void shapes/sizes, bonding of nanoparticles with matrix, and the distribution and architecture of conventional fibers and nanofibers.

The classes of nanocomposites considered are single-walled nanotubes (SWNTs), multi-walled nanotubes (MWNTs), platelets in epoxy short fibers, and spherical inclusions in polymer matrix used in (i) glass/epoxy/silica nanoparticles, (ii) talc platelets in polypropylene, (iii) alumina short fibers in aluminum matrix, (iv) spherical silica inclusions in polymer matrix, and (v) mica platelets in epoxy.

The work in this book addresses

(a) evaluation of mechanical properties of different classes of nanocomposites—the properties being determined by physical laboratory test—and the ability to perform post-test evaluation to determine the test outliers
(b) evaluation of emerging predictive methods such as molecular dynamics (MD), and nanomechanics (NM) in generating mechanical properties
(c) validation of multiscale multiphysics material characterization and qualification methodology and software
(d) generation of additional properties by risk mitigation strategy where test data are not available or are hard to obtain; examples are A-B basis allowables that account for the uncertainties in materials or tests
(e) performance of trade studies analytically to determine the parameters of nanocomposites required to achieve an optimum performance under service loading

Emerging computational models on multiscale multiphysics from an industry perspective are not mature enough to predict material performance, assess risk mitigation, and enable structural certification. These codes are basically utilized as test duplication procedures rather than as test prediction methodology. Therefore, new and emerging nano-insertion materials with "as-built/as-is" characteristics must be characterized for (i) effect of defects (i.e., void shape/size, nanofiber waviness/brush effect), (ii) damage evolution process and fracture characteristics in service, including the aging and end-of-life properties, and (iii) design characteristics (i.e., nanofiber volume fraction, thickness, aspect ratio. and interphase thickness). In this regard, researchers use quantum mechanics, molecular dynamics (MD), nanomechanics, and finite element (FEM) to predict and design nanomaterials. MD simulation is one of the most accurate methods among these, but it cannot be used in solving problems in large length-scales because of its computational inefficiencies. On the other hand, finite element simulation with homogenized properties has been used successfully in modeling large length-scale problems, but it cannot accurately capture phenomena present at smaller length-scales.

Micromechanical methods, which are based on the work of Eshelby, are among the modeling approaches that fall in between these two approaches having the capability to capture phenomena relatively in small scales as well as being computationally efficient enough to be used in large-scale problems.

These codes are challenged by complexities of nanomaterials: (i) MD does not provide accurate prediction of bulk properties, and (ii) FEM uses homogenization and approximation technique, where homogenization/scale-up resolution need validation and damage accumulation/failure mechanism is not fully characterized.

This book focuses on obtaining valid mechanical properties of nanocomposite materials by accurate prediction and observed physical tests, as well as by evaluation of test anomalies of advanced multiscale nanocomposites containing nanoparticles of different shapes, such as chopped fiber, spherical, and platelet, in polymeric, ceramic, and metallic materials. The prediction capability covers delamination, fracture toughness, impact resistance, conductivity, and fire resistance of nanocomposites. The methodology employs a high-fidelity procedure that utilizes (i) molecular dynamics to predict nanocomposite bulk properties; (ii) constituent material characterization and qualification to predict the effect of interfacial strength on nanocomposite strength and stiffness, and conductivity; (iii) durability and damage tolerance (D&DT) by multiscale multiphysics progressive failure analysis (PFA) and finite element method (FEM) to simulate the carbon nanotube (CNT) pull-out test and calibration with limited average test data available in literature; and (iv) time-dependent reliability, which combines PFA with probabilistic analysis to virtually simulate the scatter in the test data and to consider the scatter in the simulated data by introducing variations in the aspect ratio (length/diameter), waviness, void, fiber volume ratio of the CNT, strength interface, and matrix. Using the proposed approach, a good correlation between the simulation and experimental data is established. The use of the above-mentioned algorithms will result in improving nanocomposite delamination, fracture toughness, fatigue and longevity, impact resistance, conductivity, and fire resistance.

Frank Abdi

Chapter 1

Nanostructure Bulk Property Predictions Using Molecular Mechanics

Jerry Housner and Frank Abdi
AlphaStar Corporation, 5150 E. Pacific Coast Highway, Suite 650, Long Beach, CA 90804, USA
fabdi@alphastarcorp.com

Molecular mechanics is powerful tool for developing some of the fundamental bulk properties of nanostructures. It is shown in this chapter how many researchers and engineers have successfully used molecular mechanics, and the results they achieved as well as some of the challenges they faced. The intent of this chapter is not to provide new research findings but to lay out for the reader some of concepts for developing bulk properties of nanostructures using molecular mechanics and molecular dynamics with the cited references providing the details.

1.1 Introduction

The great majority of the models that engineers use are continuum based. For example, concrete is made of particles of various sizes

Characterization of Nanocomposites: Technology and Industrial Applications
Edited by Frank Abdi and Mohit Garg

ISBN 978-981-4669-02-3 (Hardcover), 978-1-315-36489-6 (eBook)
www.panstanford.com

and in addition is often reinforced by steel rods. Yet engineers design concrete structures essentially as continua. Indeed, nearly all structures tend to be treated as a continuum. This makes it possible to use continuum mechanics, beam equations, plate and shell theory, and the finite element method.

The bulk properties of the continuum, such as its constitutive properties and its strength, have been traditionally derived from test, and standard test methods such as ASTM have been developed. Once the bulk properties are derived, they become input to computer codes such as finite element programs. The traditional testing approach has proven satisfactory for many material systems. For example, for many metals, databases of properties established by test have been established. These databases may provide mean test values as well as statistical information concerning their probabilistic scatter derived from the replicas tested. However, in the age of composite materials, the traditional approach of depending solely on test has become cumbersome. This is because composites present an essentially unlimited number of material systems, so databases are limited.

In the case of composites, instead of testing the composite for its properties, one may test the constituents for their properties and then use material laws and assumptions to derive the composite properties. However, even this approach has its difficulties because some of the constituents making up the composite can be challenging to test and these difficulties invariably translate into time and money, each of which are limited resources.

In this chapter, attention is focused on composites containing nanomaterials. The nanomaterials may take the form of carbon nanotubes (CNTs) (Fig. 1.1) or another nanostructure such as nanoplatelets (Fig. 1.2). A CNT may have polymers attached to its surface (Fig. 1.3) to enhance its bonding with a composite's resin system. The process of enhancing a CNT's ability to bond with the surrounding resin is called "functionalization" of the CNT. The functionalization may change some of the properties of the CNT for better or for worse.

As Valavala[1] points out, nanostructures are very valuable to nanocomposites, particularly if they adhere well to the surrounding medium they are in. Exploring whether this is taking place may be

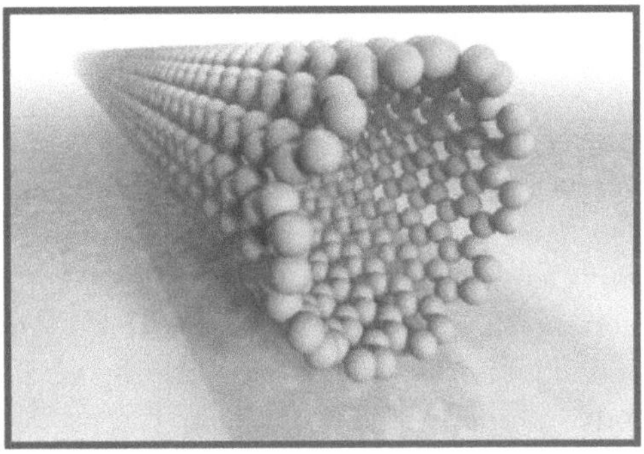

Figure 1.1 Carbon nanotube.

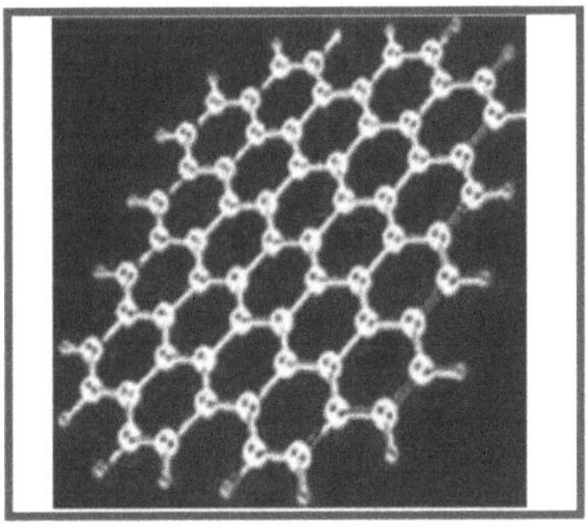

Figure 1.2 Nanoplatelet.

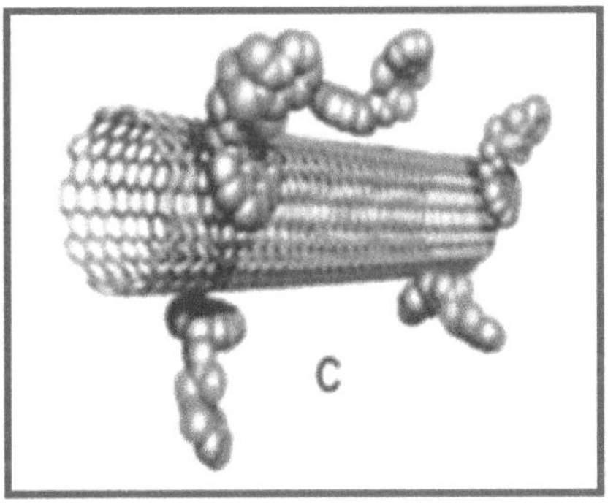

Figure 1.3 Functionalized carbon nanotube.

best left to analysis rather than test since many cases will need to be considered and testing of nanostructures can be challenging. Validation would be left to test.

Many researchers, engineers, and others have sought to derive the properties of nanomaterials from molecular mechanics (MM) and there has been a large measure of success in doing so. The purpose of this chapter is to review some of the MM methodologies that have been developed and that can be used to derive the bulk properties of nanomaterials useful to nanocomposite engineering.

The intended reader of this chapter is the practicing engineer. It is not the intent of this chapter to provide derivations of formulas presented or to present detailed explanations of molecular chemistry or any new information that has not already been published. Rather, all equations and results in the chapter have been published previously and appropriate credit is rightfully given to the authors.

1.1.1 *Modeling of the Atomistic Domain Using Molecular Mechanics and Dynamics*

The molecular nanostructure is composed of many atoms. The most accurate approach, known to date, in modeling the nanostructure atomistic domain would be to do so using quantum mechanics. Quantum mechanics involves modeling electron densities based on Schrodinger's equation. It would effectively model all the attractive and repelling forces in the nanostructure, including possible bond breaking. However, for the thousands of atoms in the nanostructure quantum mechanics, solutions tend to be computationally intensive and for engineering application purposes are in general presently beyond the capability of today's computational assets. Therefore, many researchers and practitioners have employed other methodologies.

As an approximation that is due to Max Born and Robert Oppenheimer (Oppenheimer had been Born's student), the Schrodinger equation[2] was separated into two parts: one corresponding to the motions of the atom's electrons and the other corresponding to the motions of the atom's nuclei. This approximation gives rise to molecular mechanics and dynamics, which examines the motions of the nuclei only and assumes that electrons are optimally distributed about the nuclei positions.

Molecular Dynamics (MD) adds dynamics through Newton's second law. However, MM and MD are statistical methods that averages results. Moreover, they do not account for any chemical reactions and consequently bond breaking is not accounted for. Without bond breaking, it is not possible for MM and MD to predict the strength of nanostructures, although they can predict many other useful bulk properties such as constituent properties.

The approach of many researchers has been to develop approximations for the potential energy of the nanostructure in the atomistic domain.[1,3–16] From the potential energy, useful information can be derived such as effective elastic constants. The potential energy can be computed using chemistry force fields, which describe the potential energy of a system of atoms. Chemistry force fields provide potential energy contributions whose gradients yield forces. Force fields are scalars and as such can be algebraically summed. They tend to be semi-empirical in nature, are parameterized, and depend on the atoms involved in the chemistry.

There are many different force field formulations, and the choice of the force field will often influence the numerical results achieved.

Force fields have in common that they are composed of forces due to covalent and noncovalent atomic bonding. For example, in CNTs, the covalent bonds are of the atomic form sp^2, which are very strong. They are the source of the CNTs' great strength, which has been measured to exceed 60 GPa (8700 ksi). The noncovalent bonds are due to van der Waals and electrostatic energies and are relatively much weaker. Different researchers and authors categorize these force field contributions somewhat differently. The various covalent contributions of energies involved in the force field are often represented as shown in Fig. 1.4.

In MM and MD atoms may be visualized as points and bonds as rods. Work is performed as bonds are stretched/compressed and angles are enlarged/contracted, and non-bond forces such as electrostatic forces and van der Waals forces are overcome. A review and summary of some existing force field representations

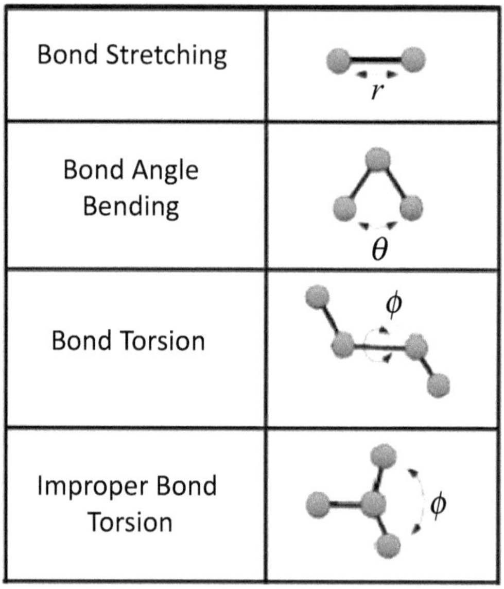

Figure 1.4 Covalent bonding states[1] contributing to potential energy.

is presented in reference 17. Presented here are two popular force fields for which computer codes can be found on the Internet.

1.1.2 *The AMBER Force Field*

A popular force field which can be used for CNTs is the AMBER force field.[18] AMBER stands for assisted model building with energy refinement. The force field potential energy can be represented as

$$\begin{aligned} E_{\text{AMBER}} = {} & (1/2)\sum_{\text{bonds}} K^{\rho}(L - L_0)^2 + (1/2)\sum_{\text{angles}} K^{\theta}(\theta - \theta_0)^2 \\ & + \sum_{\text{Torsions/Dihedrals}} V_n[1 + \cos(n\omega - \gamma)] \\ & + \sum_{j=1}^{N}\sum_{i=j+1}^{N-1} \varepsilon_{ij}\left[\left(\frac{r_{0ij}}{r_{ij}}\right)^{12} - 2\left(\frac{r_{0ij}}{r_{ij}}\right)^{6}\right] \\ & + \sum_{j=1}^{N}\sum_{i=j+1}^{N-1} q_i q_j/(4\pi\varepsilon_0 r_{ij}) \end{aligned} \quad (1.1)$$

The first two terms are recognized as belonging to the class of harmonic oscillators. The first term is equivalent to the potential energy for small stretching deformations of a linear spring joining two particles with spring constant K^{ρ}, $(L - L_0)$ being the increment of bond stretch from equilibrium. Summation is over all the bonds involved in the nanostructure chemistry being considered. The second term is similar, for a summation of angular springs with spring constant K^{θ}, $(\theta - \theta_0)$ being the increment of angular change formed by three atoms during bond angle bending as shown in Fig. 1.4. Like the first two terms, the third term is also a sum of torsions or dihedrals over all the atoms and represents a truncated Fourier series, but it does not represent a harmonic oscillator. In this term, V_n is the dihedral force constant, n the multiplicity of the function, ω the dihedral angle, and γ the phase shift. The fourth and fifth terms are employed to estimate the non-bond energies arising from van der Waals energy and electrostatic energy respectively. The first three terms are covalent energies, while the last two are not.

The fourth term estimates the van der Waals energy using the Lennard-Jones (LJ) 6–12 potential, given by

$$E(r) = 4\varepsilon\left[(\sigma/r)^{12} - (\sigma/r)^{6}\right] \tag{1.2}$$

The LJ 6–12 potential is often graphically portrayed as shown in Fig. 1.5, which indicates that the interactive strength between two atoms a distance r apart drops off to zero when the separation distance equals σ. The quantity ε represents the size of the potential energy "well" at small values of atomic separation distance. In Eq. 1.2 the term $1/r^{12}$ models the repulsion between atoms and is dominant when r is small, whereas the term $1/r^{6}$ models the attraction between atoms and becomes dominant as r gets larger. Virtually all force field representations utilize the LJ 6–12 potential representation for the non-bonded van der Waals terms.

The final term in the AMBER energy potential estimates the electrostatic energy using Coulomb's Law. Both the LJ 6–12 potential and Coulomb's law involve double sums, capturing the interaction of all the atoms in the nanostructure with all other atoms in the nanostructure. This would be a highly intensive calculation over all combinations of thousands of atoms, but fortuitously the LJ 6–12 potential and electrostatic terms drop off with atomic separation distance.

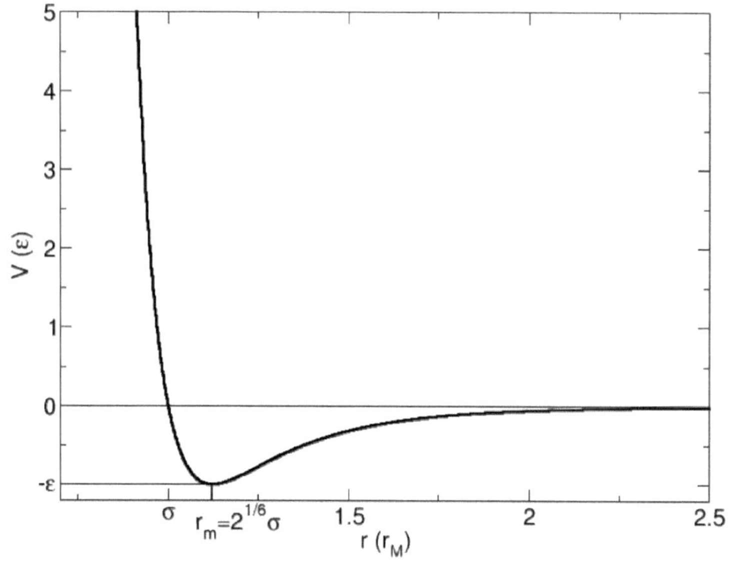

Figure 1.5 Lennard-Jones 6–12 potential.

1.1.3 *The CHARMM Force Field*

Another set of force fields comes from the CHARMM family of force fields, which stands for Chemistry at HARvard Macromolecular Mechanics. One of the versions of the CHARMM family of force fields is CHARMM22[19] as given in Eq. 1.3.

$$V = \sum_{\text{bonds}} k_b(b - b_0)^2 + \sum_{\text{angles}} k_\theta(\theta - \theta_0)^2 + \sum_{\text{dihedrals}} k_\phi[1 + \cos(n\phi - \delta)] \\ + \sum_{\text{impropers}} k_\omega(\omega - \omega_0)^2 + \sum_{\text{Urey-Bradley}} k_u(u - u_0)^2 \\ + \sum_{\text{nonbonded}} \varepsilon\left[\left(\frac{R_{\min_{ij}}}{r_{ij}}\right)^{12} - \left(\frac{R_{\min_{ij}}}{r_{ij}}\right)^{5}\right] + \frac{q_i q_j}{\varepsilon r_{ij}} \tag{1.3}$$

where the first term accounts for bond stretching, in which k_b is the bond force constant and $(b - b_0)$ is increment of bond stretch from equilibrium. The second term in the equation accounts for the bond angle bending, where k_θ is the angle force constant and $(\theta - \theta_0)$ is the increment of change from equilibrium of the angle formed by three atoms during bond angle bending as shown in Fig. 1.4. The third term is for the dihedrals (a.k.a. torsion angles), where k_ϕ is the dihedral force constant, n the multiplicity of the function, ϕ the dihedral angle, and the phase shift. The fourth term accounts for the impropers, that is, out-of-plane bending, where k_ω is the force constant and $(\omega - \omega_0)$ is the out-of-plane angle increment from equilibrium. The Urey–Bradley component (cross-term accounting for angle bending using 1,3 non-bonded interactions) comprises the fifth term, where k_u is the respective force constant and $(u - u_0)$ is the increment from equilibrium in distance between the 1,3 atoms. Non-bonded interactions between pairs of atoms (i, j) are represented by the last two terms.

Comparison of Eqs. 1.1 and 1.3 for the AMBER and CHARMM22 force fields reveals many similarities. But there are also differences. Equation 1.3 contains some additional terms denoted as "impropers" and Urey–Bradley energies, which like the bond stretching and bending terms can be modeled as harmonic oscillators.

Because many force field terms are harmonic oscillators, it is possible to represent them as mechanical elements, as was done for example in references 2, 3, and 5–8. The result is that some of the terms in the energy potential can be calculated with an appropriate finite element model. Reference 4 uses Fig. 1.6 to reveal the equivalency of the force fields and mechanical components.

However, in using force fields, it is first necessary to have the molecular structure in equilibrium. For some nanostructures it is known that their configuration is stable and in equilibrium such as a simple CNT. However, other more complex chemistries may not be in equilibrium. It is necessary to find the equilibrium state and computational molecular mechanics codes perform this necessary first step.

1.1.4 *Brenner's Equation for Interatomic Potential Energy Calculation*

An alternative to the use of force fields for calculating the potential energy of hydrocarbons is presented by Brenner's equation,[5] which can be used to provide the interatomic potential energy for carbon atoms. It is a parameterized form that approximates the

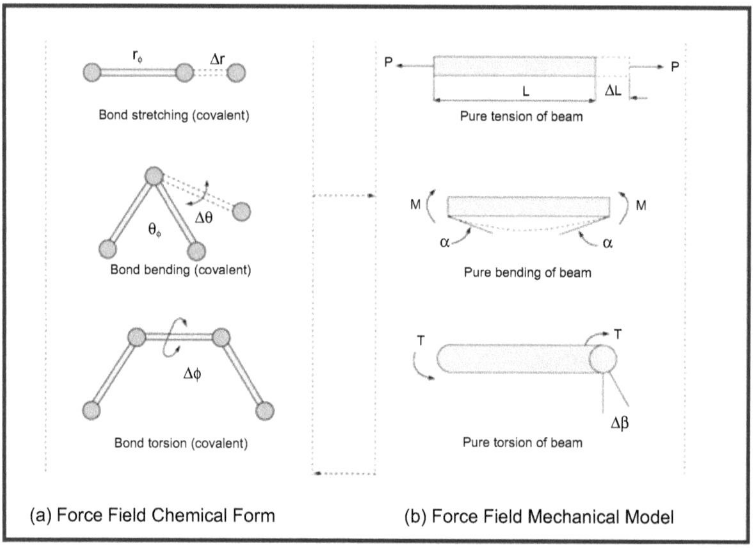

Figure 1.6 Force field chemical and mechanical representations.

atomistic bonded potential energy of molecular mechanics. Brenner expressed this energy as

$$E_{\mathrm{b}} = \frac{1}{2}\sum_{i\neq j} f_{\mathrm{c}}(r_{ij})(V^{\mathrm{R}}(r_{ij}) - b_{ij}V^{\mathrm{A}}(r_{ij}))$$
$$= \frac{1}{2}\sum_{i\neq j} f_{\mathrm{c}}(r_{ij})\left(\left(1+\frac{Q}{r}\right)A\cdot e^{-1\alpha r_{ij}} - b_{ij}\sum_{n=1}^{3} B_n\cdot e^{-\beta_n \cdot r_{ij}}\right) \quad (1.4)$$

where

$b_{ij} = b_{ij}^{\sigma-\pi} + b_{ij}^{\pi}$

$b_{ij}^{\pi} = \Pi_{ij}^{\mathrm{RC}} + b_{ij}^{\mathrm{DH}}$

E_{b}: the binding energy of the CNT system of atoms
r_{ij}: the interatomic distance between atoms i and j
$V^{\mathrm{R}}(r_{ij})$: the pairwise repulsion
$V^{\mathrm{A}}(r_{ij})$: the pairwise attraction
$f_{\mathrm{c}}(r_{ij})$: the cut-off function

A, α, B, and β are parameters for the pairwise repulsion and attraction, b_{ij} represent the bond order terms, Π_{ij}^{RC} are the radical or conjugated character in the bond order terms, and b_{ij}^{DH} the dihedral character in the bond order terms.

Wu[8] used Brenner's interatomic potential in what they called the atomistic continuum method (ACM) to derive Young's moduli and modal characteristics of SWCNTs.

1.1.5 *Developing Bulk Nanostructure Properties: Atomistic and Continuum Models*

One approach done by Odegard,[6] Gates,[3,7] Chung-Jung,[8,9] and others is to relate the molecular nanostructure to an equivalent continuum model by equating their respective potential energies for an appropriate set of displacements or forces. This is done through the use of a representative volume element (RVE) and an equivalent (or effective) continuum model (ECM).

As shown in reference 3, a three-step process may be used. In this process the same loading condition is applied to both the molecular and engineering models with the requirement that the potential energy of the molecular model and the strain energy of the continuum model have an equivalent mathematical form so that the energies will be equal for all values of the loading. These steps are expanded on here.

Step 1a: Establish a representative volume element (RVE) of the atomistic model. According to Liu,[20] RVEs were first defined by R. Hill[21] in 1963. His definition required a detailed calculation based on the microstructure of the element. Later, in 1996, Drugan and Willis[22] proposed an alternative definition for the RVE. They regard an RVE as the smallest material volume element of the heterogeneous solid for which the usual spatially constant "overall modulus" macroscopic constitutive representation is a sufficiently accurate model to represent mean constitutive response.

An example of step 1a is a graphene sheet which consists of periodic array of hexagonally arranged carbon atoms with connecting bonds. A single unit hexagon cell of atoms with connecting bonds serves as the RVE as shown in Fig. 1.7.

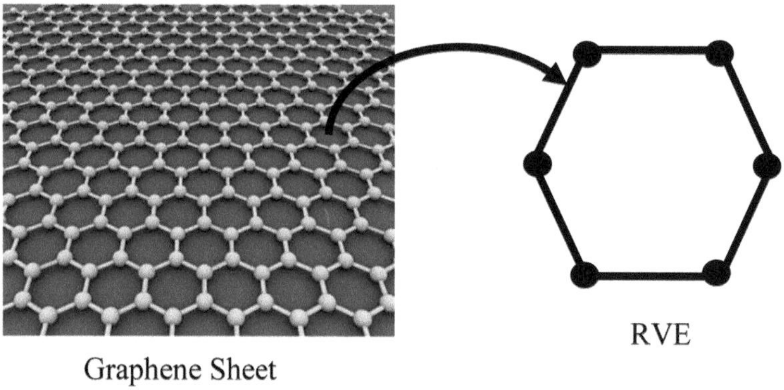

Figure 1.7 Example for Step 1a: a graphene sheet and RVE hexagonal arrangement of atoms and bonds.

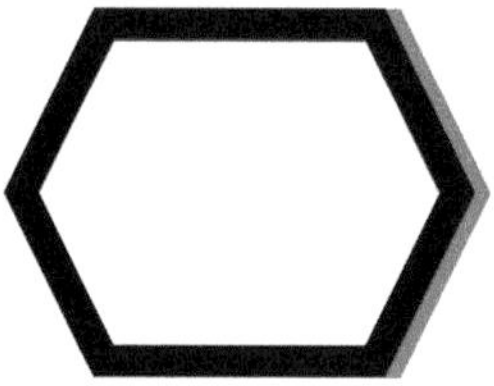

Figure 1.8 Example for Step 1b: a hexagonal plate serves as ECM.

Step 1b: Establish an equivalent continuum model (ECM) to represent the nanostructure that can be used in a finite element code to build the continuum representation and derive its overall elastic properties. The accuracy of the elastic constants derived from this procedure depends upon how the ECM is chosen.

In the example of the graphene sheet a thin hexagonal plate with the same planar dimensions as the sheet is selected to serve as the ECM. The thickness of the plate at this point is unknown.

Step 2a: Establish a constitutive relationship for the ECM. To accomplish this it must be decided if the nanostructure will behave in an isotropic, orthotropic, or anisotropic fashion, and to what degree nonlinearity is present and will be represented. In other words, the form of the constitutive law must be assumed.

This involves making some assumptions on the behavior of the nanostructure at the macroscale.

For the graphene sheet example, it is isotropic on a macroscale, and hence it is appropriate that the ECM hexagonal plate also be isotropic.

Step 3: Equate the potential energies of deformations of the RVE and ECM for a set of identical boundary conditions.

(a) Select a set of boundary conditions to be applied to the RVE and ECM. On the basis of the constitutive law, the set of boundary conditions is selected to provide a set of equations sufficient to resolve for the unknowns in the constitutive law.
(b) Select a method that will be used to calculate the energy of the RVE. This may be a force field, Brenner's equations, a corresponding computer code, or some other approach.
(c) Calculate the strain energy of the continuum under the selected set of boundary conditions.
(d) Equate the energies of the RVE and ECM, and solve for the unknowns in the constitutive law.

The coefficients of the constitutive law provide the elastic coefficients needed for finite element modeling.

For the graphene sheet example there are two elastic unknowns to be found for the ECM and so two independent boundary conditions are needed. Two boundary conditions to be applied to RVE and ECM could be a uniform radial expansion and a shearing.

The remaining question is the thickness of the isotropic plate. This type of question seems to appear in nearly all applications of continuum modeling of nanostructures. Some authors such as Leung[14] have used the interspatial experimental measurement of 0.34 nm. Other authors such as Odegard[11,12] and Yakobson[13] have calculated a thickness. The elastic properties will be highly dependent on the thickness assumed or in general, the volume of the ECM. This issue appears to account for some of the differences between analytical predictions as shown in Table 1.1.

The question concerning the assumed geometry for the ECM appears to center around the volume of an atom or perhaps more correctly, the volume to be allocated to an atom in the modeling.

Some authors[11–14] present results in terms of eV (electron volts) or eV/atom along with or instead of GPa units. Whereas this provides a means to compare modeling approaches, it does not resolve the thickness issue.

1.1.6 *Application to Carbon Nanotubes*

Because of their superior strength, stiffness, and stability, CNTs are seeing considerable use in nanocomposites. Both MM methods have been applied to CNTs to determine their bulk properties. The continuum representation methods presented can and have been applied. Before reviewing their application, a brief review of CNT geometry is provided.

1.1.6.1 Basics of carbon nanotubes

One way to visualize the CNT is to consider it as rolled from a graphene sheet of carbon atoms of infinite extent as shown in Fig. 1.9.

The axis of rotation is depicted in Fig. 1.9 and is orthogonal to a second vector called the chiral vector. The orientation of the chiral vector is given by the equation $n\mathbf{e}_1 + m\mathbf{e}_2$, where $\mathbf{e}_1$ and $\mathbf{e}_2$ are unit

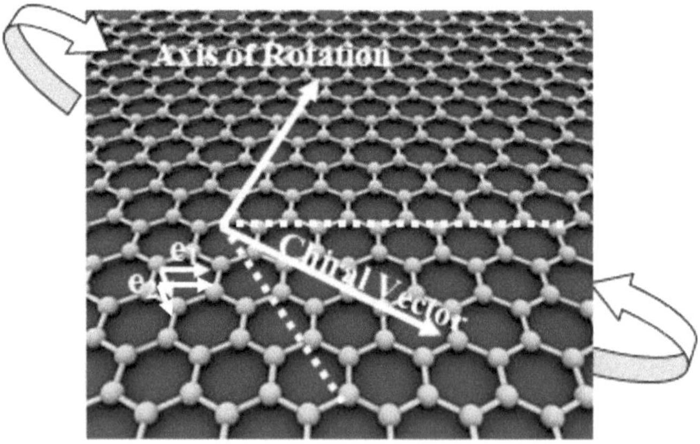

Figure 1.9 Graphene sheet with chiral vector and axis of rotation for generating CNTs.

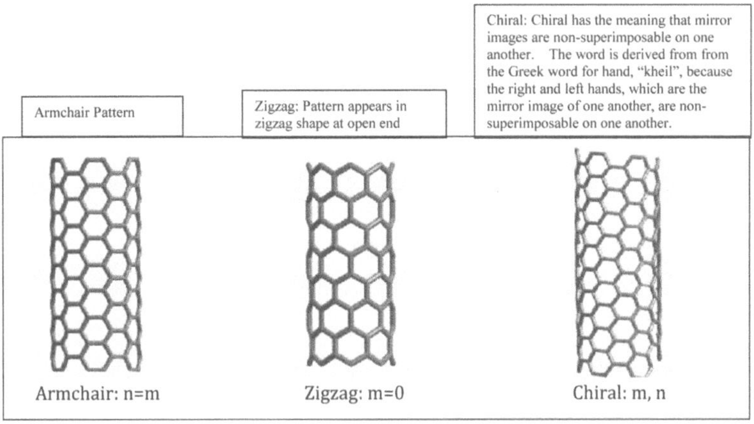

Figure 1.10 Three different types of carbon nanotubes: armchair, zigzag, and chiral.

vectors along two directions in the plane of the graphene sheet as shown in Fig. 1.9. If n and m are equal, the resulting CNT is called an armchair nanotube. If m is zero the resulting nanotube is referred to as zigzag. All other combinations of n and m are referred to as chiral CNTs. The background for the names of the CNT is suggested by the form the CNT takes as shown in Fig. 1.10.

The diameter of the CNT is fixed with the setting of n and m. The diameter is given by

$$d = (a/\pi)[n^2 + nm + m^2]^{1/2} \tag{1.5}$$

where a in nanometers is 0.246 nm.

From Eq. 1.1, the larger the values of n and m, the larger the diameter of the CNT. Typically, the diameter of the CNT is between 1 and 2 nm. The length of the CNT may be whatever is desired.

Covalent bonds exist between the carbon atoms that are sp^2 hybridized orbitals. The sp^2 bonds provide the enormous strength of CNTs. Van der Waals forces which are weak forces are found between individual CNTs and layers of CNT shells in multi-walled CNTs. The bond lengths between carbon atoms in the grapheme sheet and CNT are about 0.142 nm.

As in the graphene sheet example, the same RVE and ECM can be used for the CNT. This is the approach taken by Odegard and Gates[11,12] who examined single walled carbon nanotubes (SWNTs).

1.1.6.2 RVE-ECM method applied to SWNTs

For SWNTs subjected to axial loadings and undergoing small strains, reference 8 states "Only bond stretching and angle variation terms are significant in the system potential energy." Other authors make a similar assumption for SWCNTs or for isolated single-layer graphene sheets.

Hence, for SWNTs, Eq. 1.1 reduces to

$$E_{\text{AMBER}} = (1/2)\sum_{\text{bonds}} K^{\rho}(L - L_0)^2 + (1/2)\sum_{\text{angles}} K^{\theta}(\theta - \theta_0)^2 \quad (1.6)$$

It should be noted that for multi-walled CNTs (MWCNTs), the multiple walls are held in place by the weak van der Waals forces and hence the approximations leading to Eq. 1.6 are not valid. Nevertheless, the weak forces allow the multi-walls or layers to slide over one another rather freely. So the potential energy terms that deal with energy stored in each tube of the CNT separately may not involve van der Waals energy, but stretching between the walls of the tube and relative rotation between the walls of neighboring tubes will likely involve van der Waals contributions to potential energy. Potential energy due to van der Waals forces can be modeled using a Lennard-Jones approximation and the electrostatic energy modeled by Coulomb's law.

To aid in the calculation of the potential atomistic energy, Odegard and Gates[11] introduced an intermediate step in the form of a finite element approach for making the atomistic calculations as depicted in Fig. 1.11.

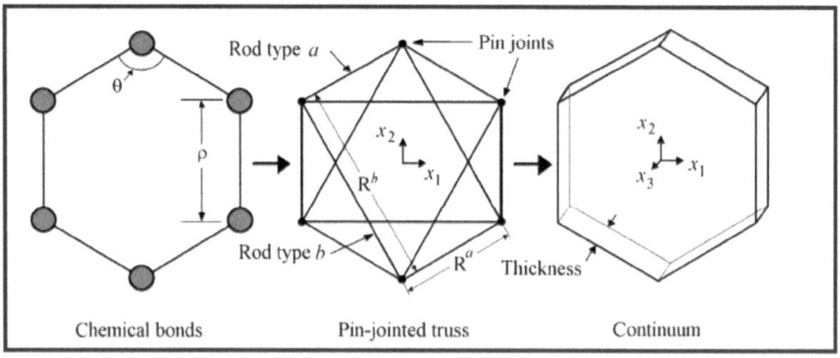

Figure 1.11 Pin-jointed truss model introduced to expedite atomistic calculations via finite element method.

Odegard and Gates used a pin-jointed truss model to represent Eq. 1.6. By proper selection of parameters the first term of Eq. 1.6 can be represented by the elongation of rods "a." However, they devised a way to represent the second term of Eq. 1.6 involving angular motion using pin jointed truss elements. This was done using the "b"-type rods. For small strains they showed that representing the truss rod moduli by the following relationship, Eq. 1.6 would be estimated by the pin-jointed truss model.

$$Y_i^a = \frac{2K_i^\rho R_i^a}{A_i^a} \qquad Y_i^b = \frac{3K_i^\theta}{2R_i^b A_i^b} \tag{1.7}$$

The consequence of this approach is that for many atomistic models, finite element truss models circumvent the need to turn to molecular mechanics codes. In addition, the value of the continuum modeling approach is that it can be extended to many nanostructures which was done by Odegard in reference 6.

Wu[8,9] used a continuum representation modeling approach which he calls atomistic continuum method (ACM) to model the nanostructure. Rather than using force fields in generating the potential energy, he used a second-generation form of Brenner's equation known as REBO (reactive empirical bond order).[23] His research showed that the Young's modulus did not vary with either the radius or the length of the CNT—that is, it was independent of the aspect ratio. Using the ACM, mode shapes and frequencies of armchair and zigzag SWNTs were also predicted.

The approach of Popov[16] is can also be thought of as an atomistic/continuum relationship approach because he derives expressions for the velocity of sound waves in SWNTs and compares these to sound wave velocities in hollow cylinders from elasticity theory. Use is made of a lattice dynamics model to accomplish this, developing the sound velocities in the SWNT using Born's perturbation technique.

There are also other methods for studying nanostructures. In general, the potential energy is employed as in atomistic/continuum relationship method. However, results are obtained by differentiating the energy expression to obtain stress tensors.[24,25] Use is made of viral, BDT (Basinski, Duesbery, and Taylor; see reference 24) and Lutsko stress tensors, and a stress tensor proposed by Shen and

Atluri[24] that is based on a smoother particle hydrodynamics (SPH) technique. The moduli in these approaches are found through the ratio of stress to strain.

Another approach used by Tu[26] ablation to involves the potential energy, but takes two derivatives to derive the stiffness parameters. Hence, in general, calculating the energy of the nanostructure is a necessary first step to deriving the moduli in all approaches.

1.1.7 *Derived Results in the Literature*

Analytically calculated values for Young's modulus of single-walled nanotubes are given in Table 1.1 from a series of references using different methodologies. (Unfortunately there are relatively few MWNT analytical results found in the literature at this time.) Similar results can be compiled for shear moduli. However, this is not presented here since the purpose of the table is to show the variation of results achieved with MM.

Most references have in common that they utilize the atomic potential energy for their calculations. Some utilize the energy to

Table 1.1 Analytical results for single-walled carbon nanotubes

Reference	Technical approach	Young's modulus
Leung [14]	Continuum equivalency	0.90 TPa (131 × 10^6 psi)
Lu [10]	Empirical lattice dynamics model	0.97 TPa (141 × 10^6 psi)
Popov [15]	Empirical lattice dynamics model	1.0 TPa (145 × 10^6 psi)
Sanchez-Portal [16]	Pseudo-potential density functional theory	0.95 to 1.10 TPa (138 to 160 × 10^6 psi)
Van Lier [27]	Ab initio multiplicative integral approach	0.75 to 1.18 TPa (109 to 171 × 10^6 psi)
Odegard [12]	Continuum equivalency	1.13 TPa (164 × 10^6 psi)
Yakobson [13]	Molecular dynamic simulations with shell theory	5.5 TPa (798 × 10^6 psi)
Wu [8]	Atomistic continuum mechanics	1.0 to 1.2 TPa (145 to 174 × 10^6 psi)

create equivalent continuum models as discussed previously, while others derive stress and/or strain forms from the potential energy equations through their derivatives or their elastic properties from their second derivative.

Table 1.1 reveals that most references predict an axial modulus ranging from 0.90 TPa to 1.18 TPa. Some researchers get higher results. As discussed earlier, it appears that the assumptions relevant to the geometry of the CNT have a strong influence on the result. Experimental results also vary. Young's modulus for SWNTs has been found to vary from 0.90 to 1.7 TPa and for MWNTs from 0.69 to 1.87 TPa over specific ranges of diameters.[16] Reference 28 states that the large variation may be due to the imperfections in MWNTs. Variations are also likely due to the difficulty in measuring behavior of both SWNTs and MWNTs. Hence, there are large variations, in both analytical and experimental predictions.

1.2 Summary

The application of the concepts of molecular mechanics to developing bulk properties of nanostructures has been reviewed. It has been shown how many researchers and engineers have used molecular mechanics to continuum modeling to predict bulk nanostructure properties. Other approaches for bulk nanostructure properties have also been used. The results derived from a variety of approaches are presented and compared to test results. Derivation of the potential energy of the atomistic representative volume of nanostructure is essential for the calculation of elastic moduli in nearly all approaches. Selection of a continuum that models the representative volume of the nanostructure can have a significant impact on the results obtained.

References

1. Valavala, P. K., and Odegard, G. M. Modeling techniques for determination of mechanical properties of polymer nanocomposites. *Adv. Mater. Sci.* **9**, 2005, 34–44.

2. Griffitks Dovid. J. *Introduction to Quantum Mechanics* (2nd ed.), Prentice Hall, 2004.
3. Gates, T. S., Odegard, G. M., Frankland, S. J. V., and Clancy, T. C. Computational materials: multi-scale modeling and simulation of nanostructured materials. *Compos. Sci. Technol.* **65**, 2005, 2416–2434.
4. Ávila, A. F., and Lacerda, G. S. R. Molecular mechanics applied to single-walled carbon nanotubes. *Mater. Res.* **11**, 2008, 325–333.
5. Brenner, D. W. Empirical potential for hydrocarbons for use in simulating the chemical vapor deposition of diamond films. *Phys. Rev. B* **42**, 1990, 9458.
6. Odegard, G. M., Frankland, S. J. V., Herzog, M. N., Gates, T. S., and Fay, C. C. Constitutive modeling of crosslinked nanotube materials. 45th AIAA/ASME/ASCE/AHS/ASC Structures, Structural Dynamics and Materials Conference, April 2004, AIAA 2004-1606.
7. Gates, T. S., and Hinkley, J. A. Computational materials: modeling and simulation of nanostructured materials and systems. NASA TM-2003-212163, March 2003.
8. Wu, C.-J., Chou, C.-Y., Han, C.-N., and Chiang, K.-N. Numerical simulation of the mechanical properties of carbon nanotube using the atomistic-continuum mechanics. ENS 2006, Paris, December 14–15, 2006.
9. Wu, C.-J., Chou, C.-Y., Han, C.-N., and Chiang, K.-N. Simulation and validation of CNT mechanical properties: the future interconnection material. Proceedings of 57th IEEE, Electronic Components and Technology Conference, 2007.
10. Lu, J. P. Elastic properties of carbon nanotubes and nanoropes. *Phys. Rev. Lett.* **79**, 1997, 1297–1300.
11. Odegard, G., Gates, T., Nicholson, L., and Wise, K. Equivalent-continuum modeling of nanostructured materials. NASA TM 210863, 2001.
12. Odegard, G., Gates, T., Nicholson, L., and Wise, K. Equivalent-continuum modeling with application to carbon nanotubes. NASA TM 211454, 2002.
13. Yakobson, B. I., Brabac, C. J., and Bernholc, J. Nanomechanics of carbon tubes: instabilities beyond linear response. *Phys. Rev. Lett.* **76**, 1996, 2511.
14. Leung, A. Y. T., Guo, X., He, X. Q., and Kittpomchai, S. A continuum model for zigzag single-walled carbon nanotubes. *Appl. Phys. Lett.* **86**, 2005, 083110.

15. Popov, V. N., Van Doren, V. E., and Balkanski, M. Elastic properties of single-walled carbon nanotubes. *Phys. Rev. B* **61**, 2000, 3078–3084.

16. Sanchez-Portal, D., Artachio, E., and Soler, J. M. Ab initio structural, elastic, and vibrational properties of carbon nanotubes. *Phys. Rev. B* **59**, 1999, 12678–12688.

17. Varma, C. K. Molecular mechanic force fields: review and critical analysis of modern day force fields with applications to protein and nucleic acid structures. *Biochemistry* **218**, Stanford University, December 2001.

18. MacKerell, A. D., Jr., Bashford, D., Bellott, M., et. al. All-atom empirical potential for molecular modeling and dynamics studies of proteins. *J. Phys. Chem. B* **102**, 1998, 3586–3616.

19. Cornell, W. D., Cieplak, P., Bayly, C. I., Gould, I. R., Merz, K. M., Jr., Ferguson, D. M., Spellmeyer, D. C., Fox, T., Caldwell, J. W., and Kollman, P. A. A second generation force field for the simulation of proteins, nucleic acids, and organic molecules. *J. Am. Chem. Soc.* **117**, 1995, 5179–5197.

20. Liu, C. On the Minimum Size of Representative Volume Element (RVE). Results of study by the U.S. Department of Energy and the U.S. Department of Defense/Office of Land Warfare & Munitions under the Joint DoD/DOE Munitions Technology Development Program. Materials Science & Technology Division, Los Alamos National Laboratory, Los Alamos, New Mexico 87545, USA.

21. Hill, R. Elastic properties of reinforced solids: some theoretical principles. *J. Mech. Phys. Solids*, **11**, 1963, 357–372.

22. Drugan, W. J., and Willis, J. R. A micromechanics-based nonlocal constitutive equation and estimates of representative volume element size for elastic composites. *J. Mech. Phys. Solids* **44**, 1996, 497–524.

23. Brenner, D. W., Shenderova, O. A., Harrison, J. A., Stuart, S. J., Ni, B., and Sinnott, S. B. A second-generation reactive empirical bond order (REBO) potential energy expression for hydrocarbons. *J. Phys.: Condens. Matter* **14**, 2002, 783.

24. Shen, S., and Atluri, S. N. Atomic-level stress calculation and continuum-molecular system equivalence. *CMES* **6**, 2002, 91–104.

25. Namilae, S., Chandra, N., and Shet, C. Mechanical behavior of functionalized nanotubes. *Chem. Phys. Lett.* **387**, 2004, 247–252.

26. Tu, Z.-C., and Ou-Yang, Z.-C. Single-walled and multiwalled nanotubes viewed as elastic tubes with the effective Young's moduli dependent on layer number. *Phys. Rev.* **65**, 2002, 233407.

27. Van Lier, G., Van Alsenoy, C., Van Doren, V., and Geerlings, P. Ab initio study of the elastic properties of single-walled carbon nanotubes and graphene. *Chem. Phys. Lett.* **326**, 2000, 181–185.
28. Coleman, J. N., Khan, U., Blau, W. J., and Gun'ko, Y. K. Small but strong: a review of the mechanical properties of carbon nanotube-polymer composites. *Carbon* **44**, 2006, 1624–1652.

Chapter 2

Obtaining Material Properties from the Bottom-Up Approach

B. Farahmand

Technical Horizon Inc., 25403 Barents Laguna Hills, California 92653, USA
bfarahmand@technicalhorizon.com

In the aerospace industry, high-risk parts must be able to handle fatigue environment during their service usage. The safe-life analysis of critical parts requires having comprehensive fracture mechanics data. Currently, these allowables are obtained through testing, which is costly and time consuming. When designers experience a lack of data prior to the service operation of components, the program can be impacted. Any method that can be implemented to eliminate the cost of testing can be helpful to the program. That is, to provide a flexible and easy-to-use method for computation that enables the simulation of the multiscale modeling, where the computation starts from the bottom of scale (the quantum mechanics) and bridge up to the continuum scale. The current virtual testing technique was originally established by the author while at the Boeing Company, and has the capability of generating fracture toughness and fatigue crack growth allowables accurately. The method utilizes only the full stress–strain curve for

Characterization of Nanocomposites: Technology and Industrial Applications
Edited by Frank Abdi and Mohit Garg

ISBN 978-981-4669-02-3 (Hardcover), 978-1-315-36489-6 (eBook)
www.panstanford.com

the material under consideration. Results of fracture data generated by this method have already been verified with test results and consistencies between analysis and testing were obtained. However, this method depends on the full stress–strain curve, which does not exist for some material. Therefore, the author recommends the use of the multiscale modeling simulation technique to overcome problems that arise from a lack of data. An attempt has been made to generate the full stress–strain curve by applying the coarse grain and molecular dynamics techniques. The proposed multi-scale modeling approach includes the use of potential energy to establish the interaction forces between atoms including the analysis of defective regions within a grain of a metal matrix. Subsequently, defects of the metal matrix, that are associated with a given grain of a given orientation, will be placed in numerous locations. Capturing the behavior of a grain through the coarse grain technique, the methodology can be then extended to numerous grains of different orientations. Finally, mechanical properties obtained through computational analysis will be compared with the actual test data and parameters will be adjusted for future analysis. *The intent of this chapter is to provide a road map to the multiscale modeling technique that will help the analyst to conduct simulation of their own in order to obtain material properties from the bottom up approach that is free from testing.*

The application of virtual testing, together with the multiscale modeling and simulation technique for the generation of the full stress–strain curve, will be of the utmost importance to the aerospace industry by greatly reducing the cost and time of testing.

2.1 Introduction

Material properties are important and essential when designing and manufacturing aerospace parts. Stress analysis tools will be able to determine the level of stress in a part. Measuring the level of stress in a body under the external load can be done experimentally or through a computational approach, such as the finite element method. The level of stress, estimated through the finite element technique, is not sufficient to allow designers to decide whether the magnitude of applied load will cause failure during the life of the part or not. Therefore, we need to have information on material properties that can be used as a reference for comparison with the maximum stress introduced in the part. That is, the maximum stress should not exceed the material allowables. Material allowables are obtainable through standard tests with stringent requirements and are costly and time consuming to the manufacturer of parts.

The failure of structural parts seldom takes place under the monotonic load condition. The in- service structural failures, in most cases, were observed to occur under a low magnitude cyclic load environment, usually below 50% of the yield of the material. But, because the applied load is cyclic, the nature of failure is totally different from monotonic load and the static allowables are not adequate to predict failure. Therefore, engineers must have data on material properties that represents the operating load environment during service usage. Having that information available allows for the accurate prediction of fatigue failure.

As was mentioned above, fatigue and fracture data must be generated in accordance with an acceptable standard for obtaining material allowables. These allowables will be useful because they allow designers to create critical parts that must survive under the operating environment. For example, in the United States, the American Society for Testing and Materials (ASTM) is used as the guideline to generate fatigue and fracture allowables.[1] ASTM E399 is used as the guideline to generate plane strain fracture toughness, K_{Ic}. The ASTM E647 and ASTM E740 have detailed instructions for obtaining the plane stress, K_c, and part through fracture toughness, K_{Ie}, values, respectively. For the aircraft and aerospace industries, this data, which is costly and labor intensive, is essential. Difficulties associated with specimen preparation, pre-cracking the notch, recording and monitoring the data, obtaining the variation of fracture toughness versus part thickness, capturing data in the threshold region of the fatigue crack growth region, and repeating the test can be a huge economic burden on the industry. The author will provide a simple example that clearly shows the cost of obtaining material allowables. Table 2.1 reflects the cost and time associated with generating fatigue and fracture allowables,

Table 2.1 The cost associated with generating fatigue and fracture allowables for a typical aluminum alloy, 2219-T6

Fracture toughness test –2219-T6	**Fatigue crack growth tests –2219-T6**
• Typically, 5 specimens per thickness are used to plot fracture toughness versus thickness. The purpose of using 5 specimens is to check for material variations. 5 different thickness will be adequate to capture the variation of fracture toughness versus thickness	• Typically, 4 specimens are needed to check for material variations
	• Usually, 3 stress ratio will be tested (R = −1, 0.1, $0.5)
	• For anisotropic materials, all 3 orientations will be tested
	• For cryogenic temperature additional tests are required
• For anisotropic materials, all 3 orientations will be tested	• Slightly above room temperature additional test are required
• For cryogenic temperature additional tests are required	Total number of specimens ~3 × 3 × 3 = 27 (Room temperature)
• For slightly above room temperature additional tests are required	Total number of specimens ~3 × 3 × 3 = 27 (Cryo temperature)
Total number of specimens (Room temperature) ~5 × 5 × 3 = 75	Total number of specimens ~3 × 3 × 3 = 27 (Above Room temperature)
Total number of specimens (Cryo temperature) ~5 × 5 × 3 = 75	Total Cost – 81 × 1500 =$121,500 (without threshold)
Total number of specimens (Above Room temperature) ~5 × 5 × 3 = 75	Total Cost – 81 × 3000 =$243,000 (with threshold)
Total Cost– 225 × 700 = $157500	*Fatigue crack growth tests ~$1500/test (threshold is not included)*
	Fatigue crack growth tests ~$3000/test (threshold is included)

Fracture toughness =$157500,

Fatigue crack growth =$243000

Total cost =$400500

for a typical aluminum alloy, 2219-T6. This alloy is currently used to manufacture aerospace parts. It can be seen that the cost of fully characterizing material properties of the 2219-T6 aluminum alloy that are essential for damage tolerance analysis of aerospace or aircraft structures can be significant. Note that, Table 2.1 did not include the cost associated with static properties, which are essential when conducting durability and damage tolerance analysis of high-risk components. Having said that, there is a need in the aerospace and other similar industries to implement an analytical (or semi-analytical) tool that can generate allowables faster and with less cost. The virtual testing technique implemented by the author is extremely helpful to provide fatigue and fracture allowables without conducting costly and time consuming ASTM testing standards.

In Section 2.3, the author will discuss a fundamental method of obtaining material allowables by studying material from an atomistic perspective (the bottom up approach) through a computational technique, such as molecular dynamics, together with finite element methods.

2.2 Virtual Testing

Obtaining fracture mechanics allowables without conducting testing through ASTM standards was first was proposed by the author.[1–6] Farahmand[2,3] utilized the extended Griffith theory to obtain a relationship between the applied stress and half a crack length by using the full stress–strain curve for the material under consideration. The extension of this work led to the estimation of material fracture toughness and later construction of the full fatigue crack growth rate curve.[4–6] A brief description of Farahmand's virtual testing theory will be given here for estimating the material fracture toughness and subsequently, the extension of this methodology to the material fatigue crack growth rate data.

Materials selected by engineers to manufacture fracture critical hardware must exhibit some amount of plastic deformation and stable slow crack growth at the crack tip region prior to their final failure. The amount of energy consumed at the crack tip for plastically deforming material is largely due to material resistance in that region prior to final failure, which is not properly accounted for in the linear elastic fracture mechanics analysis. Farahmand showed that fracture behavior can be characterized by two energy released terms representing plastic deformation at and near the crack tip, U_{U} and U_{F}. Both terms can be estimated by the full uniaxial stress–strain curve for the alloy under consideration (Fig. 2.1). The extended Griffith theory in terms of U_{U} and U_{F}, including the two new crack surfaces term, $2T$, can be written as

$$\frac{\pi\sigma^2 c}{E} = 2T + \frac{\partial U_U}{\partial c} + \frac{\partial U_F}{\partial c} \tag{2.1}$$

Energy rate associated with two crack surfaces ($2T$); Energy rate associated with uniform straining ($\partial U_U/\partial c$); Energy rate associated with non-uniform straining ($\partial U_F/\partial c$)

For ductile metals, the two terms associated with uniform and non-uniform plastic deformation are much greater than the two new crack surfaces term, $2T$.[7] If quantities to the right of Eq. 2.1 are determined, then a relationship between applied stress, σ, and half a crack length, c, at the onset of failure can be established. The derivation of the two terms, $\frac{\partial U_{\mathrm{U}}}{\partial c}$ and $\frac{\partial U_{\mathrm{F}}}{\partial c}$ are fully described in references 2 and 3. Having all quantities shown to the right of the Eq. 2.1, the half crack length, c, for a given stress level can be calculated. In brief, the critical value of stress intensity factor, K, (i.e. K_{c}), can be estimated by simply using the energy under the stress–strain curve.

The K_{c} is a useful quantity when life assessment analysis of parts must be performed. Having the K_{c} available, region I of the da/dN versus ΔK can be estimated. Other regions of the fatigue crack growth rate curve can be also estimated through the virtual testing technique proposed by Farahmand.[4–6] His approach is based on establishing each region of the da/dN curve separately and relating them through the Forman-Newman fatigue crack growth equation:[8]

$$\frac{da}{dN} = \frac{c(1-f)^n \Delta K^n \left(1 - \frac{\Delta K_{\mathrm{th}}}{\Delta K}\right)^p}{(1-R)^n \left(1 - \frac{\Delta K}{(1-R)K_c}\right)^q} \tag{2.2}$$

The traditional approach to obtaining fatigue crack growth constants shown in Eq. 2.1 (c, n, K_c, ΔKth) is through preparing a notched specimen that has been pre-cracked and conducting fatigue crack growth test in accordance with the ASTM E647 standards,[1] which are costly and labor intense. However, by using the virtual testing concept, all constants can be estimated and the *da/dN* versus ΔK, from Eq. 2.2, can be plotted. To be able to estimate region II of fatigue crack growth curve, Farahmand gathered numerous alloys from the NASGRO database.[8] The *da/dN* versus ΔK for all these alloys were plotted and as a result, two important observations were obtained that were helpful to establish the Paris region of the *da/dN* curve. The lower point in the Paris region of the fatigue curves (Fig. 2.1, point (1)) has a material independent property so that the ratio of the stress intensity factor at the lower bound point and the threshold value ($\Delta K/K$th for $R = 0$) is $\sim$1.125 for the crack growth rate per cycle, *da/dN* $\sim$*1.0E-7* inch/cycle ($\sim$2.54E-6 mm/cycle). In

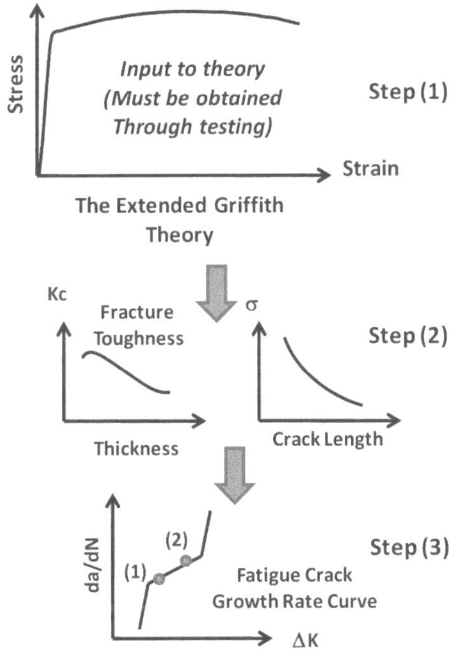

Figure 2.1 Steps for generating the *da/dN* curve.

the upper region of the da/dN curve (at the end of the Paris region, Fig. 2.1, point (2.2)), the ratio of the upper bound stress intensity factor and its critical value, Kc, ($\Delta\ K/K_c$ for $R = 0$) is found to be $\sim$0.9 for the da/dN $\sim$0.005 inch/cycle ($\sim$0.127 mm/cycle). These two points are useful to plot the entire Paris region. The da/dN curve can then be plotted using Eq. 2.1 where the fracture parameters and constants are taken from the estimated K_c, Kth, and the Paris constants C & n values for the case of $R \sim 0$. Other range of R ratios can be plotted by using the Newman closure equation, f. The threshold stress intensity factor, Kth, is a function of crack length and stress ration, R, and its true measurement is difficult to obtain. The complexity is due to plasticity and the surface roughness closure phenomenon that make the actual measurement of the threshold value difficult. The threshold stress intensity factor measurement is time consuming and sometimes it may take several days to make a few measurements of the Kth value. For these reasons the approximate value of Kth is of interest to engineers when conducting life assessment. Farahmand[9,10] was able to establish a reasonable relationship between the threshold stress intensity factor, ΔKth (for the case of stress ratio, $R \sim 0$) and the plane strain fracture toughness, K_{Ic}. The results of his observations on the threshold values of more than one hundred metallic alloys was in such a way that the quantity Kth falls between $K_{Ic}/4\pi$ to $K_{Ic}/3\pi$ (for $R = 0$), Table 2.2. Farahmand argued that materials with high K_{Ic} value also possess higher Kth value and conversely, the lower Kth value belongs to low K_{Ic}.

To verify the validity of the virtual testing approach, several fatigue crack growth data[8] were gathered for two aerospace alloys and were compared with the virtual testing results, Fig. 2.2. The da/dN data from NASGRO database[8] were prepared for 2024-T3 and Ti-6Al-4V alloys to compare and check the validity of the approach. A more detailed comparison between the virtual testing and the ASTM experimental approaches was made for numerous aerospace alloys and is available in.[10] From Fig. 2.2, it can be seen that virtual testing can be used as a reliable tool to generate the da/dN curve (da/dN data). This is of great importance to aerospace and other related industries because it can generate the fatigue crack growth data fast with practically no cost associated with it.

Table 2.2 Results of observations on the threshold values for numerous aluminum alloys where $K\text{th}$ falls between $KIc/4p$ to $KIc/3p$ (for $R = 0$)

Material	K_{Ic}	Kth ($K_{Ic}/4\pi$)	Kth (Test *)	Kth ($K_{Ic}/3\pi$)	Material	K_{Ic}	Kth ($K_{Ic}/4\pi$)	Kth (Test *)	Kth ($K_{Ic}/3\pi$)
2014-T6 (L-T)	27	2.1	2.7	3.0	2219-T87 (T-L)	27	2.1	2.9	2.9
2020-T651 (L-T)	22.5	1.8	2.2	2.4	7005-T6 & T63 (T-L)	40	3.2	3.4	4.2
2024-T3 (L-T)	33	2.6	2.9	3.5	7010-T73651	31	2.5	2.5	3.3
2024-T3 (L-T)	29	2.3	2.9	3.1	7050-T736 & T74 (T-L)	24	1.9	2.3	2.5
2024-T3 (L-T)	33	2.6	2.9	3.5	7050-T76511 (T-L)	24	1.9	2	2.5
224-T351 (L-T)	29	2.3	2.6	3.1	7075-T651 (T-L)	28	2.2	3	3.0
2024-T62 (L-T)	36	2.9	2.9	3.8	7075-T6510 (L-T)	28	2.2	3	3.0
2024-T62 (L-T)	30	2.4	2.9	3.2	7075-T6511 (T-L)	28	2.2	3	3.0
2024-T852	28	2.2	2.9	3.0	7075-T73 (T-L)	28	2.2	3	3.0
2024-T861 (L-T)	23	1.8	2.2	2.4	7075-T7351 (T-L)	29	2.3	3	3.1
2048-T851 (L-T)	35	2.8	2.7	3.7	7075-T73510 (T-L)	31	2.5	3	3.3
2048-T851 (L-T)	30	2.4	2.7	3.2	7075-T73511 (T-L)	33	2.6	3	3.5
2124-T851 (L-T)	30	2.4	3	3.2	7075-T7352 (T-L)	33	2.6	3	3.5
2124-T851 (L-T)	26	2.1	2.7	2.9	7075-T7651 (T-L)	23	1.8	2.4	2.4
2219-T62 (-320F)	30	2.4	2.9	3.2	7079-T651 (T-L)	26	2.1	2	2.8
2219-T851 (L-T)	33	2.6	3	3.5	7149-T73511 (L-T)	31	2.5	3	3.3
2219-T87 (L-T)	30	2.4	2.9	3.2	7178-T7651 (T-L)	28	2.23	3	2.97

$K_{Ic}/4\pi < \Delta K\text{th} < K_{Ic}/3\pi$

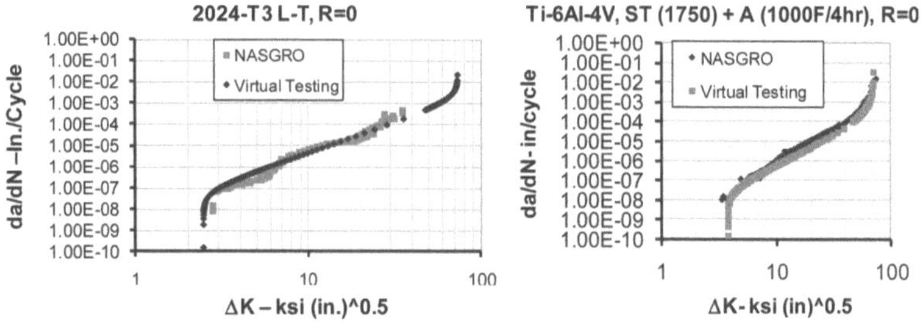

Figure 2.2 Validity of virtual testing technique versus the actual test data. Two alloys were selected for this comparison.

2.3 Virtual Testing from the Bottom-Up Approach

The virtual testing approach discussed in Section 2.2 is an innovative approach for obtaining *da/dN* data because it avoids the costs of conducting tests, which is useful to the aerospace industry when the life assessment of parts prior to their service usage must be determined. The method requires having the full stress–strain curve in order to provide a relationship between the applied stress and crack length and subsequently to generate the material fracture toughness, K_c. Also, it was mentioned that the K_c value can be used to generate region III of the *da/dN* curve. However, in many cases the full stress–strain curve is not available and must be generated in the laboratory through ASTM testing. Even though the stress–strain curve is much easier, cheaper, and less time consuming to obtain when compared with fracture properties, the ultimate goal of the aerospace industry is the ability to generate fatigue and fracture properties entirely free from testing. That is, generating material properties entirely through a computational approach. As an extension of the virtual testing discussed in the previous section, it was decided to obtain the full stress–strain curve through the multiscale modeling approach,[11,12] making the above-mentioned virtual testing technique purely by analysis. The multiscale modeling approach will be used to predict the stress–strain response of a wide range of metals. As the name implies, the multiscale modeling provides information from the bottom of the scale (that is, the atomistic simulation) and links it to the upper portion of the scale as illustrated in Fig. 2.3. Length and time scales extend many orders of magnitude when looking at various problems in computational physics, biology, and the materials sciences from the bottom up approach. For example, on the atomistic level, the typical bond distance between atoms is of the order of Ångstroms (the lattice constant), and atoms vibrate with a frequency of approximately 10^{15}/second. On the other hand, other phenomena and applications of practical interest occur on a timescale of seconds, with system sizes that can be a micrometer or larger. A grand challenge in computational physics, biology, and the material sciences is to link these vastly different time and length scales. From the material sciences point of view, the failure as the result of applied forces will be defined as the accumulation of bonds failure between atoms in a given material. Because of the structural arrangement of atoms and their properties, which is expected to be different in different material, it is likely that the stress–strain curve generated via the computational method to be also different for different material. Relating the accumulation of atomic bond failures at the bottom of scale and linking the failure phenomenon to the higher level is complicated but an attempt will be made to address it in this chapter. In Fig. 2.3, the simulation begins at the atomistic level and results must be relayed to the next level called the mesoscale. Subsequently, the continuum mechanics will take over, where finite element technique can be used in the simulation (Fig. 2.3a,b).

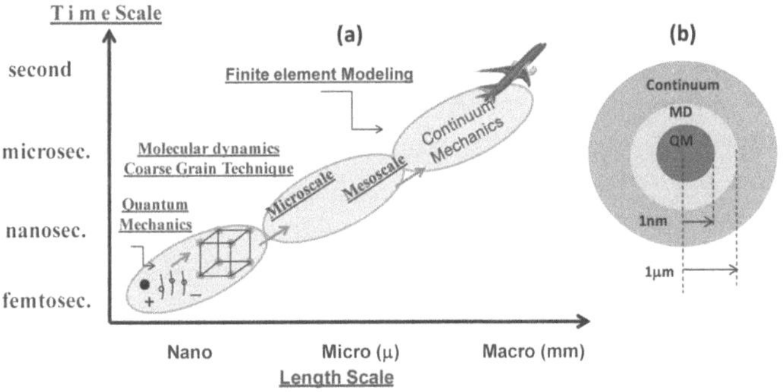

Figure 2.3 Multiscale modeling plot showing the time and length scale from the bottom up approach (quantum mechanics, mesoscale, and continuum regions are shown).

While atomistic simulations[3–16] are capable of providing a direct study of key nanoscale mechanisms associated with fracture, they are limited in both time and length scales. The most typical users are restricted to a limited number of atoms due to computational resources, which correspond to less than 100 nm^3 of material (in the case of a closely packed metal). Moreover, atomistic simulations must be performed for a short duration due to the femtosecond time step required to properly capture atomic vibrations and interaction among nearby atoms. These limitations motivated the analyst to study a localized region in the body and apply the results to the rest of the material. That is, to apply the atomistic modeling in detail to the region of interest (such as around a dislocation or grain boundary), while modeling the remainder of the simulation domain using more coarse techniques, such as the finite elements approach. The development of predictive tools that have their foundation in the fundamentals of physics is definitely of interest to the scientific community and the multiscale methods are envisioned to play an increasingly larger role in predicting bulk properties free from testing.

In conducting the atomistic and coarse grain simulation, it is important to distinguish between two regions of the simulation: (i) the region in which the nodes are in one-to-one correspondence with the atoms is called the molecular dynamics (MD) region[17–20] and (ii) the region with many atoms per node is called the coarse-grained (CG) region.[21–25] The equation of motion for the atoms in the MD region is just the usual MD equation of motion, based on Newton's laws of motion and the empirical potential describing the force of interaction between atoms. In the CG region, the equation of motion is the mass, stiffness, and displacement relationship. Figure 2.4 is the multiscale modeling of a system that contains molecular dynamics (MD) and finite element (FE) or coarse-grained molecular dynamics (CGMD) regions. A narrow handshaking region is used to couple the FE and MD regions. Another form of coarse graining is called the "united atom" (sometimes called the "extended atom") and is used in most early MD simulations in the field of biology such as proteins, lipids and nucleic acids.[26,27] For example, instead of treating all four atoms of a CH_3 methyl group explicitly

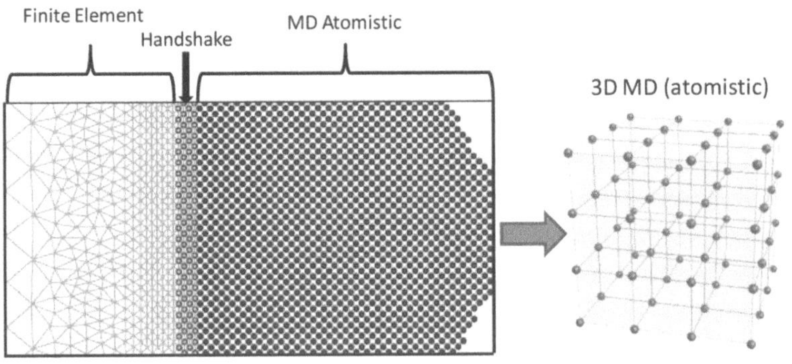

Figure 2.4 Molecular dynamics and finite element modeling. With the molecular dynamics approach nodes are in one-to-one correspondence with atoms.

(or all three atoms of CH_2 methylene group), one represents the whole group with a single pseudo-atom. This pseudo-atom must be properly parameterized so that its van der Waals interactions[28–30] with other groups have the proper distance-dependence. Similar considerations apply to the bonds, angles, and torsions in which the pseudo-atom participates. In this kind of united atom representation, one typically eliminates all explicit hydrogen atoms except those that have the capability to participate in hydrogen bonds. Figure 2.5 shows the structure of the oligopeptide sequence Glu-Ser-Ala-Tyr-Val in an all-atom representation and the equivalent CG version.[27] The CG representation is shown with beads (pseudo-atoms) in cyan, gray, blue, and in green, which their corresponding atoms are also marked in Fig. 2.5. Each bead can be analyzed separately and the resulting information will be incorporated into the assembly analysis. Figure 2.6 shows another example of coarse graining which illustrates the mapping of the atomistic model of a DNA configuration to a coarse-grained bead-spring model.[31] One bead represents one chemical repeat unit.

A number of molecular dynamics (MD) based approaches have been developed in the last decade to predict the constitutive response of metals and metal alloys. These methods have focused on predicting the mechanical response of metals on the atomic level based on applied loads. In Fig. 2.4, the critical areas are atomistically modeled using the nodes of the mesh and the equilibrium sites of

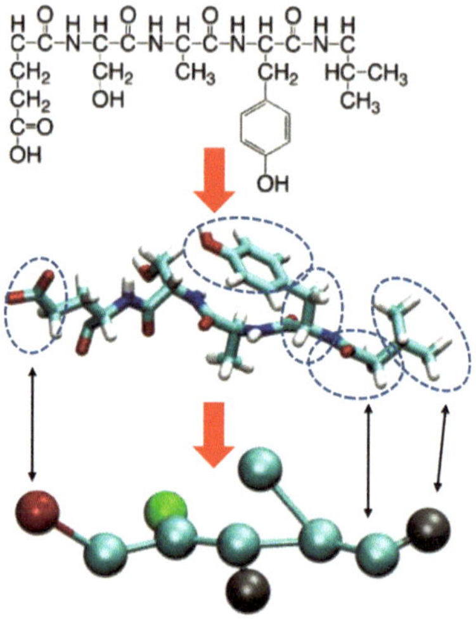

Figure 2.5 Structure of the oligopeptide sequence in an all-atom representation and the equivalent CG version. The CG representation is shown with beads in cyan, gray, blue, and in green, which their corresponding atoms are also marked.

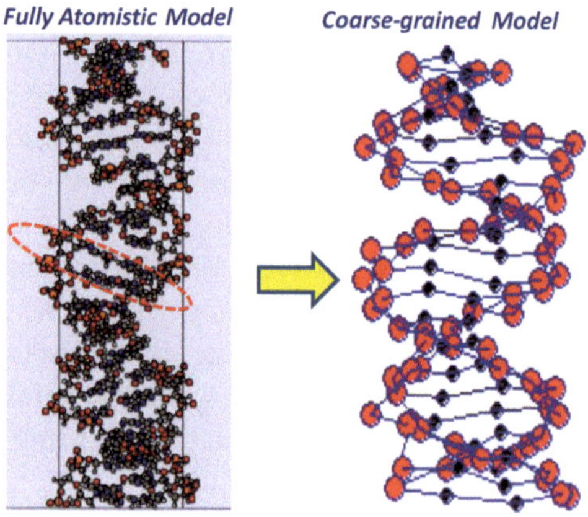

Figure 2.6 Mapping of the atomistic model of a DNA configuration to a coarse-grained bead-spring model.

the atoms coincide and are in one-to-one correspondence. Then the mesh size increases moving out further into the less critical regions, where each node of the mesh corresponds to many atoms. It is this reduction in the number of degrees of freedom that increases the efficiency of multiscale simulations. These methods can be employed to generate stress–strain curves for metallic materials that can be used as an input into the virtual testing to provide fatigue and fracture data. Computer simulations act as a bridge between microscopic length and time scales and the macroscopic world of the laboratory. It can provide a guess at the interactions between molecules and obtain accurate predictions of bulk properties including the stress–strain data. The predictions can be exact in the sense that they can be made as accurate as we like, subject to the limitations imposed by our computer budget. Accurate multiscale simulation can provide detailed bulk properties and can act as a bridge between theory and experiment. We may test a theory by conducting a simulation using the same model. We may test the model by comparing it with experimental results. We may also carry out simulations on the computer that are difficult or impossible to conduct in the laboratory (for example, working at extremes of temperature or pressure). Ultimately, we may want to make direct comparisons with experimental measurements made on specific materials, in which case a good model of molecular interactions is essential.

There are a few issues regarding the application of the MD and CG on multiscale modeling and virtual testing. The main one is to allow lattice defects to propagate from the atomistic region into the coarse-grained region (continuum) upon the application of the load. These could be point defects, line defects (dislocations) or planar defects (grain boundaries, surfaces, etc.). The need for a solution is important in the case of dislocations because they are highly mobile and in some cases tangle with each other. For example, dislocations have been observed to propagate very fast and in such a short time under driven force or thermal effects, which can make the simulation difficult to conduct.[32] In reference 32 the MD simulation of edge dislocation motion in a BCC metal, Molybdenum, at finite temperature was conducted. Therefore, a multiscale simulation incorporating atomistic and molecular dynamics techniques that can provide us material properties for virtual testing must consider the aforementioned details. The following is a brief description of MD and how it is used in multiscale modeling.

The molecular dynamics (MD) simulation is a technique by which one generates positions, velocities and accelerations of a system of N particles (atoms) by numerical integration of Newton's equation of motion, for a certain initial condition and boundary condition. It consists of the numerical step-by-step solution of the classical equations of motion, which for a simple atomic system can be written as:

$$F_i = m\frac{d^2 r_i}{dt^2} = -\nabla V_i \rightarrow F_i = m_i a_i \rightarrow a_i = \frac{\Delta v_i}{\Delta t} \rightarrow v_i = \frac{\Delta r_i}{\Delta t} \quad (2.3)$$

Here, m_i is the mass and v_i is the velocity of the ith atom. The above equation looks simple, yet it is as complicated as the famous N-body problem which cannot be solved when N is more than 2. First, we need to calculate the forces F_i acting on the atoms, and these are usually derived from a potential energy $V(r^N)$, where $r^N = (r_1; r_2; :::: \ r_N)$ represents the complete set of $3N$ atomic coordinates. In this section we focus on this function $V(r^N)$, which can provide the force value needed in the Eq. 2.1. A simplistic approach to estimate the theoretical strength between atoms of a lattice structure is to assume materials are free from defects. The theoretical strength (strength needed to break the bonds) is many times larger than the actual measured strength value reported in the literature. The calculated theoretical strength is estimated to be *E/10*, where E is the modulus of elasticity of material.[33] A more acceptable approach to evaluate the interatomic potential (also called the force field) will be to use the quantum mechanics technique, which will be used to obtain forces for the Newton's equation of motion The interatomic potential will be discussed later. To start with MD approach, we will divide the time interval of interest into many small segments, each being a time step of Δt. Given the initial position at time $r(t_0)$, the integration indicates we can advance the body by increments of Δt. The time interval, Δt, depends on the nature of the problem in consideration. For atomic simulation in metals, it is usually taken as 1fs (femtosecond).

$$\{r(t_0)\} \rightarrow \{r(t_0 + \Delta t)\} \rightarrow \{r(t_0 + 2\Delta t)\} \rightarrow \ldots \{r(t_0 + N_t \Delta t)\} \quad (2.4)$$

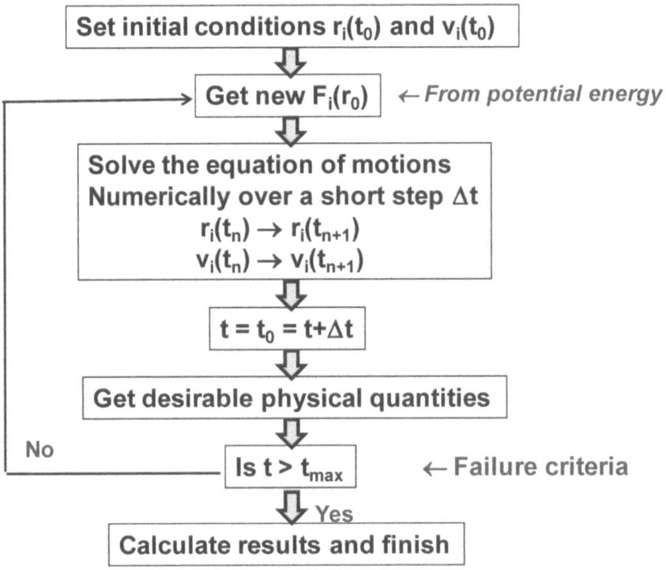

Figure 2.7 Steps for conducting molecular dynamics calculation. The simulation proceeds by calculating forces on each atom from potential and solving the equations of motion.

where N_t is the number of time steps making up the interval of integration shown in Eq. 2.1. The flowchart for a typical MD simulation is shown by Fig. 2.7. The simulation proceeds by calculating forces on each atom from potential and solving the equations of motion, where acceleration, velocity, and displacement of atoms over a short step, Δt, can be estimated. That is, (i) set initial particle position and velocities, (ii) calculate force on each particle, (iii) move particles by time step Δt, (iv) save current positions and velocities, (v) if the preset number of time steps is reached, stop, otherwise go back to (ii), (vi) analyze data and record results. To make this simulation possible and tractable, it is common to assume the inter-atomic potential energy, V, can be represented as the sum of pairwise interactions. Therefore, to calculate a trajectory (positions, velocities and accelerations), one only needs the initial positions of the atoms, an initial distribution of velocities and the acceleration, which the acceleration $\left(\frac{F}{m}\right)$ is determined by the gradient of the potential energy function, V (Eq. 2.3). A well-known molecular dynamics algorithm is called the *Leapfrog algorithm* and starts by defining velocity, v at half time steps $\Delta t/2$.[34,35]

$$r\,(t+\Delta t) = r\,(t) + v\left(t+\frac{1}{2}\Delta t\right)\Delta t \tag{2.5}$$

$$v\left(t+\frac{1}{2}\Delta t\right) = v\left(t-\frac{1}{2}\Delta t\right) + a(t)\,\Delta t \tag{2.6}$$

In this algorithm, the velocities are first calculated at time $t+\frac{1}{2}\Delta t$; these are used to calculate the positions, r, at time $t+\Delta t$. In this way, the velocities leap over the positions, then the positions leap over the velocities. The advantage of this algorithm is that the velocities are explicitly calculated from the force and potential. However, the disadvantage is that they are not calculated at the same time as the positions. The velocities at time t can be approximated by the relationship

$$v\,(t) = \frac{1}{2}\left[v\left(t-\frac{1}{2}\Delta t\right) + v\left(t+\frac{1}{2}\Delta t\right)\right] \tag{2.7}$$

The precision of MD simulation depends on the accuracy of the interatomic potential energy. The true calculation of interatomic potential can be performed through quantum mechanics where electrons play a major role in determining the bonding properties of the system. It is a system of interacting atoms that is made up of nuclei and electrons which interact with each other. In the case of metals, it involves the delocalized sharing of free electrons among a lattice of metal atoms. The bond typically contains a high number of electrons surrounding a giant lattice of positive ions. The fundamental assumption for the MD simulation is that the equation for the electronic problem considers the nuclei to be fixed. Otherwise, the simulation of individual atoms within an assembly of N atoms requires massive computer resources, and poses severe limits on the maximum size of the system and simulation time will be limited to only a few hundred atoms. To avoid the quantum mechanics techniques, the phenomenological potentials are obtained by using a mathematical function and fitting its unknown parameters to various experimentally determined properties of the system, such as its lattice constants. The interatomic potential must be able to model the dynamics of nanostructures and be able to describe the properties of the model system.

2.4 Interatomic Potential

Constructing a potential involves two steps: (i) Selection of an analytical form for the potential. This is a sum of pairwise terms, with the energy of a pair depending on their relative distance. Currently, the attempt is shifted to capture as much information on the characteristics of the bond as is possible. A typical analytical form is constituted by a number of functions, depending on geometrical quantities such as distances, angles, or on intermediate variables such as atom coordination. (ii) Finding an actual parametrization for the functions that constitutes the analytical form that analyst has chosen. This step is very important and can be technically elaborate. A variety of techniques have been utilized to correctly implement the potential. Some groups have tried to start from a *first-principle* description (that is, where the electronic structure is kept into account, the ab initio MD simulation), and obtain an expression for the energy as a function of the nuclei position by means of successive approximations. Other groups chose to fit the potential on the experimental data, giving maximum priority to practicality rather than to connections with first-principles. In all cases, potentials are designed with a range of applicability in mind. The ability of a potential to work properly in different environments is called *transferability*. When using a potential, the simulator should always be familiar with its transferability properties, and look critically at results obtained in unusual conditions such as very high temperature or very high pressure.

As mentioned previously, the heart of MD simulations is the inter-atomic potential. In classical simulations the atoms are most often represented by point-like mass, which interact through many-body interactions potential. In that way the highly complex description of quantum mechanics is neglected and effective picture is implemented. Herein, the atoms and degrees of freedom are modeled by a set of parameters and analytical functions, which depend on the mutual positions of the atoms in the configuration. These parameters and functions give complete information about system energy, as well as about the forces acting on each particle. The best choice of a potential for simulations of metals is a many-body potential. Pair-wise potentials, like Lenard-Jones (LJ) potentials,[28] do not give an adequate description of all the properties of metals. For the interaction potential between a pair of atoms, the LJ potential can be written as:

$$V(r) = 4\varepsilon\left[\left(\frac{\sigma}{r}\right)^{12} - \left(\frac{\sigma}{r}\right)^{6}\right] \tag{2.8}$$

where ε = depth of the well shown in Fig. 2.8 and σ is the distance at which the interatomic force is zero. Another potential equation similar to the LJ is called the Morse potential. The Morse potential can be written as

$$V(r) = D\left[e^{-2\alpha(r-r_e)} - 2e^{-\alpha(r-r_e)}\right] \tag{2.9}$$

where r is the distance between the two atoms, D the depth of the well r_e the position of minimum, and α the measure of curvature near minimum. At long range, as $r \to \infty$, the potential energy becomes a constant (since the interaction between the atoms becomes negligible) which is chosen to be almost zero on the energy scale. This energy is often referred to as the dissociation limit since the diatomic molecule will dissociate into two separate atoms if the energy of the system is greater than this value.

The LJ potential has an attractive tail at large r (far distance), it reaches a minimum around 1.122σ, and it is strongly repulsive at close distance, passing through 0 at $r = \sigma$ and increasing steeply as r is decreased further. The term $\frac{1}{r^{12}}$, dominating at short distance, models the repulsion between atoms when atoms are brought very close to each other. Its physical origin is related to the Pauli principle[36] when the electronic clouds surrounding the atoms start to overlap, causing the energy of the system to increase abruptly. The exponent 12 was chosen entirely on a practical basis. Equation 2.5 is relatively simple to compute. In fact, on physical grounds an exponential behavior would be more appropriate. The term $\frac{1}{r^{6}}$, dominating at large distance, constitutes the attractive part. This is the term which gives cohesion to the system. A $1/r^6$ attraction is originated by van der Waals dispersion forces, originated by dipole-dipole interactions in turn due to fluctuating dipoles. The plot of energy versus atomic distance is shown in Fig. 2.8. The depth of the well can be calculated through the quantum mechanics technique. The accuracy of the potential energy can be checked

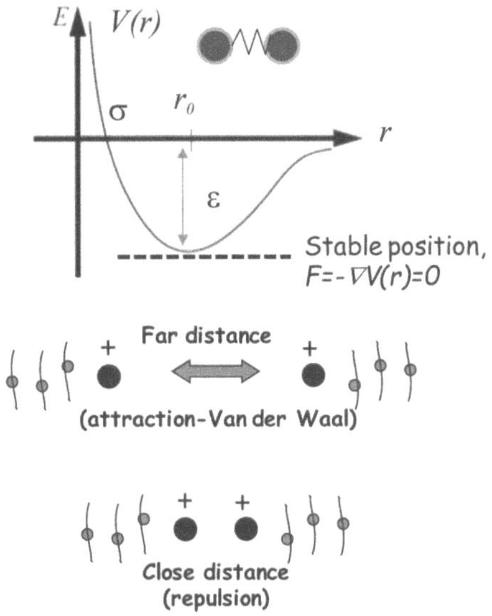

Figure 2.8 The depth of the well and the distance at which the interatomic force is zero. Repulsive and attractive interatomic forces are also illustrated here.

against test data. In solids, there are attractive forces pulling the atoms together and repulsive forces that prevent the atoms from getting too close. If the repulsive force were not present, then solids would collapse in on themselves. To describe the forces between atoms, we need a potential energy function that gives a potential well with both attractive and repulsive terms.

The van der Waals forces are rather weak interactions, which dominate the bonding character of closed-shell systems, that is, rare gases such as Ar or Kr. Therefore, these are the materials that a LJ potential could mimic fairly well. The parameters ε and σ are chosen to fit the physical properties of the material. For instance, in the case of the interaction between the carbon atoms of nanotubes and the polymer (–CH2– units), the LJ potential was parametrized with $\varepsilon = 0.4492$ kJ = mol and $\sigma = 0.3825$ nm.[37]

On the other hand, a LJ potential is not adequate to model situations with open shells, where strong localized bonds may form (as in covalent systems), or where there is a delocalized situation as in metals. In these systems the two-body interactions scheme itself fails very badly. However, regardless of how well it is able to model actual materials, the exponent of LJ (12 and 6 of equation) potential is an important model system. There is a vast body of work that investigated the behavior of atoms interacting via LJ on a variety of different geometries (solids, liquids, surfaces, clusters, two-dimensional systems, etc.). One could say that LJ is the standard potential to use for all the investigations where the focus is on fundamental issues, rather than studying the properties of a specific material. The simulation work done on LJ systems helped us to understand basic points in many areas of condensed matter physics, and for this reason the importance of LJ cannot be underestimated.

It is also worthy to mention that there are alternative methods of estimating the inter-atomic potential and forces (both positive and negative). One experimental method is through the atomic force microscope (AFM) and the scanning tunneling microscope (STM). The following is a brief description of the AFM and STM methods.[38–40]

2.5 Measuring Interatomic Forces through the AFM and STM

Currently, several powerful surface imaging techniques are available that can measure the inter-atomic forces for metallic material. Results of these experimental measurements are useful for the MD analysis. This technique relies on a very sharp probe positioned within a few nanometers above the surface of the material (Fig. 2.9). When the probe moves laterally relative to the specimen, any change in the height of the surface causes the detected probe signal to change. This can be done through a laser beam and the detector shown in the figure. In general, if the probe signal decreases, this means that the point on the surface directly beneath the probe is further from the probe than the previous point was. Conversely, if the probe signal increases, then the point on the surface is closer to the probe than the previous point. In the Scanning Tunnel Microscope, STM, and the Atomic Force Microscopy, AFM, the probe signals depend so strongly on the probe–substrate interaction in

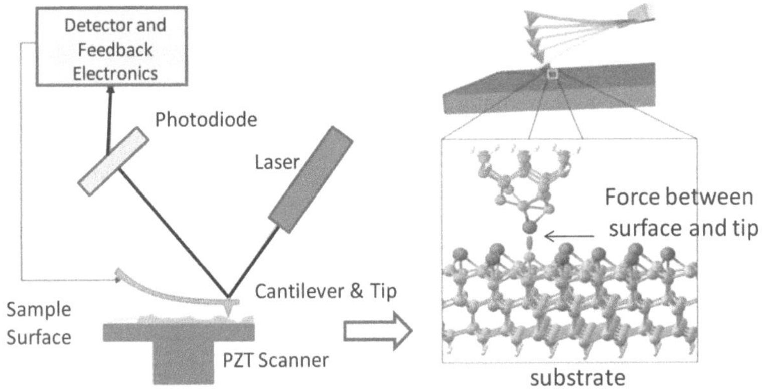

Figure 2.9 AFM is one of the most powerful methods for imaging, measuring, and manipulating surfaces at nanoscale. The size of the deflection is measured by a laser and this information is used to generate an image of the surface.

such a way that slight change in substrate height of as little as 0.01 nm can be detected. In addition, the STM and AFM probes can interact with regions of the substrate that are of atomic-scale lateral dimensions, which allows the substrate height to be measured with sub-nanometer lateral resolution as well.

STM image information is derived from measurements of the electron current that can flow when two electrodes, one a sharp metal tip, and the other a relatively flat surface, are brought within 1 nm of each other. When the two electrodes are that close together (a few atomic radii), electrons can pass from one electrode to the other by tunneling through the potential energy barrier that normally confines them inside each electrode. The magnitude of the tunneling current is a very strong function of the distance between the probe tip and the sample. In fact, the current is so sensitive to the probe/sample spacing that, under normal operating conditions, it changes by about a factor of ten for a change of only 0.1 nm in the separation distance (smaller than the radius of a single atom). Assuming the probe tip is made of atoms, and, if the tip is sharp, chances are good that one of these atoms sticks out a little farther than the others (Fig. 2.9). Therefore the tunneling barrier is thinnest right below the atom that sticks out, so it is much easier for electrons

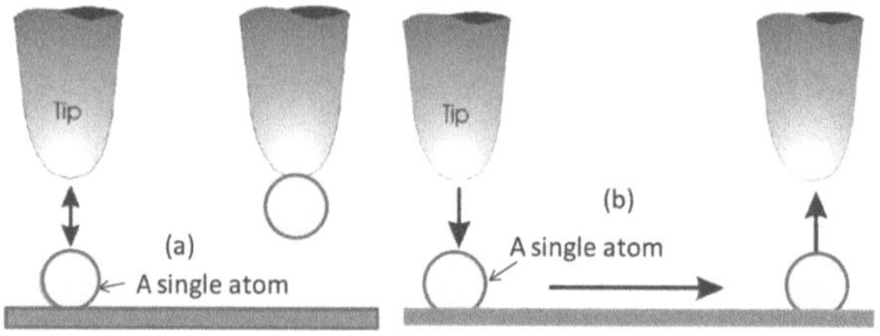

Figure 2.10 Vertical (a) and lateral (b) manipulation of a single atom on the surface by the STM technique.

to tunnel between the tip and the sample at this point than anywhere else. Using STM manipulation techniques, quantum structures can be constructed on an atom-by-atom basis. Single molecules can be manufactured on a one-molecule-at-a-time basis and detailed physical and chemical properties of atoms or molecules, which are vague to other experimental measurements, can be accessed at an atomic level. Figure 2.10 shows vertical and lateral manipulation of a single atom on the surface. Moreover, STM is an instrument not only used to see individual atoms by imaging, but also used to touch and take the atoms or to hear their vibration by means of manipulation. However, the AFM technique works by scanning the surface of a sample with a sharp tip, normally by a silicon or silicon nitride probe with a radius on the nanometer scale. As the probe scans the surface, forces (such as mechanical contact, van der Waals, chemical bonding, magnetic, and electrostatic) cause the probe to deflect. In manipulating the atom, the interatomic force (the force necessary to separate the atom) can be measured and can contribute to MD analysis.

2.6 Molecular Dynamic and *N*-Body (Atoms) Assessment

As was mentioned previously, molecular dynamics is a form of computer simulation in which atoms and molecules are allowed to interact for a period of time and their motion can be described through the classical laws of motion. Molecular dynamics simulation is used extensively in the field of biology, including the study of protein, as well as in the material sciences. Sometimes the molecular dynamics technique is called the "virtual microscope," because it allows scientists look closely into the motion of individual atoms in a way which is not possible in laboratory experiments. As indicated before, the potential energy functions (also called force fields) must be available and accurate to allow molecular dynamics analysis to be meaningful. The quantum mechanics approaches are rigorous but have limitations due to computer capability which can assess no more than a few hundred atoms at most. This approach, however, serves two important purposes. First of all, it provides direct information on the response of materials to external force and temperature. Secondly, it can generate a database of properties that can be used to construct effective (empirical) interatomic potentials. To determine the properties of an assembly of atoms that cannot be handled by computational quantum mechanics, the description of the atomic interactions must be approximated. With effective interatomic potentials, the MD method is used to study the properties of material volumes containing billions of atoms. The basic idea is to eliminate all electronic degrees of freedom and to assume that the electrons are part of the nuclei. Thus, the interaction between two atoms is simply represented by a potential function that depends on the atomic configuration (relative displacement) and the local environment (electrons). Based on the electronic structure database, or using experimental measurements of specific properties, approximate effective potentials can be constructed. As mentioned before, using classical Newtonian mechanics, the dynamic evolution of all atoms can be fully determined by numerical integration. In principle, once the positions and velocities of atoms in the finite ensemble within the simulation cell are known, all thermodynamic properties can be readily extracted. The implementation of MD simulations is more involved than the conceptual description illustrated here. As discussed before, a successful simulation depends on three major factors:

- The computational capability of the MD method
- The construction of accurate inter-atomic potentials
- The analysis/interpretation of massive data resulting from computer simulations

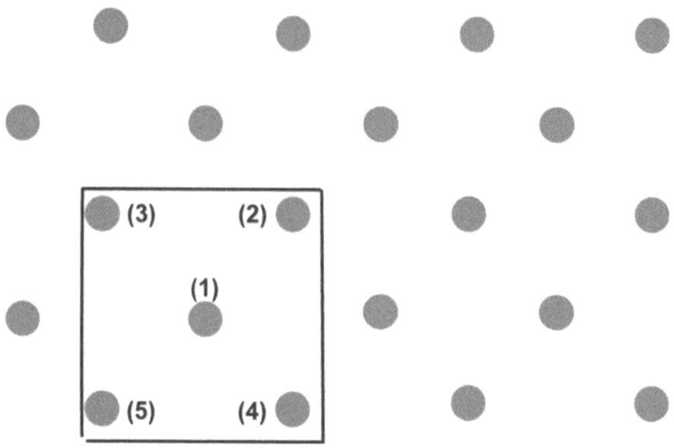

Figure 2.11 The N-atom interaction. Here the atom (1) will interact with the surrounding atoms (atoms 2, 3, 4, and 5).

In general, an atom of mass m_i moves as a rigid particle at the velocity v_i in the effective potential V of other particles. The atomic force F_i is obtained as the negative gradient of the effective potential (Eq. 2.3). Solving these second-order ordinary differential equations for all the atoms in a simulation cell can be performed in a short amount of time. For example, the simulations for 10^9 atoms, applying a simple inter-atomic potential such as LJ, will take less than 10 nanoseconds as a result of short integration time steps in the femtosecond range. In general, there is a compromise between the potential accuracy and the computational Multiscale modeling of nanomechanics and micromechanics efficiency. For high computational efficiency, pair potentials, such as the Lennard-Jones and the Morse[41] potentials, are used.

Below is a simple example of molecular dynamics analysis of N body interaction. First, the interaction of atom (1) with the surrounding masses (mass 2, 3, 4 and 5) will be assessed (Fig. 2.11). In Fig. 2.12, at time t, the distances r_{ij} and hence forces F_{ij} between nearby atoms are calculated. From these forces we can solve the equations of motions and hence, get new positions and velocities. The leapfrog Algorithm mentioned previously (Eqs. 2.5–2.7) together with LJ or Morse potentials can be used to perform MD analysis of N-body interaction shown in Fig. 2.11. The displacement over a time step Δt is denoted with Δr. The Δr has to be much

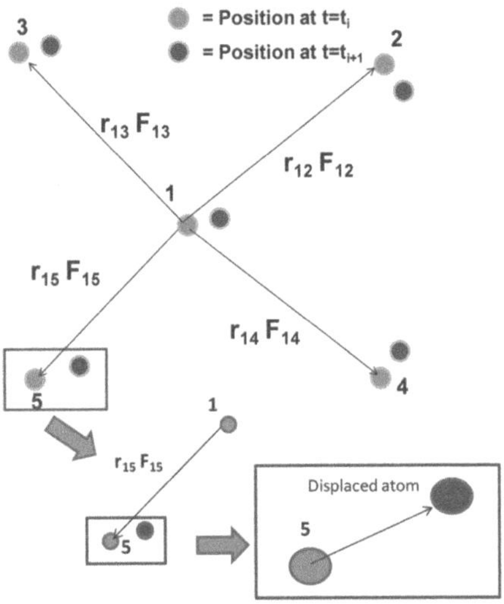

Figure 2.12 Interaction of atom (1) with other atoms in the vicinity. The interaction caused the displacement of atoms as indicated in the figure. *Note*: $\Delta r << r_{54}$, r_{53}, or r_{51}.

smaller than the distance between nearby atoms. Figure 2.12 shows the detailed MD analysis of body (1) and (5). The time interval, Δt, is usually taken to be around 10^{-12} to 10^{-15} (femtosecond) seconds. The quantity r in the FCC structure is around $2.4A^0$, the mass of aluminum atom is $(26.98\ \text{g/mol.})/(6.02 * 10^{23}) = 4.48 * 10^{-23}$ grams, and at the equilibrium position $F = -\nabla V(r) = 0$. The change in energy between its current position and its position a small distance away can provide the force on the atom. This can be recognized as the derivative of the energy with respect to the change in the atom's position (2.10). Knowledge of the atomic forces and masses can then be used to solve for the positions of each atom along a series of extremely small time steps (on the order of femtoseconds $= 10^{-15}$ seconds).

$$\Delta r = \frac{1}{2} a\, \Delta t^2 \quad \text{where} \quad a = F/m \quad \Rightarrow \quad F_i = -\nabla V \qquad (2.10)$$

In Fig. 2.11, if all N atoms interact with all atoms, one has to calculate N^2 interactions. This is largely impossible for $N =$ millions of atoms. Fortunately in practice most atomic interactions decrease rapidly (as $r \to \infty$) and it is enough to calculate only interactions to

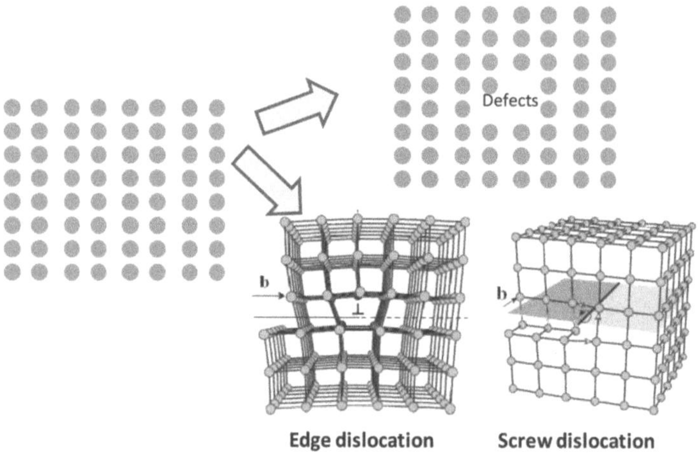

Figure 2.13 In obtaining material properties of metals, the interatomic potential in the vicinity of dislocation/defects must be taken into consideration.

nearby atoms. In metals, atoms more than ~5 Å apart can usually be neglected. That translates into about 80 atoms interactions in 3D apace. Finally, when defects are present in the structure, it must be taken into consideration for the MD analysis (Fig. 2.13). The potential energy around that locality changes and must be reevaluated. The same is true for the case of dislocation when modeling a metal lattice structure. Using the molecular dynamics approach and modeling atoms around the area of interest and using the finite element technique in the continuum region, the MD simulation of a crack tip will be carried out as shown in Fig. 2.14.[42]

In Fig. 2.14, four regions have been defined: (i) the inner MD domain simulates fundamental material processes and each atom is taken into consideration, (ii) the FEM region simulates material as a continuum with small deformation gradients, (iii) the interface MD region performs primary interfacing between MD and FEM, and (iv) the surface MD region corrects bogus forces that are primarily due to the effect of free MD surface. The purpose of this work was to study crack propagation through the grain boundaries using the MD approach in a material free from dislocation. Figure 2.15 illustrates the detailed analysis conducted by reference 42. Using the same

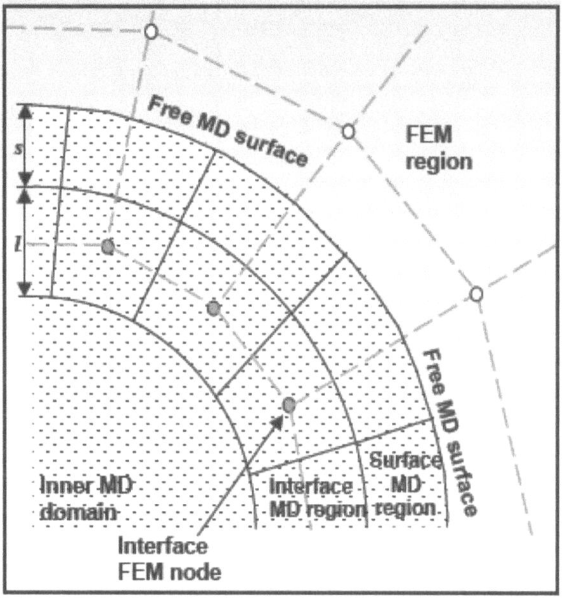

Figure 2.14 Molecular dynamics modeling of a metallic material with the presence of crack where atoms at the crack tip were used for the simulation.[42]

approach but eliminating the presence of crack and replacing it with dislocations in several regions of a grain, it is possible to capture the material properties including the stress–strain curve. Also, the propagation of the dislocation from the point of their nucleation at a grain boundary, through their glide across the grain to the point that they are absorbed into the grain boundary, can be observed when conducting MD simulation.

Another example of simulating material failure for ductile metals, currently used in the aerospace industry, was conducted by modeling up to 1 billion atoms with the help of the world's fastest computer.[43] The White computer at the Lawrence Livermore National Laboratory was used to perform molecular dynamics of ductile failure where the complexity of the formation and interaction of hundreds of dislocations were exposed. In this simulation, the simple inter-atomic force law was adapted. The goal was to investigate the generic features of a particular many-body problem

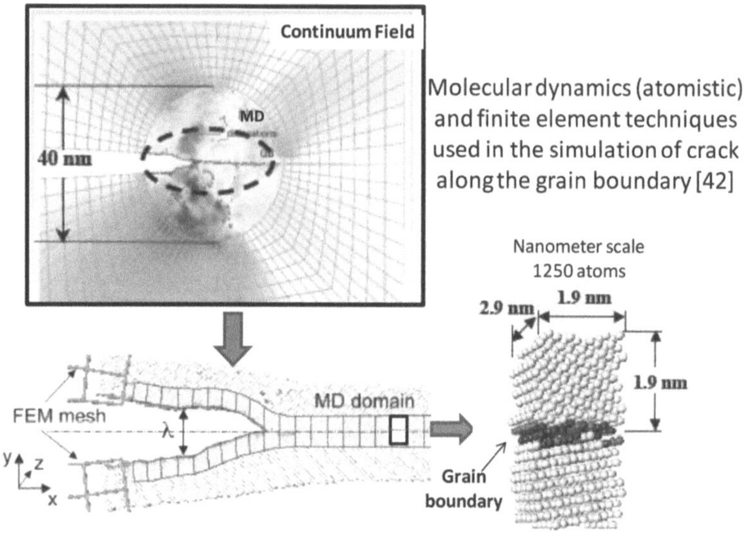

Figure 2.15 Molecular dynamics simulation at the crack tip where crack is propagating between two grain boundaries.

common to a large class of real physical systems and not governed by the particular complexities of a unique molecular interaction. Simulation time was of concern in the billion atoms simulation. If the goal was to use a suitable inter-atomic potential describing the real interaction (e.g., the embedded atom potential, EAM, describing the energy between two atoms for metals), a factor of 3 increases in execution time could result in an excessive computational burden when simulating a large number of atoms. Using the embedded atom potential[44] for the billion atoms simulation required more than one million computer hours and 4 days to complete. Therefore, a simple inter-atomic potential such as the Lennard-Jones 12:6 potential[28] (Eq. 2.8) was used for this simulation. Figure 2.16 shows early stages of the propagating dislocations, growing, and the subsequent collision of the dislocations from the opposing notches (two opposing surface cracks on opposite faces of a three dimensional FCC solid cube that have been applied with mode I loading). The reduced times in a clockwise sequence are 0, 22.5, 45 and 67.5 (reduced time units). Only atoms with a potential

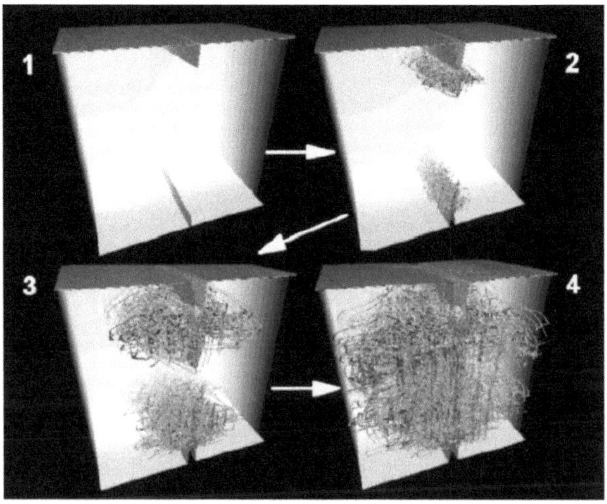

Figure 2.16 Temporal sequence of snapshot pictures for the early time evolution of dislocation dynamics. From two virgin cracks (case 1) subjected to mode I loading, a spaghetti-like network of atomic strings flying from the vertices of the two opposing crack edges. This is simply a large number of mobile dislocations being created at each crack edge, rapidly flowing through the stretched solid in a wiggly manner, and eventually colliding with intersecting dislocations from the opposite edge.

energy less than 97% of the bulk value are displayed, resulting in the selected visualization of atoms neighboring surfaces and dislocations. The total simulation time for this study is 200,000 time steps. It takes 1.7 seconds per time step for a 4,096-node spatial decomposition simulation on the White computer that translates into approximately 4 days of total simulation time. For the simple FCC solid, dislocations are easily created at the apex of the two micocracks where the stress is maximized and easily flows through the solid, giving rise to the ductility of the solid. The appearance of dislocation as described by[43] can be viewed much like a ripple in a rug that is being moved across a floor. In this case, the ripple is only a few atomic spacings in width, and at any instant in time those atoms associated with the ripple are distorted from their lowest energy lattice site. The distortion and associated energy increase are sufficient to be captured by the energy filtering.

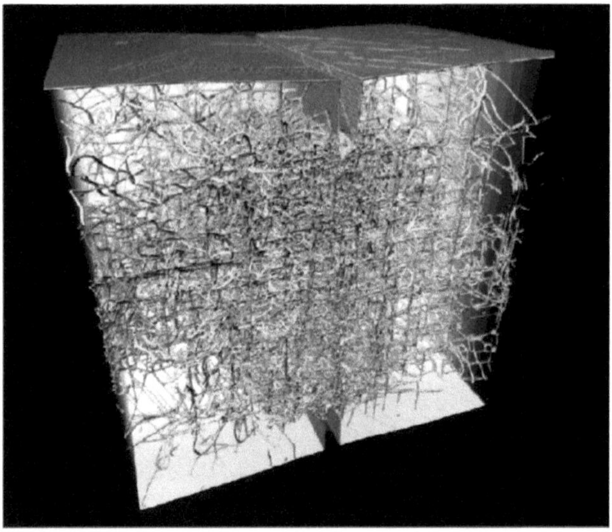

Figure 2.17 Late-time snapshot picture of the dislocation network of the billion atom slab.[43]

Figure 2.17 shows a late-time snapshot of the dislocation distribution in the billion atom slab. A coarse grain cubical skeleton of rigid junctions (sessile dislocations) becomes apparent from a distant view. The molecular simulation also revealed the tangling of dislocation which gives rise to work-hardening phenomenon. It should be emphasized that while a general understanding of dislocation intertwining (obstacles to further dislocation mobility) has been known for quite some time, it is presently impossible to observe the rich microscopic details of the many-body dislocation dynamics in the laboratory. However, using proper molecular dynamics and the ability to compute billions of atoms quickly makes this a reality.

2.7 Summary

The method of obtaining fracture properties from a traditional approach was briefly presented in Section 2.2 Virtual testing using Farahmand's approach estimates fracture properties of metallic material used in the aerospace industry.[1–6] As an input it requires to have data from the full stress–strain curve for the material under consideration. In this chapter, the author suggests that the full stress–strain curve should be generated through the molecular dynamics approach and multiscale modeling technique, which allows virtual testing forego ASTM testing. Achieving this goal will eliminate the cost and time of conducting fracture related tests that are costly to the parts manufacturer (Table 2.1). An attempt was made here to provide a road map to engineers to use the bottom up approach (Fig. 2.3) by studying material properties from the atomistic perspective using the molecular dynamics technique, together with continuum mechanics, to generate the full stress–strain curve. For atomistic modeling to work at the bottom of the scale requires a valid interatomic potential in order to generate a meaningful simulation. Despite today's enhanced computational platforms and sophisticated quantum mechanics techniques, the nanostructures that can be studied are limited at most to a few hundred atoms. A balance must be made between accuracy and feasibility in order to simulate a much larger number of atoms. As was mentioned in Section 2.6, the molecular dynamics simulation of billion atoms is much easier to perform by using the effective interatomic potential such as LJ or Morse potential, rather than the embedded atom potential technique.[44] However, the billion atoms simulation still represents a very small solid, only 0.3 μm per side. Another obstacle is the time interval that has been chosen for atomic simulation. As mentioned previously, the time interval, Δt, is usually taken to be around 10^{-15} seconds (femtosecond), which is close to what it should be. For that reason, the molecular dynamics simulation is difficult to perform. Using larger Δt, will be helpful but may not provide for accuracy in the final computational results. With future computers, hopefully, larger number of atoms and shorter time interval, Δt, will make it possible to simulate material properties through atomistic simulation. Finally, by achieving the ability to generate the stress–strain curve, results will be of the utmost help to the aerospace industry, where the cost and time of testing can greatly be reduced.

References

1. Annual Book of ASTM Standards, Metals Test Methods and Analytical Procedures, Volume 3.01.
2. B. Farahmand. *Fatigue and Fracture Mechanics of High Risk Parts.* Chapman & Hall, 1997, Chapter 5.
3. B. Farahmand. *Fracture Mechanics of Metals, Composites, Welds, and Bolted Joints.* Kluwer Academic, 2000, Chapter 6.
4. B. Farahmand. Predicting fracture and fatigue crack growth properties using tensile properties. *Eng. Fract. Mech.* **75**, 2144–2155 (2008).
5. B. Farahmand, De Xie, and F. Abdi. Estimation of fatigue and fracture allowables for metallic materials under cyclic loading. AIAA-2007.
6. B. Farahmand. Multiscaling of Fatigue: *Application of Virtual Testing for Obtaining Fracture Allowables of Aerospace and Aircraft Materials.* Springer, 2008. Book chapter by G. Sih.
7. E. Orowan. Fracture and strength of solids. *Rep. Prog. Phys.* **12**, 185–232 (1949).
8. Fatigue Crack Growth Computer Program "NASGRO 4.0." JSC, SRI, ESA, and FAA, 2002.
9. B. Farahmand. Virtual testing for estimating material fracture properties (reducing time & cost of testing). 11th International Conference on Fracture (ICF11), Turin Italy, March 2005.
10. B. Farahmand. *Virtual Testing and Predictive Modeling.* Springer, 2009.
11. R. E. Rudd. Plasticity and failure in nanocrystalline BCC metals via molecular dynamics simulation. 16th US National Congress on Theoretical and Applied Mechanics State College, PA, United States, June 27, 2010 through July 2, 2010.
12. A. M. Krivtsov and M. Wiercigroch. Molecular dynamics simulation of mechanical properties for polycrystal materials. *Mater. Phys. Mech.* **3**, 45–51 (2001).
13. H. A. Wu, G. R. Liu, X. Han, and X. X. Wang. An atomistic simulation method combining molecular dynamics with finite element technique. *Chaos, Solitons Fractals* **30**, 791–796 (2006).
14. D. Alfe. Ab initio molecular dynamics, a simple algorithm for charge extrapolation. *Comp. Phys. Commun.* **118**, 31–33 (1999).
15. M. T. Dove. Aluminum: an introduction to atomistic simulation methods. Department of Earth Sciences, University of Cambridge, Downing Street, Cambridge, UK.

16. H. C. Andersen. Molecular dynamics simulations at constant pressure and/or temperature. *J. Chem. Phys.* **72**, 2384 (1980).
17. A. Arkhipov, P. L. Freddolino, and K. Schulten. Stability and dynamics of virus capsids described by coarse-grained modeling. *Structure* **14**, 1767–1777 (2006).
18. A. Y. Shih, P. L. Freddolino, A. Arkhipov, and K. Schulten. Assembly of lipoprotein particles revealed by coarse-grained molecular dynamics simulations. *J. Struct. Biol.* **157**, 579–592 (2007).
19. M. P. Allen. *Introduction to Molecular Dynamics Simulation*. John von Neumann Institute for Computing (NIC), NIC Series, Vol. 23, pp. 1–28 (2004).
20. S. Chatterjee, B. Gersten, S. Thakur, and A. Burin. Molecular dynamics simulation of a single stranded (ss) DNA. *Technical Proceedings of the 2006 NSTI Nanotechnology Conference and Trade Show*, Vol. 2, pp. 305–308 (2006).
21. A. Y. Shih, P. L. Freddolino, A. Arkhipov, and K. Schulten. Coarse grained protein-lipid model with application to lipoprotein particles. *J. Phys. Chem. B* **110**, 3674–3684 (2006).
22. P. D. Dans, A. Zeida, M. R. Machado, and S. Pantano. A coarse grained model for atomic-detailed DNA simulations with explicit electrostatics. *J. Chem. Theory Comput.* **6**, 1711–1725 (2010).
23. A. Arkhipov, Y. Yin, and K. Schulten. Four-scale description of membrane sculpting by BAR domains. *Biophys. J.* **95**, 2806–2821 (2008).
24. S. O. Nielsen, C. F. Lopez, G. Srinivas, and M. L. Klein. A coarse grain model for n-alkanes parameterized from surface tension data. *J. Chem. Phys.* **119**, 7043 (2003).
25. R. E. Rudd and J. Q. Broughton. Coarse-grained molecular dynamics and the atomic limit of finite elements. *Phys. Rev. B* **58**, R5993(R) (1998).
26. A. Y. Shih, P. L. Freddolino, A. Arkhipov, and K. Schulten. A coarse grained protein-lipid model with application to lipoprotein particles. *J. Phys. Chem. B.* Author manuscript, available in PMC 2008, September 15.
27. T. Knotts IV, N. Rathore, and J. de Pablo. A united atom model for simulation of DNA from angstroms to microns in length. American Physical Society, APS March Meeting, March 13–17, 2006.
28. S. Tsuzuki and H. P. Lüthi. Interaction energies of van der Waals and hydrogen bonded systems calculated using density functional theory: assessing the PW91 model. *J. Chem. Phys.* **114**, 3949 (2001).

29. B. J. Alder and C. E. Hecht. Studies in molecular dynamics. VII. Hard, sphere distribution functions and an augmented van der Waals theory. *J. Chem. Phys.* **50**, 2032 (1969).
30. B. J. Alder, D. A. Young, and M. A. Mark. Studies in molecular dynamics. X. Corrections to the augmented van der Waals theory for the square well fluid. *J. Chem. Phys.* **56**, 3013 (1972).
31. T. Knotts, N. Rathore, D. C. Schwartz, and J. de Pablo. A coarse grain model for DNA. *J. Chem. Phys.* **126**, 084901 (2007).
32. J. Chang, W. Cai, V. V. Bulatov, and S. Yip. Dislocation motion in BCC metals by molecular dynamics. *Mater. Sci. Eng.* **309–310**, 160–163 (2001).
33. J. T. Norton, T. Alfrey, Jr., and J. J. Gilman. *Theoretical Strength of Materials*. National Academy of Sciences, National Research Council Washington, D.C., August 1966.
34. M. A. Cuendet and W. F. van Gunsteren. On the calculation of velocity-dependent properties in molecular dynamics simulations using the leap-frog integration algorithm. Laboratory of Physical Chemistry, Swiss Federal, November 14, 2006.
35. W. F. van Gunsteren and H. J. C. Berendsen. A Leap-frog algorithm for stochastic dynamics. *Mol. Simul.* **1**, 173–185 (1988).
36. J. E. Lennard-Jones. Cohesion. *Proc. Phys. Soc.* **43**, 461–482 (1931).
37. S. J. V. Frankland, V. M. Harik, G. M. Odegard, D. W. Brenner, and T. S. Gates. The stress–strain behavior of polymer–nanotube composites from molecular dynamics simulation. *Compos. Sci. Technol.* **63**, 1655–1661 (2003).
38. Y. Sugimoto, P. Jelinek, P. Pou, M. Abel, S. Morital, R. Perez, and O. Custance. Mechanism for room-temperature single-atom lateral manipulations on semiconductors using dynamic force microscopy. *Phys. Rev. Lett.* **98**, 106104 (2007).
39. G. Binnig and C. F. Quate. *Atomic Force Microscope*. The American Physical Society, 1986.
40. S. Morita, R. Wiesendanger, and E. Meyer. *Noncontact Atomic Force Microscopy*. Springer, 1986.
41. L. A. Girifalco and V. G. Weizer. Application of the Morse potential function to cubic metals. *Phys. Rev.* **114**, 687–690 (1959).
42. E. Saether, V. Yamakov, and E. H. Glaessgen. Multiscale modeling of damage processes in fcc aluminum: from atoms to grains. 11th Joint NASA/FAA/DoDConference on Aging Aircraft Phoenix, AZ, April 21–24, 2008.

43. F. F. Abraham, R. Walkup, H. Gao, M. Duchaineau, T. D. De La Rubia, and M. Seager. Simulating materials failure by using up to one billion atoms and the world's fastest computer: work hardening. Communicated by L. B. Freund, Brown University, Providence, RI, January 30, 2002 (received for review September 3, 2001).
44. M. S. Daw and M. Baskes. Embedded-atom method: derivation and application to impurities, surfaces, and other defects in metals. *Phys. Rev. B* 1984.

Chapter 3

Fiber–Matrix Interphase Effects on Damage Progression in Composite Structures

Levon Minnetyan,[a] Xiaofeng Su,[b] and Frank Abdi[b]

[a]*Civil and Environmental Engineering, Mail Stop 5710, Clarkson University, Potsdam, NY 13699-5710, USA*

[b]*AlphaStar Corporation, 5150 E. Pacific Coast Highway, Suite 650, Long Beach, CA 90804, USA*

Fiber–matrix interphase properties affect the composite response and the progressive fracture characteristics, including the appearance of matrix microcracks and evolution to structural fractures. A comprehensive life estimation methodology is developed simulating nonlinear structural degradation of ceramic matrix and other composites with fiber–matrix interphase regions. Hierarchical progressive failure analysis (HPFA) methodology and associated software capable of predicting accurate life of structures under thermomechanical loading is enhanced to accommodate the effects of interphase thickness and properties.

Nomenclature

E_{f11} = fiber longitudinal modulus

Characterization of Nanocomposites: Technology and Industrial Applications
Edited by Frank Abdi and Mohit Garg

ISBN 978-981-4669-02-3 (Hardcover), 978-1-315-36489-6 (eBook)
www.panstanford.com

E_{f22} = fiber transverse modulus
E_m = matrix modulus
ν = Poisson's ratio
S_{f11T} = fiber longitudinal tensile strength
S_{f11C} = fiber longitudinal compressive strength
S_{mT} = matrix tensile strength
S_{mC} = matrix compressive strength
S_{mS} = matrix shear strength

3.1 Introduction

Interphase regions between the fiber and matrix phases are present in several types of composite structures. For example in SiC/SiC ceramic matrix composites, fibers are usually coated with a weak interphase to prevent the propagation of matrix cracks into fiber fractures. Another example is reinforced concrete bridge decks where steel reinforcement bars are coated with epoxy resin to prevent oxidation of steel bars because of the seepage of deicing chemicals. In both cases the effect of an interphase is to modify the fiber–matrix bond properties. Interphase properties affect the progressive fracture characteristics, including the appearance of matrix microcracks and evolution to structural fractures. A comprehensive life estimation methodology is developed simulating nonlinear structural degradation of ceramic matrix and other composites with fiber–matrix interphase regions. Hierarchical progressive failure analysis (HPFA) methodology and associated software capable of predicting accurate life of structures under thermomechanical loading is enhanced to accommodate the effects of interphase thickness and properties. The enhanced software takes into account (1) the effect of thermal expansion and contraction, (2) creep and shrinkage, (3) degradation of a structure represented by the appearance of matrix cracks, (4) mechanical loading, (5) cyclic fatigue, (6) the state of stress/strain in composite structures, (7) interlaminar micro stress, (8) strain energy release rates, (9) quantitative predictions of damage initiation/propagation, progression to fracture and residual strength, and (10) sensitivity of composite failure response to material properties, spectral loading, and geometry.

A hierarchical progressive failure analysis (HPFA) methodology is used to simulate the (1) locations of cracking and failure distribution, (2) life cycle predictions, (3) degradation of a composite structure represented by the changes in natural frequency, (4) instabilities associated with thermal buckling, (5) degradation response of composite structural materials, and (6) cumulative tracking of quantitative predictions of damage initiation, damage progression, propagation to fracture and residual strength. HPFA modeling of composite structures considers the physics, chemistry, and mechanics of materials and structure by integration of (1) a hierarchical multilevel macroscale (structural configuration) and microscale (material constituent) and (2) multidisciplinary analysis, which determines the effects of all structural and loading variables based on the ranges of the energy release rate. Comprehensive durability/life of composite structures is estimated subjected to service thermomechanical loads at high temperatures and extended harsh environments taking into account the frequency response modes to rapidly and cost effectively evaluate service life based on evaluation of cumulative environment effects on durability.

Under thermal/structural loading the solution procedure takes into account the (1) degradation of material properties due to matrix cracking, (2) nonlinear buckling analysis as the change in the component geometry due to critical temperature changes takes place, (3) thermal gradients, and (4) progressive failure mechanisms of high-temperature composite materials (e.g., ceramic matrix composites). The software-based methodology utilizes (1) an incremental layering FEM concept and (2) composite micromechanics in which each ply, ply orientation, fiber volume ratio, void volume ratio, and fiber weave architecture is further divided into sublayers "unit cells" representing the fiber, matrix and interface.

Computational simulation is implemented via the integration of three modules: (1) composite mechanics, (2) finite element analysis, and (3) damage progression tracking. The composite mechanics module is designed to analyze fiber composite structures with an updated composite mechanics theory. Its main function is to calculate ply and composite properties of laminates from the fiber and matrix constituent characteristics and the composite layup. Prior to each finite element analysis, the composite mechanics module utilizes a resident data bank that contains the typical fiber and matrix constituent properties, computes the composite properties and synthesizes the laminate generalized force–displacement relations according to the composite layup.[1] Additionally, the composite mechanics module determines the composite structural response and ply stresses from the FEM analysis results. In simulation, the composite mechanics module is called before and after each finite element analysis. The finite element analysis module is capable of linear and nonlinear static and dynamic analysis. Four-node anisotropic thick-shell elements with six degrees of freedom per node and eight-node three-dimensional solid elements with three degrees of freedom per node are used to model composites. The finite element analysis module accepts laminate properties from the composite mechanics module and performs the structural analysis at each load increment. After a finite element analysis, the computed generalized nodal force and moment time histories are supplied to the composite analysis module that evaluates the nature and amount of local damage, if any, in the plies of the composite laminate. The evaluation of local damage due to multidisciplinary loading environment is based on simplified mathematical models embedded in the composite mechanics module.[1]

The generalized stress–strain relationships are revised locally according to the composite damage evaluated after each finite element analysis. The model is automatically updated with a new finite element mesh having reconstituted properties, and the structure is reanalyzed for further deformation and damage. If there is no damage after a cyclic load increment, the structure is considered to be in equilibrium and an additional number of cycles are applied leading to possible damage growth, accumulation, or propagation. Simulation under multidisciplinary loading is continued until structural fracture.

3.2 Effect of Interphase on Composite Mechanics

Assume a square-packed array of fibers in a unidirectional lamina. Each fiber will have a square region of matrix with side a surrounding it, as shown in Fig. 3.1. Considering a unit depth into

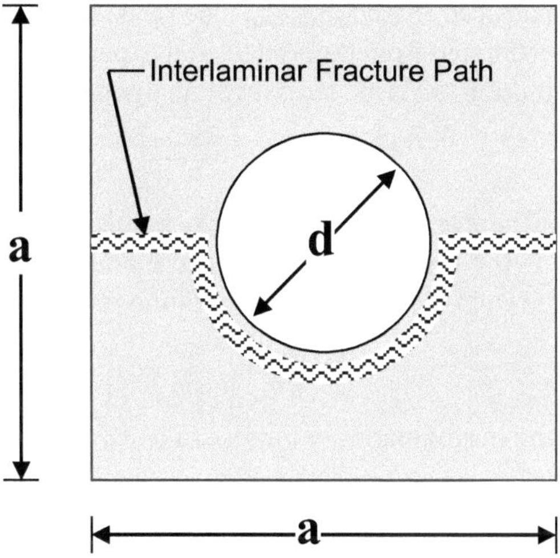

Figure 3.1 Interlaminar fracture path for square-packed unidirectional composite.

the figure, the fiber volume ratio is written as

$$k_f = \frac{\pi d^2}{4a^2} \quad \text{or} \quad \frac{d}{a} = 2\sqrt{\frac{k_f}{\pi}} \tag{3.1}$$

An interlaminar weak layer will be formed due to the fracture path passing through the weak fiber–matrix interphase as shown in Fig. 3.1. The length of the fracture path through the matrix is a minus d. Substituting for d in terms of a, we obtain

$$a - d = a - a\left(2\sqrt{\frac{k_f}{\pi}}\right) = \left(1 - 2\sqrt{\frac{k_f}{\pi}}\right)a \tag{3.2}$$

The length of the fracture path through the fiber–matrix weak interphase is

$$\frac{\pi d}{2} = \frac{\pi}{2}a(2)\sqrt{\frac{k_f}{\pi}} = a\sqrt{\pi k_f} \tag{3.3}$$

Interlaminar strength will be the sum of matrix in situ strength (taking into account the stress concentration effects) when the fracture line goes through the matrix plus the fiber–matrix interphase strength when the fracture line goes through the fiber–matrix interphase. Stress concentrations will also play a significant role on the in situ matrix strength $S_{\mathrm{mT(in\ situ)}}$. The stress concentration factor, S_{cc}, can be estimated from theoretical and experimental assessment of the distribution of elastic stresses at an open hole[2] as

$$S_{\mathrm{cc}} = \frac{\sigma_{\max}}{\sigma_n} = \frac{3\kappa - 1}{\kappa + 0.3} \tag{3.4}$$

where σ_n is the average stress over the weakened cross-sectional area and κ is the ratio of the width a to diameter d, as shown in. κ can be written in terms of the fiber volume ratio k_{f} as

$$\kappa = \frac{a}{d} = \sqrt{\frac{\pi}{4k_{\mathrm{f}}}} \tag{3.5}$$

Therefore, the in situ matrix strength can be written as

$$S_{\mathrm{mT(in\ situ\ )}} = S_{\mathrm{mT}}/S_{\mathrm{cc}} \tag{3.6}$$

where S_{mT} is the matrix tensile strength. Designating the fiber–matrix interface (fiber coating) tensile strength as S_{coatingT}, the total interlaminar force, F_a, that can be transmitted over the distance a in with unit depth is given by the sum

$$F_{\mathrm{a}} = S_{\mathrm{mT(in\ situ)}} \left(1 - 2\sqrt{\frac{k_{\mathrm{f}}}{\pi}}\right) a + S_{\mathrm{coatingT}} a \sqrt{\pi k_{\mathrm{f}}} \tag{3.7}$$

The composite interlaminar tensile strength S_{ilT} is obtained by dividing F_a by the area $1 \times a$. Therefore,

$$S_{\mathrm{ilT}} = S_{\mathrm{mT(in\ situ)}} \left(1 - 2\sqrt{\frac{k_{\mathrm{f}}}{\pi}}\right) + S_{\mathrm{coatingT}} \sqrt{k_{\mathrm{f}}} \tag{3.8}$$

We note that composite micromechanics also takes into account the stress concentration effects in computing transverse tensile strength S_{l22T} of a lamina as

$$S_{\mathrm{l22T}} = \frac{S_{\mathrm{mT}}}{\left[1 - \sqrt{k_{\mathrm{f}}}\left(1 - \frac{E_{\mathrm{m}}}{E_{\mathrm{f22}}}\right)\right] \sqrt{1 + \varphi(\varphi - 1) + \frac{1}{3}(\varphi - 1)^2}} \tag{3.9}$$

where

$$\varphi = \frac{\sqrt{\frac{\pi}{4k_{\mathrm{f}}}} - \frac{E_{\mathrm{m}}}{E_{\mathrm{f22}}\left[1 - \sqrt{k_{\mathrm{f}}}\left(1 - \frac{E_{\mathrm{m}}}{E_{\mathrm{f22}}}\right)\right]}}{\sqrt{\frac{\pi}{4k_{\mathrm{f}}}} - 1} \tag{3.10}$$

If we substitute S_{l22T} instead of $S_{\mathrm{m(in\ situ)}}$ in Eq. 3.8, the calculated interlaminar tensile strength S_{ilT} should be similar.

The interlaminar shear strength is computed by micromechanics as

$$S_{l12} = \frac{\left[\left(\sqrt{\frac{\pi}{4k_f}} - 1 + {}^{G_m}/_{G_{f12}}\right)\left(1 - \sqrt{\frac{4k_v}{\pi k_m}}\right) G_{l12} S_{ms}\right]}{G_m \sqrt{\frac{\pi}{4k_f}}} \tag{3.11}$$

Similar to Eq. 3.1, we can write the interlaminar shear strength S_{ilS}, including the weak fiber coating effect as

$$S_{ilS} = S_{l12}\left(1 - 2\sqrt{\frac{k_f}{\pi}}\right) + S_{coatingS}\sqrt{k_f} \tag{3.12}$$

3.3 Uniaxial Composite Results

Carbon fiber reinforced Epon laminate is used to demonstrate the effects of relative interphase properties on the uniaxial laminate strengths (transverse tensile, in-plane shear, interlaminar tensile and interlaminar shear strengths). The nominal carbon fiber properties are given as follows: longitudinal modulus $E_{f11} = 186$ GPa (27 Msi), transverse modulus $E_{f22} = 12.4$ GPa (1.8 Msi), Poisson's ratios $\nu_{f12} = 0.22$, $\nu_{f23} = 0.20$, tensile strength $S_{f11T} = 2.14$ GPa (0.31 Msi), compressive strength $S_{f11C} = 1.24$ GPa (0.18 Msi). The Epon matrix has an elastic modulus of $E_m = 3.1$ GPa (0.45 Msi) and a Poisson's ratio of $\nu_m = 0.4$.

Figures 3.2–3.6 show the different ply strengths due to the variation of interphase properties. Figure 3.2 shows the ply transverse tensile strength as a linear function of the interphase tensile strength.

Next, we look at a graphite fiber-reinforced Epon triaxial composite laminate. This example is a triaxial polymer composite with linear constituents. The objective is to demonstrate the effect of interphase defect or debonding on the mechanical response of a composite system with complex fiber architecture. The fiber architecture is [45/90/−45]s fabric. Tension is selected as the 0° direction in simulation in order not to conceal the interphase effect by the fiber response. The graphite fiber properties are given as follows: longitudinal modulus $E_{f11} = 241$ GPa (35 Msi), transverse modulus $E_{f22} = 3.5$ GPa (0.5 Msi), Poisson's ratios $\nu_{f12} = 0.4$,

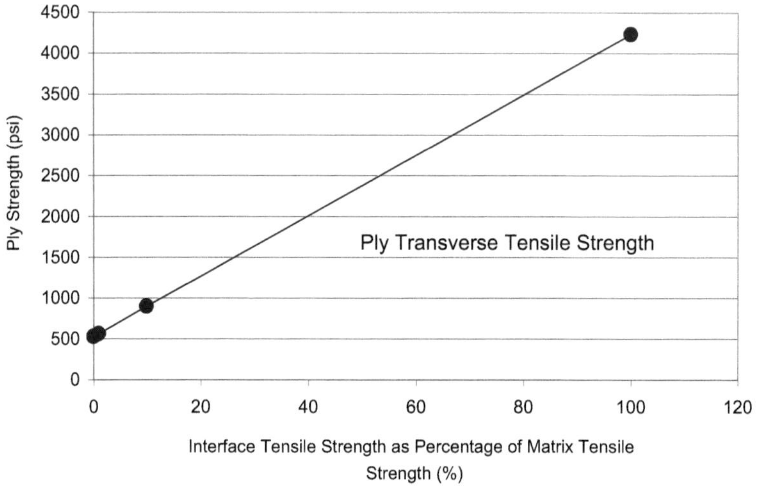

Figure 3.2 Effect of Interphase tensile strength on Akzo-Epon uniaxial laminate (ply) transverse tensile strength (1.0 psi = 6894.757 Pa).

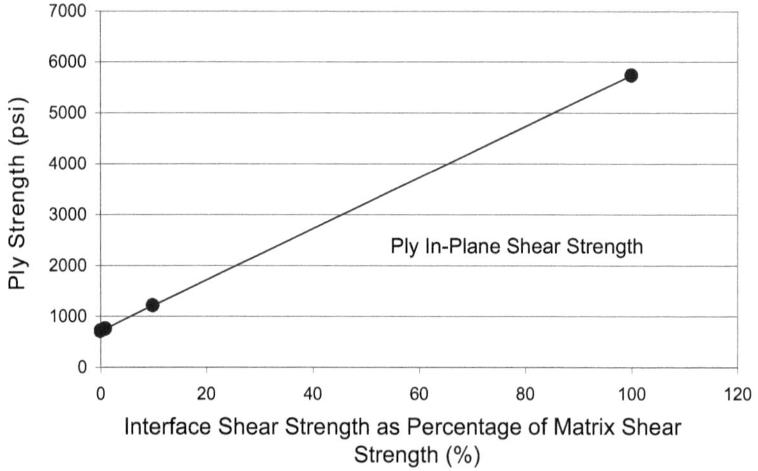

Figure 3.3 Effect of Interface shear strength on Akzo-Epon uniaxial laminate (ply) in-plane shear strength (1.0 psi = 6894.757 Pa).

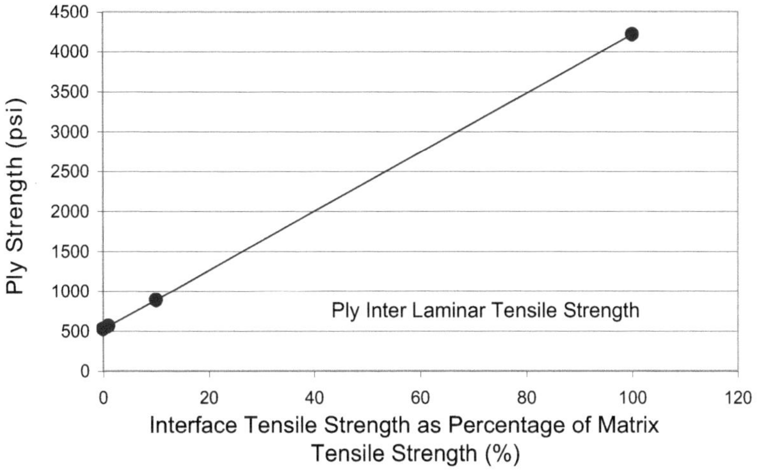

Figure 3.4 Effect of Interphase tensile strength on Akzo-Epon interlaminar tensile strength (1.0 psi = 6894.757 Pa).

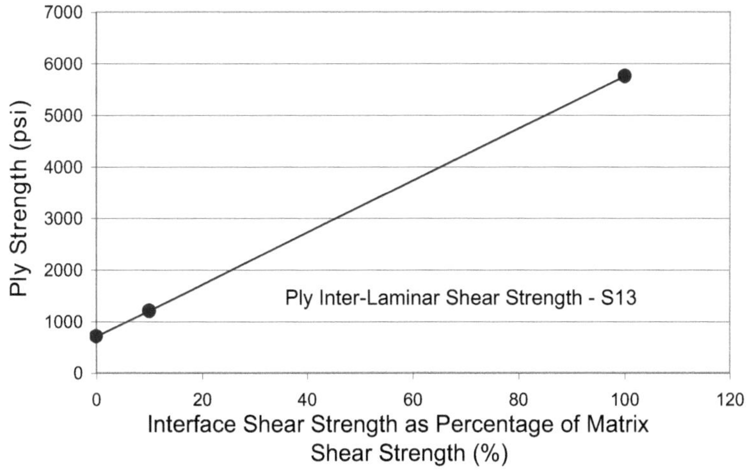

Figure 3.5 Effect of Interphase shear strength on Akzo-Epon inter laminar shear strength (S13 with respect to ply longitudinal and normal directions) (1.0 psi = 6894.757 Pa).

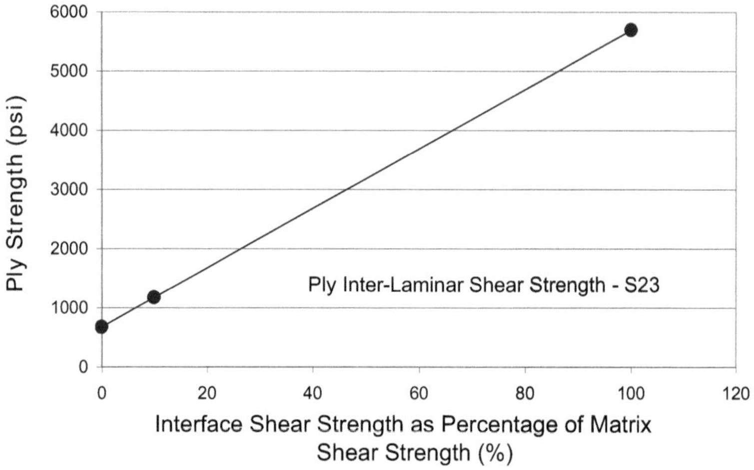

Figure 3.6 Effect of Interphase shear strength on Akzo-Epon interlaminar shear strength (S23 with respect to ply transverse and normal directions) (1.0 psi = 6894.757 Pa).

$\nu_{f23} = 0.3$, tensile strength $S_{f11T} = 3.24$ GPa (0.47 Msi), compressive strength $S_{f11C} = 2.1$ GPa (0.3 Msi). The matrix properties are identical with the matrix in the first example.

Figure 3.7 illustrates the effects of interphase tensile and shear strengths on the triaxial composite ([45/90/−45]s) tensile strength along 0° direction. For the convenience of description in this paper, we define the 0° as the laminate transverse direction and 90° as its longitudinal direction. It can be seen that the interface tensile strength has more significant effect on the transverse tensile strength of this triaxial composite laminate than its shear strength.

Next we look at the response of E-glass fiber reinforced triaxial composite laminate with nonlinear Dion resin matrix.[3] The E-glass fiber properties are longitudinal or transverse modulus $E_{f11} = E_{f22} = 89.6$ GPa (13 Msi), Poisson's ratios $\nu_{f12} = \nu_{f23} = 0.20$, tensile strength $S_{f11T} = 2.03$ GPa (0.295 Msi), compressive strength $S_{f11C} = 1.45$ GPa (0.21 Msi). The Dion resin has an elastic modulus of $E_m = 3.21$ GPa (0.466 Msi), a Poisson's ratio of $\nu_m = 0.42$, a tensile strength S_{mT} of 76.4 MPa (11.1 Ksi), compressive strength S_{mC} of 241 MPa (35.0 Ksi), and a shear strength S_{mS} of 89.6 MPa (13.0 Ksi).

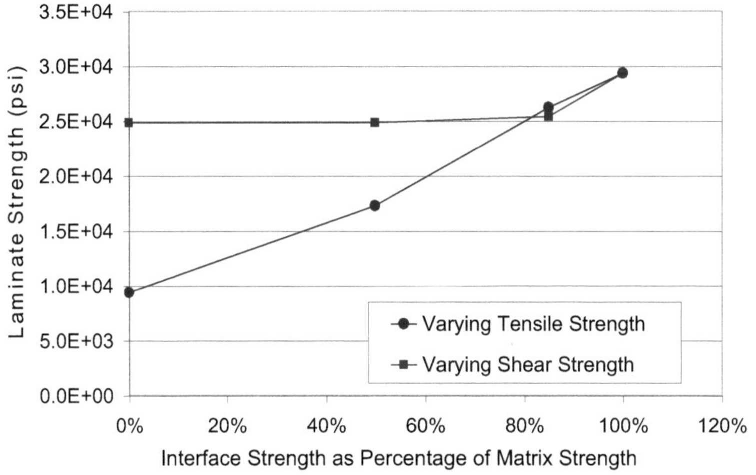

Figure 3.7 Effects of interface tensile and shear strengths on graphite fiber reinforced Epon triaxial composite ([45/90/−45]s) tensile strength. Load is along 0° direction. (1.0 psi = 6894.757 Pa).

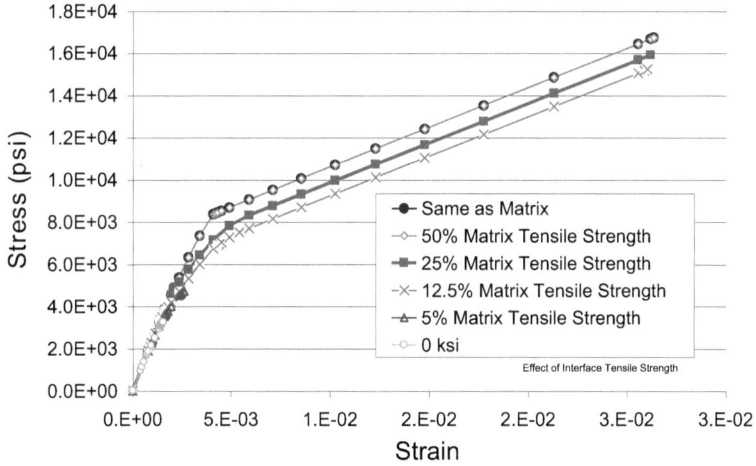

Figure 3.8 Effect of Interphase tensile strength on glass fiber reinforced Dion triaxial composite ([45/90/−45]s) tensile behavior. Load is along 0° direction. (1.0 psi = 6894.757 Pa).

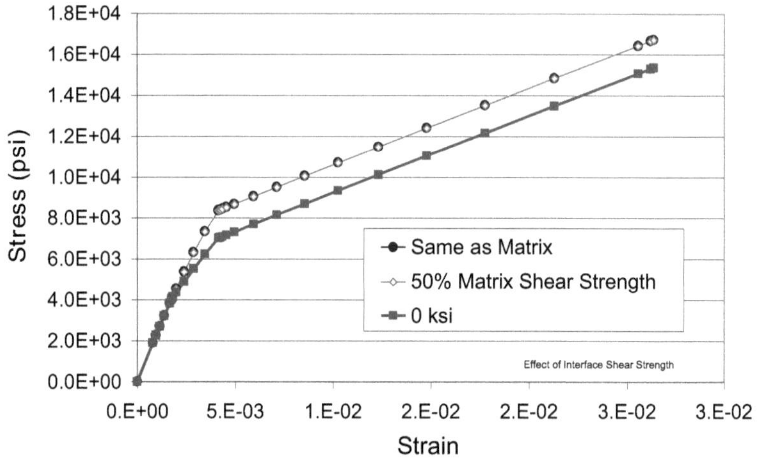

Figure 3.9 Effect of Interphase shear strength on glass fiber reinforced Dion triaxial composite ([45/90/−45]s) tensile behavior. Load is along 0° direction. (1.0 psi = 6894.757 Pa).

The fiber architecture for the E-glass reinforced Dion resin is triaxial ([45/90/−45]s), identical to the configuration of the graphite fiber–Epon composite in the above examples. The simulated tension direction was along the 0° direction. A parametric study was performed to understand the effect of varying the interfacial tensile or shear strength on the triaxial composite nonlinear tensile response.

As illustrated in Figs. 3.8 and 3.9, the proportional limit and ultimate strength of the composite decreased as the interphase tensile or shear strength was reduced. However, the weak interphase did not change the slope of each stress–strain curve at the work hardening stage.

3.4 Conclusions

Theory and representative unidirectional and multidirectional composite results were discussed for assessing the effect of fiber–matrix interphase properties on composite response under loading. The results can be summarized as follows:

1. The interphase properties mainly affect the composite tensile and shear strengths perpendicular to the fiber direction, i.e., transverse tensile strength, in-plane shear strength, interlaminar tensile strength, and interlaminar shear strengths.
2. Interphase properties need be separately defined in addition to the fiber and matrix properties and represented in the composite mechanics module. Material properties of the interphase need to be provided in the databank file.
3. For a nonlinear composite with complex fiber architecture, e.g., triaxial, the proportional limit and ultimate strength decrease as the interfacial tensile or shear strength decreases. However, the weak interface does not affect the slope of each stress–strain curve at the work hardening stage.

The approach described here is inclusive in that it integrates composite mechanics (for composite behavior) with finite element analysis (for global structural response) and incorporates progressive fracture algorithms to simulate composite structural fracture. The effect on the composite structural damage of the interphase properties is accounted for at all composite scales. The methods/codes are demonstrated by application to uniaxially and multiaxially reinforced composite specimens with a fiber–matrix interphase region Examples are subjected to monotonically increasing loading.

References

1. Murthy, P. L. N., and Chamis, C. C., *Integrated Composite Analyzer (ICAN): Users and Programmers Manual*, 1986, NASA Technical Paper 2515.
2. Boresi, A. P., Schmidth, R. J., and Sidebottom, O. M., *Advanced Mechanics of Materials*, 5th ed., John Wiley & Sons, 1993, p. 566.
3. Lee, M., Short, M., Abdi, F., and Qian, J., "A Math Based Methodology for Longevity Fatigue Prediction of 3D Woven Fiber Glass Reinforced Vynleester Composite." SAE, 2006, Detroit, MI. 06M29-M28 Fatigue.

Chapter 4

Composite Nanomechanics: A Mechanistic Properties Prediction

Christos C. Chamis*

Former Senior Aerospace Scientist, NASA Glenn Research Center, Cleveland, OH 44135, USA

christos.c.chamis@aol.com

A unique mechanistic theory is described to predict the properties of nanocomposites. The theory is based on composite micromechanics with progressive substructuring down to a nanoscale slice of a nanofiber where all the governing equations are formulated. These equations have been programmed in a computer code. That computer code is used to predict 25 properties of a mononanofiber laminate. The results are presented graphically and discussed with respect to their practical significance. Most of the results show smooth distributions. Results for matrix-dependent properties show bimodal through-the-thickness distribution with discontinuous changes from mode to mode.

4.1 Introduction

The research in the nanoscale technology has exploded over the recent past. An indication of this explosion is that the

*Retired.

Characterization of Nanocomposites: Technology and Industrial Applications
Edited by Frank Abdi and Mohit Garg

ISBN 978-981-4669-02-3 (Hardcover), 978-1-315-36489-6 (eBook)
www.panstanford.com

SAMPE (Society of Aerospace Material and Processing Engineers) Conference is devoting four sessions of about six papers each in the last three years. These papers cover practically all current research activities. The majority of the research is devoted to processing because of the difficulties involved in making a useful material.[1] A few investigators have been fortunate to make some testing samples, which they subsequently tested to obtain limited data.[2] A few other investigators researched the characterization of fatigue[3] and creep.[4] A couple of papers explored the construction of nanocomposites for rocket ablative material[5] and for carbon nanotubes for adaptive structures.[6] One paper ventured to describe a computer simulation of macroscopic properties of carbon nanotubes polymer composites.[7] However, there are no results of what special macroscopic properties are included. Reference 7 shows one stress strain curve and citation of several references. One recent article[8] describes multiscale modeling and simulation of nanostructural materials from atomistic to micromechanics. This article does not include information on nanocomposites, but it mentions that mechanistic models will be needed in the end. It is becoming abundantly clear that no holistic approach has been used to investigate the mechanistic prediction of all nanocomposite uniaxial properties: fabrication parameters (3), physical (10), mechanical moduli (6), and uniaxial strengths (6),[9] which is the objective of the research reported herein.

Herein a unique mechanistic method is described, which is used to predict all 25 nanocomposite properties. The method is unique because it is based on composite mechanics that are reduced down to nanoscale by progressive substructuring. At the nanoscale the same assumptions are made that are consistent with those made in the composite mechanics. The method is illustrated by applying it to an assumed nanofiber aligned laminate of one nanofiber diameter ($2.756 \times 10^{m-6}$ in.). The properties are then predicted by all the composite micromechanics equations, which are formulated on a thin slice of the substructured nanofiber. The substructuring and the equations are programmed in an in-house computer code called ICAN/JAVA.[10] A logic block diagram of ICAN/JAVA for application to nanocomposites is shown in Fig. 4.1. It is assumed that the nanofiber is a graphite fiber (Pyrograf II, Applied Sciences Incorporated)

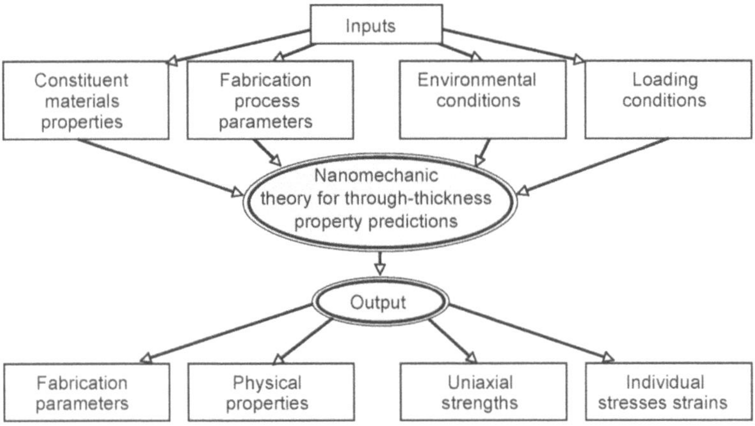

Figure 4.1 Mechanistic approach to nanocomposites mechanics.

with modulus 1.0×10^9 psi and a tensile strength of 0.8 million psi. The remaining properties are estimated to be those from a Thornel 300 fiber (Table 4.1) and Intermediate-Modulus High Tensile Strength Epoxy Matrix (Table 4.2). The fiber volume ratio is about 0.05, and is about the same as that for the Pyrograf (private communication).

It is important that the reader keeps in mind that the emphasis herein is on the method used. The properties obtained are for illustrative purposes only and will depend on the specific input (Tables 4.1 and 4.2). The simplified equations used[11,12] are summarized before each corresponding set of results presented.

4.2 Fundamentals

The fiber alignment with uniform dispersion is not met in nanocomposites. It is assumed herein that the fibers are aligned only for predicting "point" through-the-thickness properties. The fussiness can be simulated by estimating the angle of single fibers through the thickness. Therefore, it is assumed that an aligned unidirectional typical section of a nanocomposite is as illustrated schematically in Fig. 4.2 on the left. A nanoply is schematically shown in Fig. 4.2 on the right. It is interesting to note that the substructuring into

Table 4.1 T300 graphite nanofiber (Pyrograf Ii) properties

Description	Symbol	Value	Units
Number of fibers per end	Nf	1.0	number
Filament equivalent diameter	df	2.756×10^{-6}	in.
Weight density	Rhof	0.064	lb/in.**3
Normal moduli (11)	Ef11	1.0×10^{9}	psi
Normal moduli (22)	Ef22	7.0×10^{7}	psi
Poisson's ratio (12)	Nuf12	0.2	Nondimensional
Poisson's ratio (23)	Nuf23	0.25	Nondimensional
Shear moduli (12)	Gf12	5.0×10^{7}	psi
Shear moduli (23)	Gf23	3.5×10^{7}	psi
Thermal expansion coefficient (11)	A1faf11	-5.5×10^{-7}	in./in./°F
Thermal expansion coefficient (22)	Alfaf22	5.6×10^{-6}	in./in./°F
Heat conductivity (11)	Kf11	444.0	Btu/hr/ft^2/°F/in.
Heat conductivity (22)	Kf22	4.0	Btu/hr/ft^2/°F/in.
Heat capacity	Cf	0.22	Btu/lb/°F
Dielectric strength (11)	KeF11	0.0	V/in.
Dielectric strength (22)	Kef22	0.0	V/in.
Dielectric constant (11)	Gamma11	0.0	in./V
Dielectric constant (22)	Gamma22	0.0	in./V
Capacitance	Cef	0.0	V
Resistivity	Ref	0.0	Ω–in.
Tensile strength	SfT	8.0×10^{5}	psi
Compressive strength	SiC	6.0×10^{5}	psi
Shear strength	SfS	4.0×10^{5}	psi
Normal damping capacity (11)	psi11f	0.38	%Energy
Normal damping capacity (22)	psi22f	6.3	%Energy
Shear damping capacity (12)	psi12f	3.34	%Energy
Shear damping capacity (23)	psi23f	6.3	%Energy
Melting temperature	TMf	6000.0	°F

slices the monofiber nanoply is not constrained by the maximum fiber volume ratio, even though the monofiber was assumed to be in a square array with a limiting fiber volume ratio of about 0.78. A block diagram depicting a mechanistic approach to nanoscale mechanics is shown in Fig. 4.1, as was mentioned previously. This diagram shows the three major parts of nanoscale mechanics: (1) input, (2) mechanics theory for through-the-thickness predictions of properties, and (3) output. The input includes the constituent material properties, the fabrication parameters, environmental,

Table 4.2 Intermediate-Modulus High-Strength Matrix (EPOXY)

Description	Symbol	Value	Units
Weight density	Rhom	0.044	lb/in.**3
Normal modulus	Em	500000.0	psi
Poisson's ratio	Num	0.35	Nondimensional
Thermal expansion coefficient	Alfa m	3.6×10^{-5}	in./in./°F
Heat conductivity	Km	0.008681	Btu/hr/ft^2/°F/in.
Heat capacity	Cm	0.25	Btu/lb/°F
Dielectric strength	Kem	0.0	V/in.
Dielectric constant	Gammam	0.0	in./V
Capacitance	Cem	0.0	V
Resistivity	Rem	0.0	Ω−in.
Moisture expansion coefficient	Betam	0.0033	in./in./%moisture
Diffusivity	Dm	2.16×10^{-7}	in.**2/hr
Saturation	Mm	0.0	%moisture
Tensile strength	SmT	15000.0	psi
Compressive strength	SmC	35000.0	psi
Shear strength	SmS	13000.0	psi
Allowable tensile strain	eps mT	0.02	in./in.
Allowable compression strain	eps mC	0.05	in./in.
Allowable shear strain	eps mS	0.035	in./in.
Allowable torsional strain	eps mTOR	0.035	in./in.
Normal damping capacity	psiNM	6.6	%energy
Shear damping capacity	psiSm	6.9	%energy
Void heat conductivity	Kv	0.0012	Btu/hr/in./°F
Glass transition temperature	Tgdr	420.0	°F
Melting temperature	TMm	0.0	°F

and the loading conditions. The nanomechanics theory includes all the equations that are required to predict the output. The output includes the fabrication parameters as are present in the nanocomposite, the physical properties, the mechanical properties, and individual uniaxial strengths.

The properties prediction is expedited by the following geometric diagrams: (1) A typical section of the nanocomposite shown in Fig. 4.2(a) and a nanoply in Fig. 4.2(b). (2) An exploded view of nanoscale isolation of a typical part is shown in Fig. 4.3 with nanoscale dimensions. (3) A single nanofiber schematic with substructuring is shown in Fig. 4.4(a), and a typical subslice is shown in Fig. 4.4(b). A nanosubply with its corresponding stresses

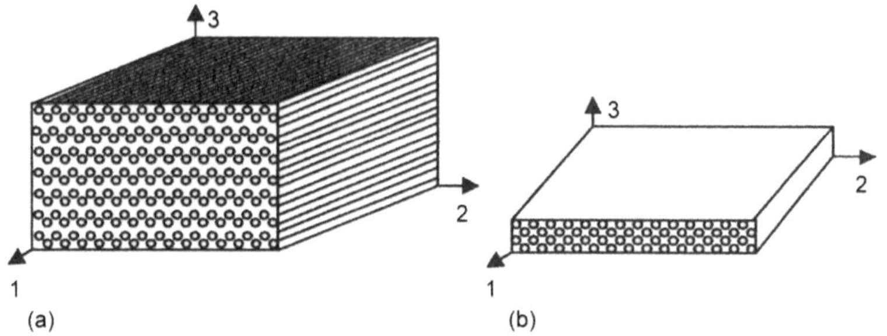

Figure 4.2 Unidirectional nanocomposite typical section. (a) Nanocomposites. (b) Nanoply.

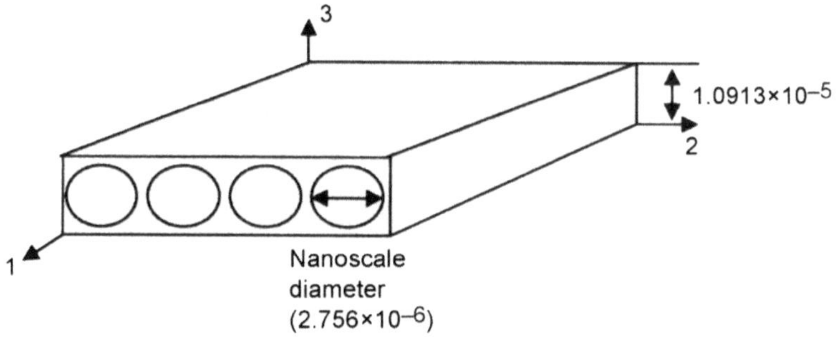

Figure 4.3 Nanoscale isolation of a typical part (units are in inches).

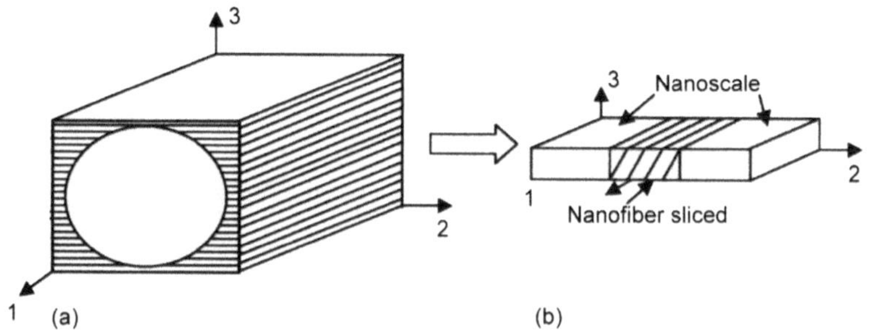

Figure 4.4 Nanofiber substructuring. (a) Several slices through the thickness. (b) Nanofiber sliced.

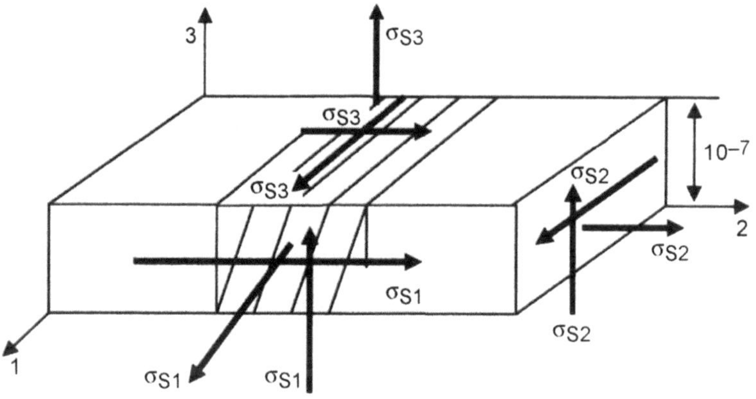

Figure 4.5 Nanostresses on a nanosubply (units are in inches).

is shown in Fig. 4.5. All the nanomechanics predictive equations are derived by using Fig. 4.5. The equations used are extensive; they are all programmed in ICAN/JAVA.[10] Details are not explicitly shown here because the emphasis is on the results obtained and their practical significance. A simplified form of the equation is summarized prior to predicted results. The equations are for (1) In situ nanofabrication parameters—fiber volume ratio, matrix volume ratio, and void volume ratio; (2) Nanophysical properties—density, heat capacity, heat conductivities, diffusivities, moisture expansion coefficients, and thermal expansion coefficients; (3) Mechanical properties—normal moduli, shear moduli, and Poisson's ratios; and (4) Individual uniaxial strengths, as shown in Fig. 4.5. As previously mentioned, all these equations are programmed in ICAN/JAVA and are available for obtaining the results that are subsequently described. This approach has the unique advantage that it can be used directly to predict nano, micro, macro, and structural properties of composites as required in item 7 of Ref. 12.

Prior to describing the results obtained, it is instructive to describe the interface and how it is modeled. The schematics in Fig. 4.6 show a vertical slice with unit thickness of the nanocomposite with a single fiber in it. As can be seen in this slice,

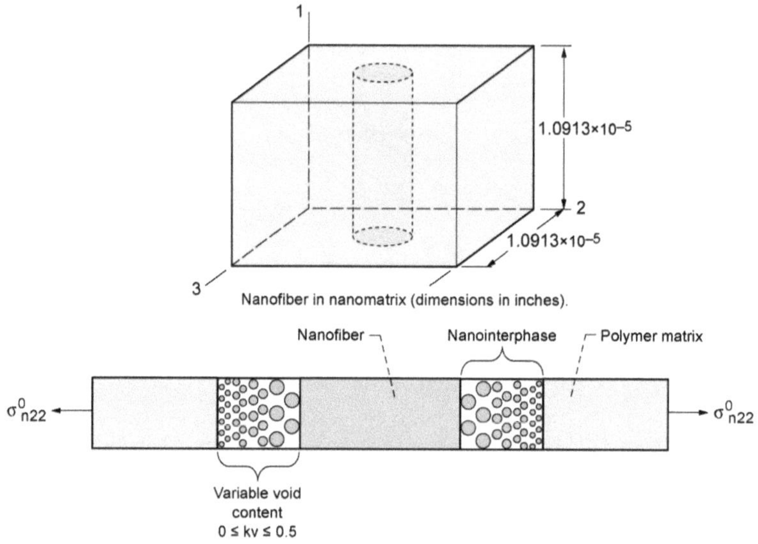

Figure 4.6 Vertical section of a composite nanocell through nanofiber center.

the fiber interface is represented by a series of progressively larger volume voids starting with the smallest near the matrix interface and ending with the fiber interface. It can be visualized by cutting a vertical section through and part in the interface that the resisting force will be equal to the applied force. However, the stress in the matrix will be magnified because of the voids. This magnification is shown in Fig. 4.7 for a specific nanocomposite with 0.05-fiber volume ratio and with void volume ratio varying from 0.05 to 0.4.

The results that will be presented are based on a typical composite for structural application. That is, the fiber volume ratio is assumed to be 0.05 and a monofiber ply thickness of 7×10^{-5} in. thick. The number of nanofibers of diameter of 2.756×10^{-6} in. to fill a ply thickness of 0.005 in. thick by 1 in. wide with a volume ratio of 0.05 (typical of fiber nanocomposites) is about 42×10^{6}, a large number. However, when those numbers become practical, the herein described procedure to predict the respective properties will be available.

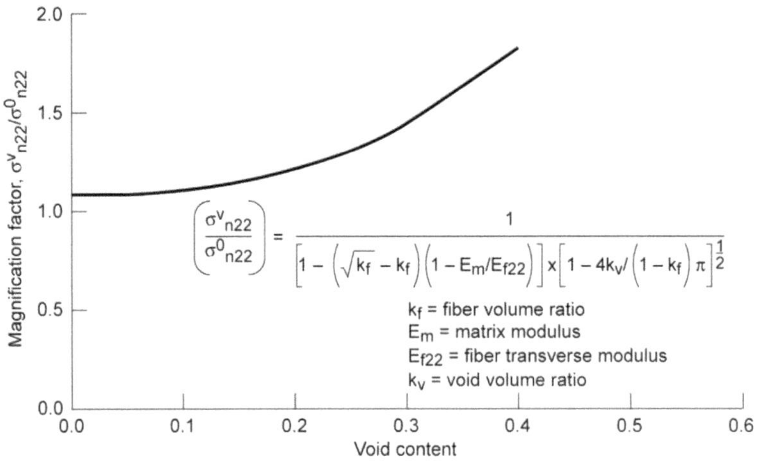

Figure 4.7 Nanocomposite magnification factor.

4.3 Results and Discussion

The results obtained are by assuming there is no interphase and are discussed in the following order: (1) in situ fabrication parameters; (2) physical variables; (3) mechanical variables; and (4) individual uniaxial strengths.

4.3.1 *In situ Fabrication Parameters*

The in situ fabrication parameters include the fiber volume ratio, the matrix volume ratio, and the voids volume ratio. The equations used for these parameters are summarized in Fig. 4.8. The in situ fiber volume ratio is graphically plotted through the nanoply thickness in Fig. 4.9. As can be seen, it starts at zero and rises up to 0.25. The fiber volume ratio distribution illustrated in Fig. 4.9 is a result of the substructuring of the single fiber into 10 subslices. The practical significance of that distribution is that the damage will most probably initiate at the center of the laminate. Figure 4.9 is also instructive in interpreting the input fiber volume ratio of 0.05, which shows that the in situ fiber volume ratio will be a weighted average compared to its through-the-thickness distribution.

Partial volumes: $k_f + k_m + k_v = 1$

Ply density: $\rho_\ell = k_f \rho_f + k_m \rho_m$

Resin volume ratio: $k_m = (1 - k_v)/[1 + (\rho_m/\rho_f)(1/\lambda_m - 1)]$

Fiber volume ratio: $k_f = (1 - k_v)/[1 + (\rho_f/\rho_m)(1/\lambda_f - 1)]$

Weight ratios: $\lambda_f + \lambda_m = 1$

Ply thickness (S.A.): $t_\ell = 1/2\, N_f\, d_f \sqrt{\pi/k_f}$

Interply thickness: $\delta_\ell = 1/2\left[\sqrt{\pi/k_f} - 2\right] d_f$

Interfiber spacing (S.A.): $\delta_s = \delta_\ell$

Contiguous fibers (S.A.): $k_f = \pi/4 \sim 0.785$

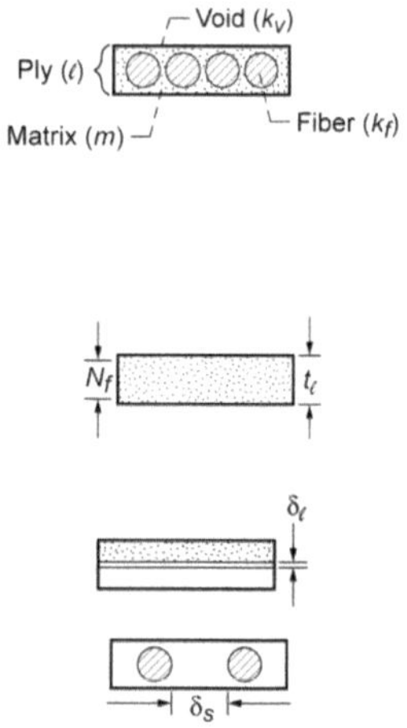

Figure 4.8 Micromechanics and geometric relationships.

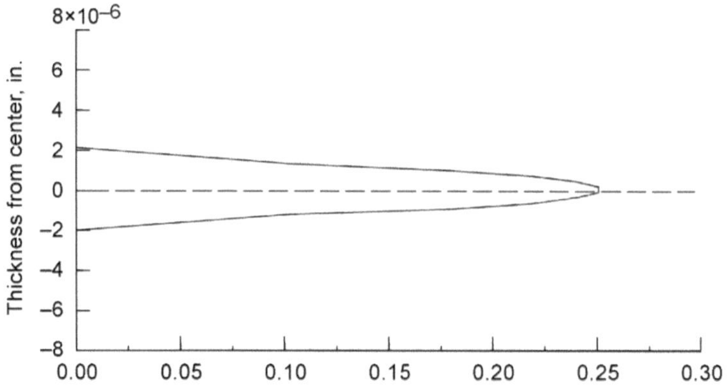

Figure 4.9 Nanoscale in situ fabrication parameter, k_f.

The in situ matrix volume ratio is shown graphically in Fig. 4.10. Its distribution through the nanolaminate thickness is the complement of the in situ fiber volume ratio. It starts at 1.0, at the matrix region, and decreases progressively to about 0.75 at the center. This very small amount of matrix volume ratio will definitely

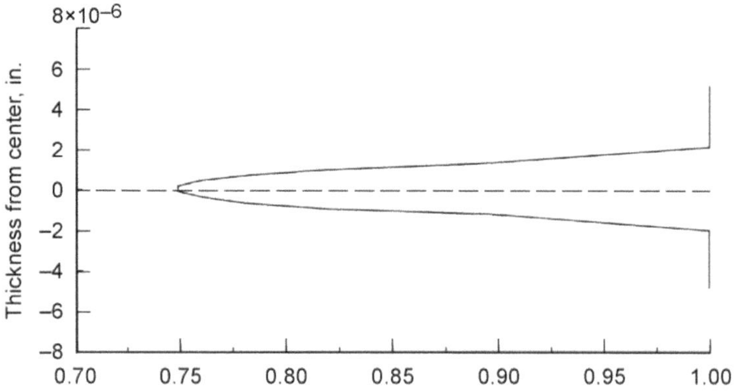

Figure 4.10 Nanoscale in situ fabrication parameter, k_m.

cause stress risers at this point and damage to initiate. It is also interesting to note that the average matrix volume ratio of 0.95 is a weighted average through the thickness as can be deduced from Fig. 4.10. Because the two distributions are complementary, the sum of the two must equal one at any one-fiber/matrix volume ratio in the absence of voids. These plots are presented here to illustrate their variation through the thickness of a single-ply nanocomposite.

The void volume ratio is plotted in Fig. 4.11. Note that the void volume ratio is constant through the thickness. It is constant because it is assumed in the theoretical development that in each slice the

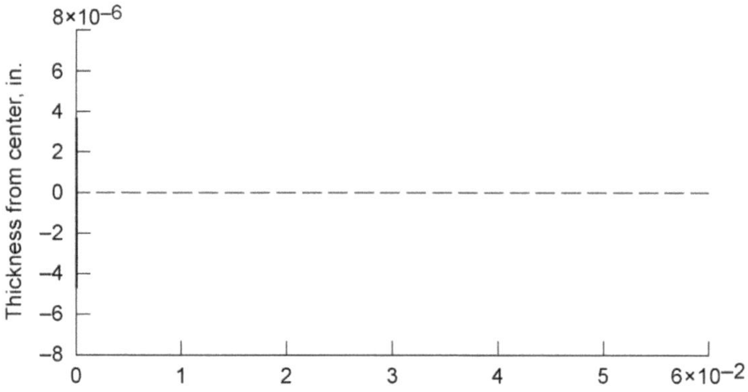

Figure 4.11 Nanoscale in situ fabrication parameter, k_v.

Heat capacity: $$C_\ell = \frac{1}{\rho_\ell}\left(k_f\, \rho_f\, C_f + k_m\, \rho_m\, C_m\right)$$

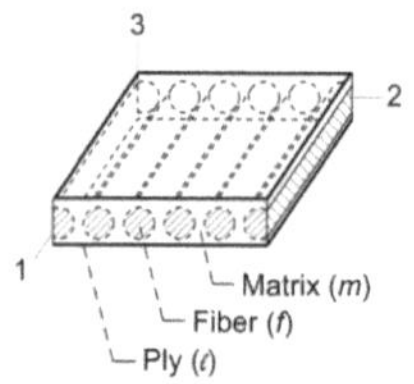

Longitudinal conductivity: $$K_{\ell 11} = k_f K_{f11} + k_m K_m$$

Transverse conductivity: $$K_{\ell 22} = \left(1 - \sqrt{k_f}\right)K_m + \frac{K_m \sqrt{k_f}}{1 - \sqrt{k_f}\left(1 - K_m/K_{f22}\right)} = K_{\ell 33}$$

For voids: $$K_m = \left(1 - \sqrt{k_v}\right)K_m + \frac{K_m \sqrt{k_v}}{1 - \sqrt{k_v}\left(1 - K_m/K_v\right)}$$

Longitudinal thermal expansion coefficient: $$\alpha_{\ell 11} = \frac{k_f\, \alpha_{f11}\, E_{f11} + k_m\, \alpha_m\, E_m}{E_{\ell 11}}$$

Transverse thermal expansion coefficient: $$\alpha_{\ell 22} = \alpha_{f22}\sqrt{k_f} + \left(1 - \sqrt{k_f}\right)\left(1 + k_f\, \nu_m\, E_{f11}/E_{\ell 11}\right)\alpha_m = \alpha_{\ell 33}$$

Figure 4.12 Composite micromechanics and thermal properties.

void volume ratio is a constant, which is the input value and that the properties are predicted for one void volume ratio which is "0."

4.3.2 *Physical Properties*

The equations for predicting these properties are summarized in Fig. 4.12 except the equation for the weight density, which is the second equation in Fig. 4.8. The first physical property of a nanocomposite is the weight density whose distribution through the thickness is plotted in Fig. 4.13. It is noted that this physical property has analogous distribution as that for the fiber volume ratio. It starts at the weight density of the matrix and progresses to its highest value at midheight. The second physical property investigated is the heat capacity, which is plotted in Fig. 4.14 through the thickness. This property behaves opposite to that of the density—starts at the heat capacity of the resin and decreases progressively at midheight. This is so because the heat capacity of the resin is greater than that of the fiber (0.25 to 0.23 Btu/lb-°F). This is the last of the scalar properties. The subsequent ones are directional. The major observation to be made about the scalar properties is that they all have about the same shape except the voids where their distribution is constant through the thickness of the nanolaminate.

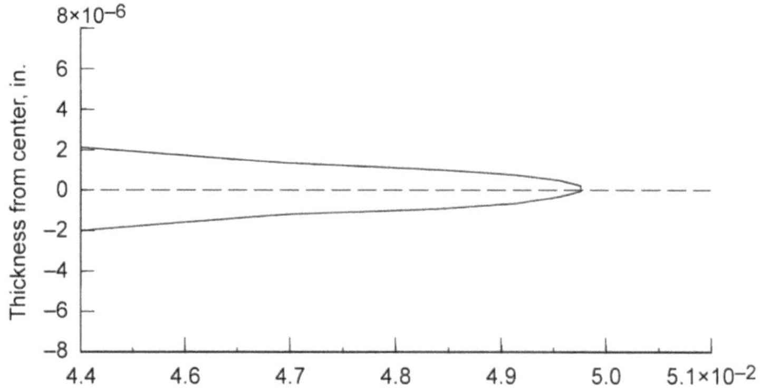

Figure 4.13 Nanoscale physical parameter density, ρ_ℓ (lb/in.^{3}).

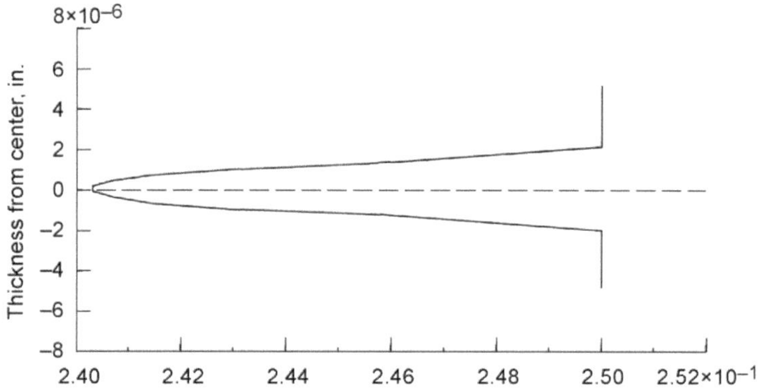

Figure 4.14 Nanoscale physical parameter heat capacity, H_C (Btu/lb-°F).

4.3.3 *Heat Conductivities*

The longitudinal heat conductivity is plotted in Fig. 4.15. The variation is similar to the scalar properties. It starts at the matrix conductivity and it increases progressively to the midheight of the nanolaminate, and then decreases symmetrically to the conductivity of the matrix. The transverse heat conductivity is shown graphically in Fig. 4.16. It can be seen that this one has the same shape as the longitudinal, but much smaller scale. The heat conductivity in the third direction has the same distribution as that in the transverse.

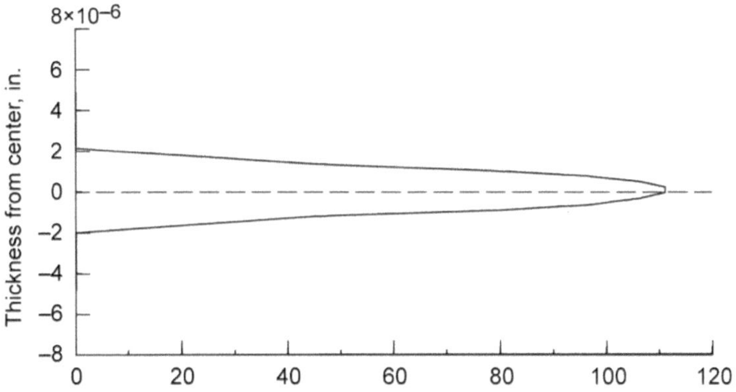

Figure 4.15 Nanoscale physical variable longitudinal heat conductivity, $K_{\ell 11}$ (Btu/hr/ft^2/°F/in.).

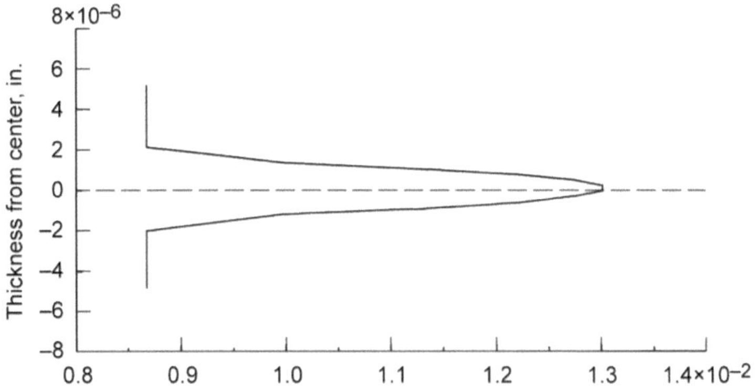

Figure 4.16 Nanoscale physical variable transverse heat conductivity, $k_{\ell 22}$ (Btu/hr/ft^2/°F/in.).

Diffusivities

The equations used for these properties are summarized in Fig. 4.17. The longitudinal moisture diffusivity for the nanocomposite is plotted in Fig. 4.18. It can be seen in the figure that the longitudinal diffusivity curve is analogous to the matrix volume ratio curve, which is at a maximum in the matrix region and decreases continuously to the midheight where it attains its least value. This is because moisture does not diffuse through the fibers. The transverse

Longitudinal diffusivity: $D_{\ell 11} = (1 - k_f)D_m$

Transverse diffusivity: $D_{\ell 22} = \left(1 - \sqrt{k_f}\right)D_m = D_{\ell 33}$

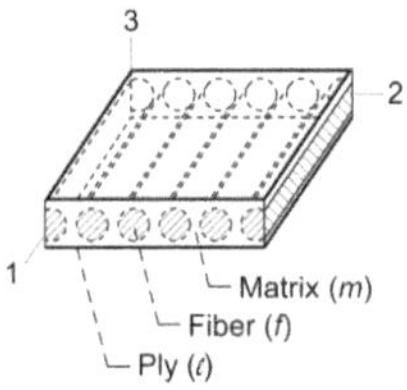

Longitudinal moisture expansion coefficient $\beta_{\ell 11} = \beta_m(1 - k_f)E_m/E_{\ell 11}$

Transverse moisture expansion coefficient: $\beta_{\ell 22} = \beta_m\left(1 - \sqrt{k_f}\right)\left[1 + \frac{\sqrt{k_f}\left(1 - \sqrt{k_f}\right)E_m}{\sqrt{k_f}E_{\ell 22} + \left(1 - \sqrt{k_f}\right)E_m}\right] = \beta_{\ell 33}$

For incompressible matrix:

$$\beta_{\ell 11} = 0$$
$$\beta_{\ell 22} = \beta_m\rho_\ell/2\rho_m = \beta_{\ell 33}$$

Figure 4.17 Composite micromechanics and hygro properties.

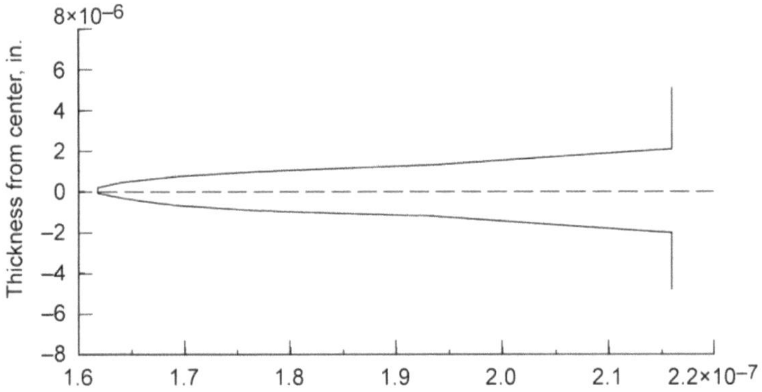

Figure 4.18 Nanoscale physical variable longitudinal moisture diffusivity, $D_{\ell 11}$ (in.^{2}/hr).

diffusivity is plotted in Fig. 4.19. It has approximately the same shape as the longitudinal, but it is longer and reaches lower value at midheight about 0.9×10^{-7} to about 2.2×10^{-7}. Through-the-thickness diffusivity is the same as the transverse.

4.3.4 *Moisture Expansion*

The longitudinal moisture expansion coefficient is shown graphically in Fig. 4.20. As can be seen in Fig. 4.20, it starts at a relatively high matrix expansion and decreases rapidly to almost zero values. The reason for this rapid decrease is that the fibers do not expand in

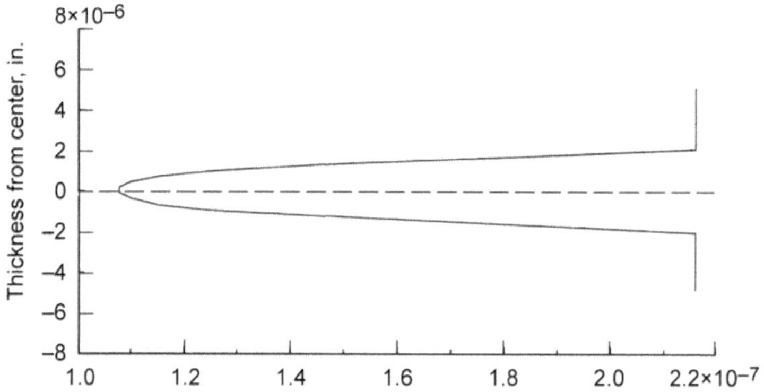

Figure 4.19 Nanoscale physical variable transverse moisture diffusivity, $D_{\ell 22}$ (in.^{2}/hr).

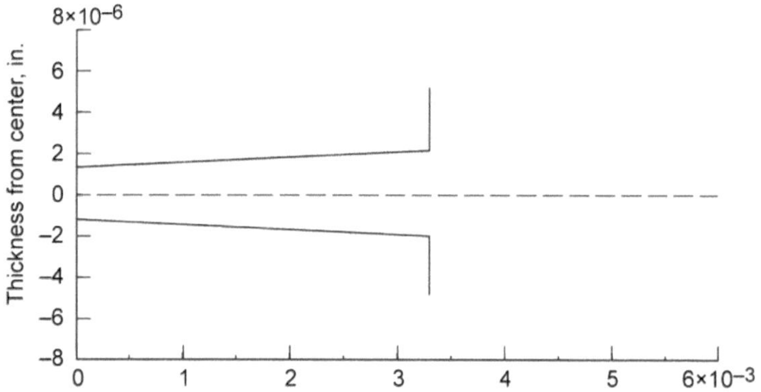

Figure 4.20 Nanoscale physical variable longitudinal moisture expansion, $\beta_{\ell 11}$ (in./in./%M).

moisture fields. The weighted average of the longitudinal moisture expansion is near zero, thus exhibiting a huge stress concentration in the bounding matrix. The practical significance of this large discrepancy between the matrix and fiber moisture expansion is that moisture will induce nanoscale crazes in the matrix in that region.

The transverse moisture expansion is plotted in Fig. 4.21. It can be seen in this figure that its behavior is analogous to the longitudinal diffusivity.

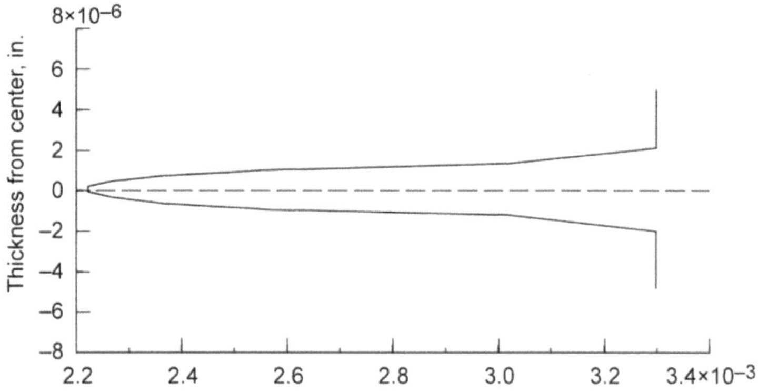

Figure 4.21 Nanoscale physical variable transverse moisture expansion, $\beta_{\ell 22}$ (in./in./%M).

4.3.5 *Thermal Expansion*

The longitudinal nanothermal expansion coefficient is shown graphically in Fig. 4.22. This coefficient shows the same behavior as the longitudinal moisture expansion but at different scales. It starts with the thermal expansion of the matrix and rapidly decreases to near zero. The reason for this rapid decrease from the matrix to the fiber is that the fibers are assumed to have a near-negative longitudinal expansion. The comments noted in the

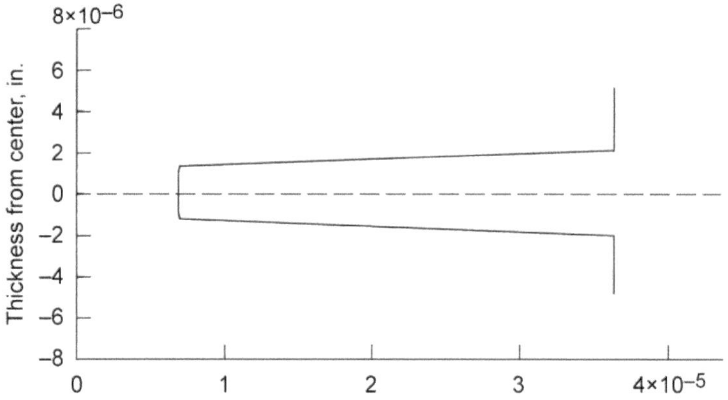

Figure 4.22 Nanoscale physical variable longitudinal thermal expansion, $\alpha_{\ell 11}$ (in./in/.°F).

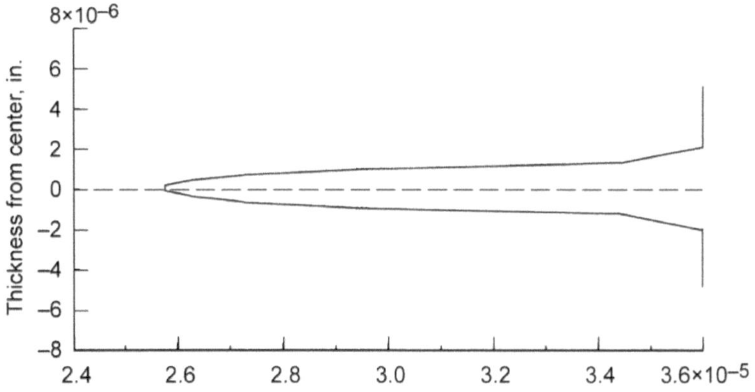

Figure 4.23 Nanoscale physical variable transverse thermal expansion, $\alpha_{\ell 22}$ (in./in./°F).

discussion of the longitudinal moisture expansion, with respect to stress concentration, apply to the thermal longitudinal expansion as well. The practical significance is that crazes in this matrix region will occur during the cooling down of the nanocomposite.

The transverse thermal expansion is plotted in Fig. 4.23. It shows similar behavior as that for the transverse moisture expansion in Fig. 4.21. As can be seen, it exhibits a gradual decrease from the matrix region to the fiber reaching a minimum at midheight. It is noted here, as a reminder, that the distribution shown in these figures are the direct result of the progressive monofiber laminate substructuring. It is also noted that the through-the-thickness thermal expansion is the same as that for the transverse. This property is the last one of the physical properties of the nanounidirectional laminate.

4.3.6 *Mechanical Properties*

These properties include all the information that is needed to perform a stress and or displacement analyses. Included are moduli, Poisson's ratio and uniaxial strengths as described below.

Longitudinal modulus: $$E_{\ell 11} = k_f\, E_{f11} + k_m\, E_m$$

Transverse modulus: $$E_{\ell 22} = \frac{E_m}{1 - \sqrt{k_f}\left(1 - E_m/E_{f22}\right)} = E_{\ell 33}$$

Shear modulus: $$G_{\ell 12} = \frac{G_m}{1 - \sqrt{k_f}\left(1 - G_m/G_{f12}\right)} = G_{\ell 13}$$

Shear modulus: $$G_{\ell 23} = \frac{G_m}{1 - \sqrt{k_f}\left(1 - G_m/G_{f23}\right)}$$

Poisson's ratio: $$\nu_{\ell 12} = k_f\, \nu_{f12} + k_m\, \nu_m = \nu_{\ell 13}$$

Poisson's ratio: $$\nu_{\ell 23} = \frac{E_{\ell 22}}{2G_{\ell 23}} - 1$$

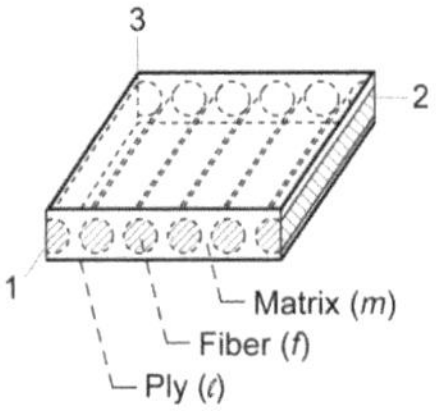

Figure 4.24 Composite micromechanics and mechanical properties.

4.3.6.1 Moduli

The equations used for these properties are summarized in Fig. 4.24. The longitudinal modulus is shown graphically in Fig. 4.25. The graph starts with the matrix modulus, which is relatively low in the scale of the abscissa. It increases gradually to its maximum value at midheight and it is symmetric as would be expected. Note that the highest value is close to 2.5×10^8 psi, which is at about 0.25 of the fiber volume ratio for a fiber whose modulus is 1.0×10^9 psi.

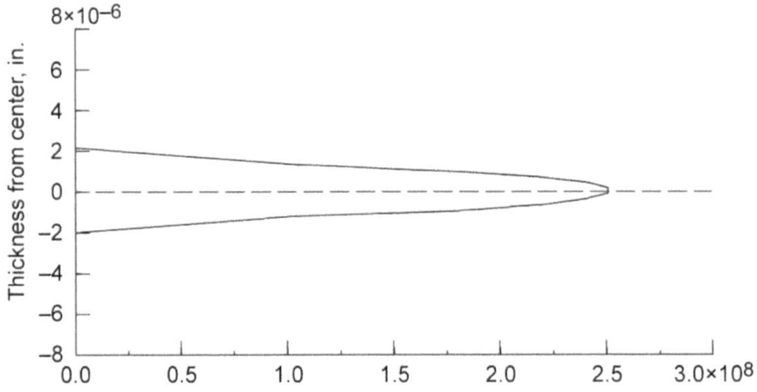

Figure 4.25 Nanoscale mechanical property variable longitudinal modulus, $E_{\ell 11}$ (lb/in.^{2}).

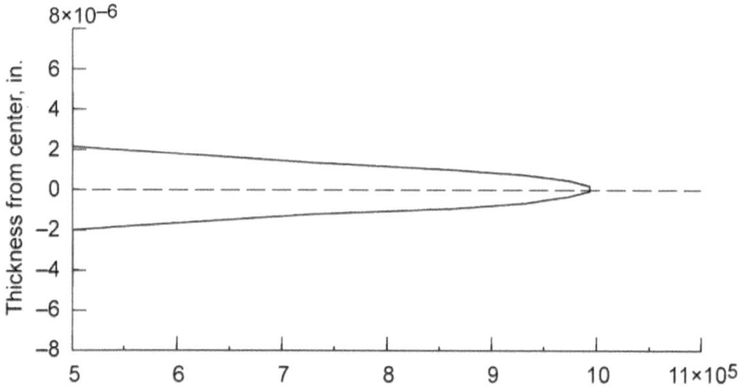

Figure 4.26 Nanoscale mechanical property variable transverse modulus, $E_{\ell 11}$ (lb/in.^{2}).

The transverse modulus is plotted in Fig. 4.26. This value starts at the matrix value and increases gradually to its maximum value at midheight. Note that the maximum value for this modulus is about 1.0×10^6 psi. This low value results from the very low fiber volume ratio (0.05). The important point here is that this nanocomposite is highly anisotropic at about 250/1.

The in-plane shear modulus is plotted graphically in Fig. 4.27. This plot is analogous to the transverse modulus but one-third the scale in the abscissa. It reaches its maximum value at midheight by increasing gradually from the matrix shear modulus to that value.

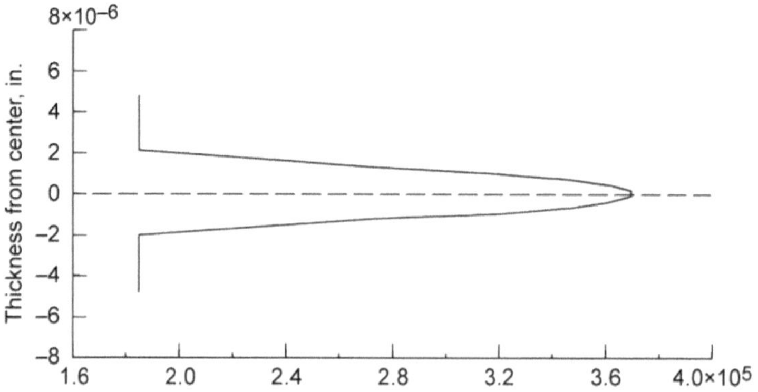

Figure 4.27 Nanoscale mechanical property shear modulus, $G_{\ell 12}$ (lb/in.^{2}).

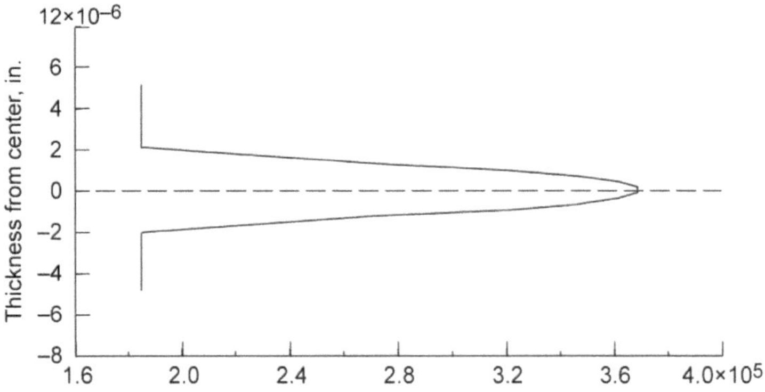

Figure 4.28 Nanoscale mechanical property, $G_{\ell 23}$ (lb/in.^{2}).

One observation is that the transverse and shear moduli are matrix-controlled quantities and their respective nanocomposite values are expected to be low.

The through-the-thickness shear modulus is plotted graphically in Fig. 4.28. It is the same as the in-plane shear modulus. The reason is that they are both matrix dependent and the fiber shear moduli are relatively low.

4.3.6.2 Poisson's ratios

The major Poisson's ratio is plotted in Fig. 4.29. It can be seen that it starts at its maximum value (the average value at 0.05 fiber volume ratio) and decreases gradually reaching its minimum value of about 0.29 at midheight. It can also be observed that the graph is a well-behaved function.

The through-the-thickness Poisson's ratio is plotted graphically in Fig. 4.30. Observe that this is a peculiar plot. It starts from an average value of the Poisson's ratio of the matrix, rapidly increases to point 0.651 and then progressively decreases to a lower value of about 0.58 at midheight, and it is symmetric. The only explanation at this time is that the restraints of the substructured layers at the nanoscale cause this bimodal behavior. This is the last moduli property of the mechanical properties of a unidirectional nanolaminate.

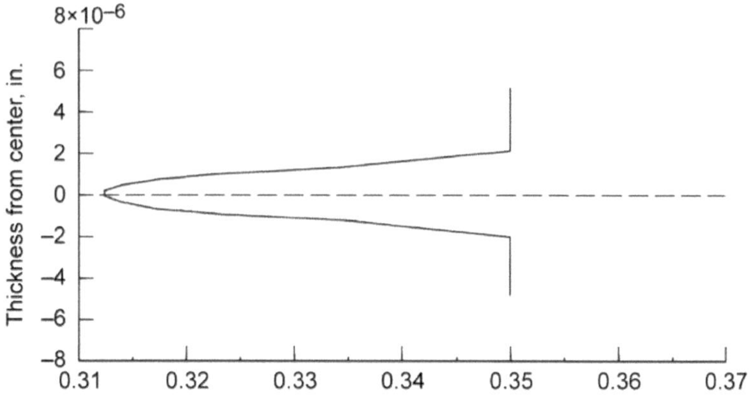

Figure 4.29 Nanoscale mechanical property Poisson's ratio, $\nu_{\ell 12}$ (in./in.).

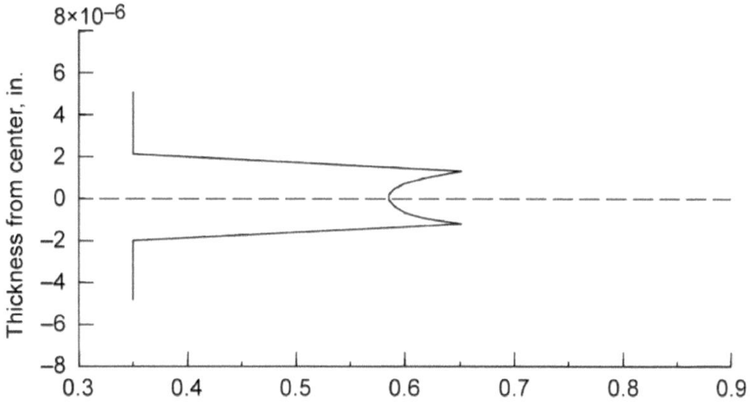

Figure 4.30 Nanoscale mechanical property Poisson's ratio, $\nu_{\ell 23}$ (in./in.).

4.3.6.3 Uniaxial nanocomposite strengths

The equations used for predicting these properties are summarized in Fig. 4.31. The uniaxial strengths of interest in nanocomposite designs are longitudinal tension and compression, transverse tension and compression, in-plane shear, and through-the-thickness shear. Below each of these is discussed in some detail:

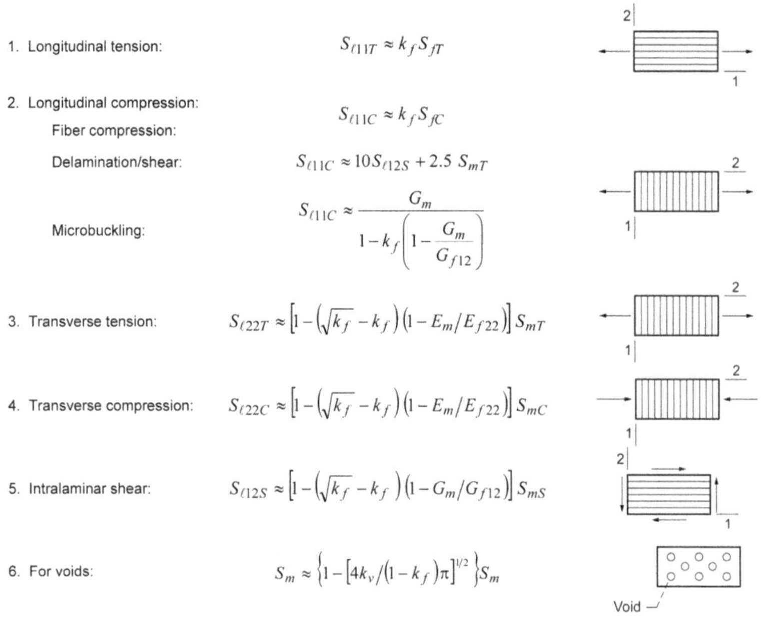

Figure 4.31 Composite micromechanics, uniaxial strengths, in-plane.

4.3.7 *Longitudinal Tension*

The longitudinal tensile strength is plotted graphically in Fig. 4.32. The plot is symmetric about the midheight and it is analogous to that of the longitudinal modulus (Fig. 4.25). The fiber strength of course is about three orders of magnitude less. It is observed that this strength behaves normally. It starts at a strength compared to matrix (about 15 ksi) and increases gradually to its highest value (about 2.0 $\times$ 10^5 ksi) at midheight. This very high value for tensile strength results from an input value of 8 $\times$ 10^5 psi for the nanofiber.

4.3.8 *Longitudinal Compression*

The longitudinal compression strength is plotted graphically in Fig. 4.33. It starts at the matrix compressive strength (about 35 ksi) and increases rapidly to its highest value, which is about 160 ksi at the midheight of the nanolaminate. Then it decreases symmetrically to about the compressive strength of the matrix.

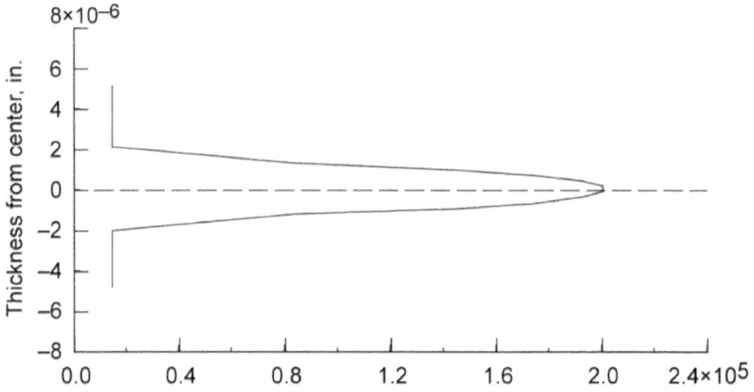

Figure 4.32 Nanoscale uniaxial longitudinal tensile strength, $S_{\ell 11T}$ (lb/in.^{2}).

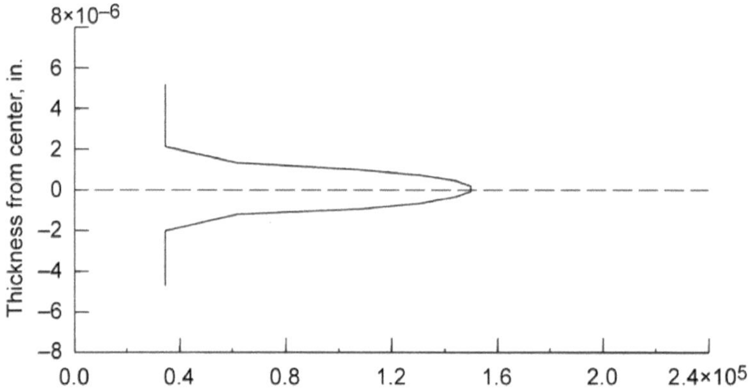

Figure 4.33 Nanoscale uniaxial longitudinal compressive strength, $S_{\ell 11C}$ (lb/in.^{2}).

4.3.9 *Transverse Tensile Strength*

The nanocomposite transverse strength of a unidirectional laminate is plotted graphically in Fig. 4.34. As can be seen in the figure, it starts at the transverse tensile strength of the matrix, about 15 000 psi and decreases very rapidly to about 11 300 psi at the midheight of the nanolaminate.

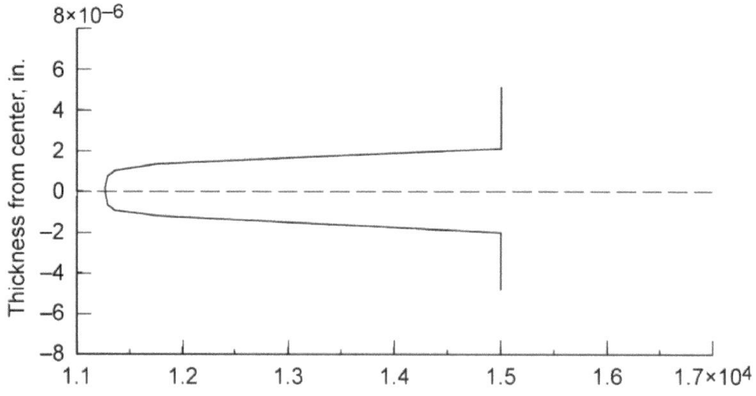

Figure 4.34 Nanoscale uniaxial transverse tensile strength, $S_{\ell 22T}$ (lb/in.^{2}).

4.3.10 *Transverse Compressive Strength*

The transverse compressive strength is shown graphically in Fig. 4.35. The behavior of this strength is comparable to the transverse tensile strength (Fig. 4.34). The comments made in that figure apply to this figure as well.

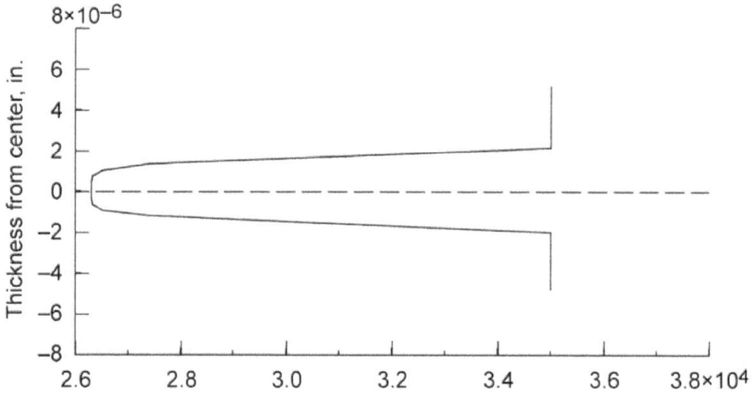

Figure 4.35 Nanoscale uniaxial transverse compressive strength, $S_{\ell 22C}$ (lb/in.^{2}).

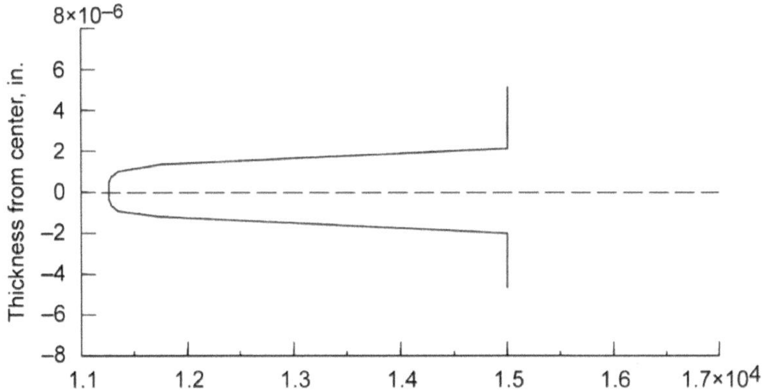

Figure 4.36 Nanoscale uniaxial in-plane shear strength, $S_{\ell 12S}$ (lb/in.^{2}).

4.3.11 *Intralaminar Shear Strength*

The intralaminar shear strength is plotted graphically in Fig. 4.36. The behavior of this strength is comparable to the two transverse strengths except that scale is a bit lower. The comments made for those two strengths apply to the intralaminar shear strength as well.

4.3.12 *Interlaminar Shear Strength*

The interlaminar (through-the-thickness) shear strength is plotted graphically in Fig. 4.37. This strength has analogous behavior as the intralaminar shear strength except that it reaches about 10 000 psi. Then it reverses symmetrically. The practical significance of this behavior is that the matrix-bounding interface needs to be very thin otherwise the unidirectional nanolaminate will have very little transverse and shear strengths.

This property is the last of the mechanical properties. All the fabrication parameters, the physical variables and the mechanical properties moduli and uniaxial strengths of the unidirectional (fiber aligned) nanolaminate have been completely characterized computationally.

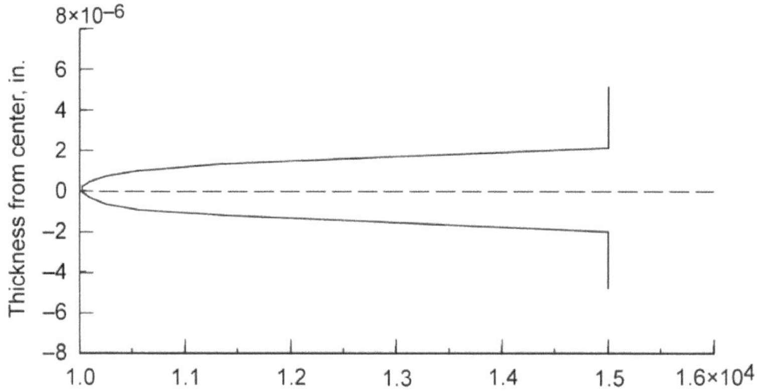

Figure 4.37 Nanoscale uniaxial transverse tensile strength, $S_{\ell 23T}$ (lb/in.^{2}).

4.4 Concluding Remarks

The salient remarks from an investigation to characterize an aligned monofiber nanolaminate are as follows:

1. The characterization for the nanolaminate (composite) was based on a series of progressive substructuring down a sliced single-diameter fiber where all the equations are based.
2. The theoretical development and all the equations are included in a computer code called ICAN/JAVA.
3. The characterization includes the following 25 properties: in situ fabrication parameters (3), physical variables (10), and mechanical properties (12) (6 moduli and 6 uniaxial strengths).
4. These properties are plotted graphically versus the nanolaminate thickness as ordinate.
5. The nanolaminate investigated consists of single nanofiber laminate with 0.05 weighted fiber volume ratio.
6. The nanofiber diameter is 2.756×10^{-6} in.
7. All the figures are symmetric about the midheight; most of them exhibit continuous behavior as would be expected. There is one exception: the through-the-thickness Poisson's ratio exhibits bimodal symmetric behavior.
8. The formulation and results are possible where the fiber diameter is a variable in the formulation and including the fiber,

matrix, and void volume ratios input quantities into the computer code.

References

1. Jose, M., Tyner, J., Nyairo, E., and Dean, D. "Synthesis and Processing of Aligned Carbon Nanotube Based Fibers," presented at the 49th International SAMPE Symposium and Exhibition, (CD). May 16–20, 2004.
2. Ayalasomayajuala, G., Garg, A., Kapila, S., Chandrashekhara, K., and Flanigan, V. "Fabrication and Evaluation of Rice Hull Derived Nano Silica Composites," presented at the 49th International SAMPE Symposium and Exhibition (CD). May 16–20, 2004.
3. Karaki, T., Killgore, J. P., and Seferis, J. C. "Characterization of Fatigue Behavior of Polynanomeric Matrix Composites," presented at the 49th International SAMPE Symposium and Exhibition (CD). May 16–20, 2004.
4. Ranade, A., D'Souza, N. A., Nayak, K., Gnade, B., and Fairbrother, D. 2003. Correlation Between Creep-Recovery, Crystallization and Dispersion of Linear Low Density Polyethylene Nanocomposite Films, 48th International Symposium and Exhibition, vol. 48, book 1 of 2, pp. 2164–2176.
5. Koo, J. H., Stretz, H., Weispfenning, J., Luo, Z. P., and Wootan, W. "Nanocomposite Rocket Ablative Materials: Subscale Ablation Test," presented at the 49th International SAMPE Symposium and Exhibition (CD). May 16–20, 2004.
6. Muhle, S., Monner, H. P., and Wiersch, P. 2003. Carbon-Nanotubes for Adaptive Structures, 48th International SAMPE Symposium and Exhibition, vol. 48, book 1 of 2, pp. 1181–1190.
7. Srivastava, D. and Wei, C. 2003. Computer Simulations of Macroscopic Properties of Carbon-Nanotube Polymer Composites, 48th International SAMPE Symposium and Exhibition, vol. 48, book 1 of 2, pp. 2153–2163.
8. Gates, T. S., et al. "Computational Materials: Multi-scale Modeling and Simulation of Nanostructural Materials." *Composites Science and Technology* **65** (2005), pp. 2416–2434.
9. Naiwa, H. S. (ed.) *Nanostructured Materials and Nanotechnology*. Academic Press, San Diego, CA. 2002.

10. Handler, L. M., and Chamis, C.C. ICAN/JAVA Computer Code.
11. Chamis, C. C. "Simplified Composite Micromechanics Equations for Hygral, Thermal and Mechanical Properties," NASA TM–83320, February 1983.
12. Chamis, C. C. "Simplified Composite Equations for Strength, Fracture, Toughness, Impact Resistance and Environmental Effects," NASA TM–83696, January 1984.
13. Freund, M. M., Pomrenke, G. S., Brown, G. J., and Vaia, R. A. "Nanoscience and Technology for the Air Force." *AFRL Technology Horizons* (2005), pp. 9–13.

Chapter 5

Analyzing Interlaminar Shear Strength of Multiscale Composites via Combined Finite Element and Progressive Failure Analysis Approach*

Mohit Garg,[a] Frank Abdi,[a] and Stuart McHugh[b]

[a] *AlphaStar Corporation, 5150 E. Pacific Coast Highway, Suite 650, Long Beach, CA 90804, USA*
[b] *Lockheed Martin Corporation, 3251 Hanover St., Palo Alto, CA 94304, USA*
stuart.mchugh@lmco.com

Newly developed multiscale composites use nanoparticles in fiber-reinforced composites to dramatically increase strength and stiffness in the out-of-plane direction. The small weight percentage of nanoparticles significantly improves the delamination resistance. Analyzing multiscale composites is a challenging issue, requiring analysts to consider the mechanics from nanoscale to macroscale. The approach developed in this chapter demonstrates a powerful, easy, and time-efficient methodology that combines analytical and numerical analyses with progressive failure analysis. First,

---

**SAMPE Fall Technical Conference on Multifunctional Materials.* Co-located with American Society for Composites Annual Technical Conference and ASTM D-30 Committee Meetings. September 8–11, 2008. Memphis, Tennessee.

Characterization of Nanocomposites: Technology and Industrial Applications
Edited by Frank Abdi and Mohit Garg

ISBN 978-981-4669-02-3 (Hardcover), 978-1-315-36489-6 (eBook)
www.panstanford.com

the effective anisotropic properties for the enhanced matrix are calculated that contains aligned nanoparticles of known aspect ratio, strength, and stiffness. Second, the isotropic enhanced matrix properties are calculated using micromechanics. The enhanced matrix is assumed to be same as interphase around the conventional fibers in the neat matrix. Then, progressive failure analysis is performed to characterize/simulate the multiscale composites and structural components. Simulation results indicate a 21.5% increase in interlaminar shear strength for 1 wt% multiwalled carbon nanotubes in the advanced material. The results obtained compare reasonably well with similar test data (18.2% increase in interlaminar shear strength for 1 wt% multiwalled carbon nanotubes).

5.1 Introduction

Laminated fiber-reinforced polymer (FRP) composites have found wide use for a range of structural and functional applications in industries such as aerospace, automobile, marine, and infrastructure. FRP composites are high-performance materials, which have high stiffness and strength-to-weight ratios compared to metals. FRP composites exhibit high corrosion resistance and significant reduction in maintenance costs and, most importantly, expand a designer's horizons to design a material to meet the limit load requirements. The fibers in FRP composites reinforce mostly the in-plane properties. The weaker matrix interface dominates the out-of-plane properties, such as delamination resistance. Delamination along ply interfaces (interlaminar shear strength) due to fiber/matrix debonding is one of the major failure modes in laminated FRP composites when subjected to transverse and compressive loading.

Reinforcement of the matrix of laminated composites by stitching, weaving, or braiding fibers or by adding toughening particles is a widely used approach to improve delamination resistance. The approach, however, can degrade other FRP properties. For example, using toughening particles with lower stiffness than the matrix can reduce the overall stiffness of a laminate.

Improving interlaminar shear strength (ILSS) by reinforcing the laminate interface with nanoparticles such as double-walled [1] or multiwalled [2] carbon nanotubes (CNTs) and nano-clay [3] have recently been considered [4]. The nanoparticles used had high aspect ratios and high stiffness/strength and were nano-size. A low weight percentage (<15 wt%) of nanoparticles in neat matrix can increase both the strength and stiffness of the matrix. For example, Yeh et al. (2007) reported an increase of 64.6% in the tensile strength and 64.8% in Young's modulus with only 2 wt% addition of MWCNTs in a phenolic matrix. Other experimentalists have reported similar improvements [6, 7].

Literature indicates that the nanoparticle-related data has a wide scatter: conducting experiments at the nanoscale is an extremely difficult and expensive process. To overcome this obstacle, researchers simulate the nanoparticles using molecular dynamic (MD) simulation techniques.

Using carbon nanotubes, the ILSS of the laminated composites is usually improved either by dispersing the nanoparticles in the matrix [8] or by growing CNTs around the conventional fibers [2]. Researchers have reported that adding less than 5 wt% of CNTs in the matrix material or growing CNTs around the conventional fibers can improve the ILSS for advance hybrid multiscale composites by approximately 45% [8] and 69% [2], respectively. Growing CNTs on conventional fibers allows larger weight percentages than mixing the CNTs in the matrix for using vacuum-assisted resin transfer molding (VARTM) manufacturing. The VARTM approach increases the viscosity during the laminate fabrication, which in turn adversely affects mixing of the enhanced matrix containing high weight percent nanotubes. It is easier for the matrix to exhibit capillary behavior in between the closely packed nanotubes, allowing a higher weight percentage of CNTs. Both approaches are affected because of the bonding between the CNTs (or nanoparticles) and the matrix. Functionalization can help increase the bond (interfacial shear strength [ISS]) between the CNTs (or nanoparticles) and the matrix. Note that sometimes ISS is also referred to as IFSS in the literature.

The lack of test data, the nanoscale size involved, and the range of unknown variables make analyzing hybrid multiscale composite materials extremely complex. For industrial applications, analyzing these materials with finite element (FE) or MD simulation alone is next to impossible. There is a need for an analytical/numerical approach that can easily and rapidly analyze the complete deformation response of large-scale components made of multiscale composites.

5.2 Methodology and Approach

In the suggested approach, the effective nanocomposite (or enhanced matrix) material properties, where CNTs of known weight fraction are aligned in a neat matrix, are determined first analytically. The analysis approach uses the well-known Mori–Tanaka formulation [9, 10] for calculating the anisotropic nanocomposite properties from the isotropic matrix and nanoparticle properties (stiffness, aspect ratio, and volume fraction). In an independent review, Tucker and Liang (1999) found that Mori–Tanaka-type models give the best results for large-aspect-ratio fillers (nanoparticles). For a composite material reinforced with aligned fiber-like particles, the Tandon and Weng (1984) prediction of the moduli E_{11} (aligned particle direction), E_{22} (transverse to the aligned particle direction), the in-plane shear modulus G_{12}, and the out-of-plane shear modulus G_{23} of the composite are

$$E_{11} = {}^{E_m}\!/\left(1 + f_p\left(A1 + 2\upsilon_m A2\right)/A\right) \tag{5.1}$$

$$E_{22} = {}^{E_m}\!/\left(1 + f_p\left(-2\upsilon_m A3 + \left(1-\upsilon_m\right)A4 + \left(1+\upsilon_m\right)A5^{*}A\right)/\left(2A\right)\right) \tag{5.2}$$

$$G_{12} = {}^{G_m}\!/\left(1 + f_p\left(\mu_m/\left(\mu_p - \mu_m\right)\right) + 2\left(1 - f_p\right)H_{1212}\right) \tag{5.3}$$

$$G_{12} = {}^{G_m}\!/\left(1 + f_p\left(\mu_m/\left(\mu_p - \mu_m\right)\right) + 2\left(1 - f_p\right)H_{2323}\right) \tag{5.4}$$

where A and A_i are constants depending on the components of the Eshelby tensor and the matrix/nanoparticle properties, and H_{ijkl} are the Cartesian components of the Eshelby tensor [12].

In this chapter, we used a closed-form analytical solution for the complete set of anisotropic elastic properties of the composite derived by Tandon and Weng (1984) by combining the Eshelby theory [12] and the Mori–Tanaka model [9, 10], as shown in Eqs. 5.1–5.4.

The effective tensile and shear strengths for the unidirectional composite are calculated using the CNT aspect ratio and the ISS between the CNTs and the matrix material using shear lag theory [13]. The ISS is obtained either from the tests [14], MD simulation [15], or effective continuum model [16]. At this stage we have calculated the effective material properties of enhanced matrix where chopped fiber-like nanoparticles are aligned and conventional fiber is not yet used.

An FE-based coupon model is then subjected to progressive failure analysis (PFA). This simulates an ASTM D3039 tensile test assigned with quasi-isotropic $[45/0/-45/90]_S$ lay-up. Each ply is assigned unidirectional enhanced matrix properties (no conventional fibers) that were previously derived analytically. The combined FE and PFA simulation of a quasi-isotropic ply lay-up gives equivalent effective random (isotropic) stiffness, strength, and Poisson's ratio of the enhanced matrix (nanocomposite). Note that the derived random enhanced in-plane properties are also assumed for the out-of-plane direction for full randomness.

The random enhanced matrix properties are used as interphase properties between conventional fiber and the neat matrix. The random enhanced matrix and interphase here refers to the phase where CNTs are grown with random orientation in the matrix, as shown in Fig. 5.1a. Note that the CNTs are grown around the conventional fibers. Although not done in this chapter, the random enhanced matrix properties could also be used to simulate the multiscale composite where conventional fibers are used with enhanced matrix, as shown in Fig. 5.1b.

Radially aligned CNTs, grown outward from conventional fibers (Fig. 5.2a,b), can be simulated by taking the interphase properties directly from the unidirectional ply where the aligned direction is the radial direction. For the aligned case (not done in this chapter), the analytical results from the Mori–Tanaka model can be used directly to define the interphase. GENOA uses continuum interphase formulation [19] developed assuming that the crack propagates around the continuous fiber and through the interphase or the enhanced matrix.

Next, the FE and PFA are combined to simulate a short-beam strength (SBS) test, similar to ASTM D 2344/D 2344M. For this

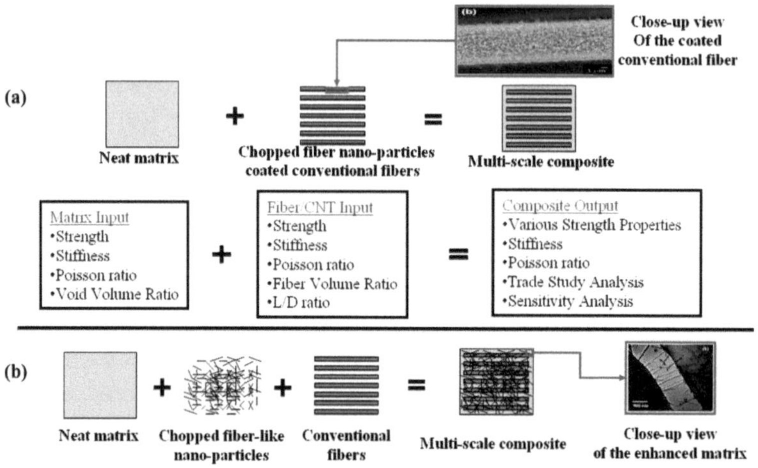

Figure 5.1 Micrographs of (a) coated conventional fiber [17] and (b) enhanced matrix [18].

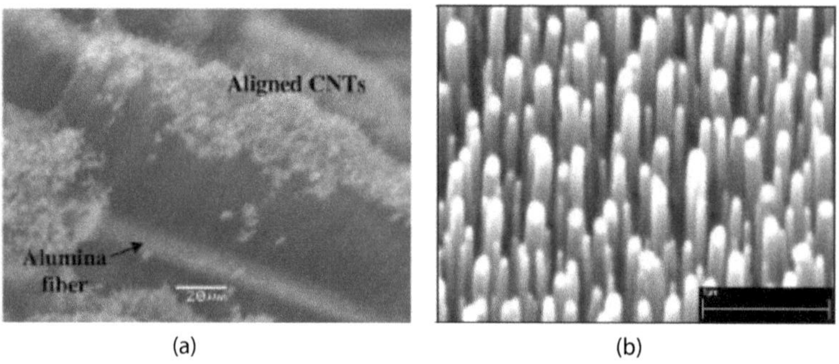

Figure 5.2 (a) Radially aligned carbon nanotubes around conventional fiber [2] and (b) vertically aligned carbon nanotubes [20] obtained via scanning electron microscope.

simulation, multiscale composite properties are used with a mixed ply lay-up ($[0/45/-45]_s$), conventional fiber volume ratio of 60%, and 0% void volume ratio. Progressive failure analysis plays a vital role simulating the SBS strength test. In PFA, individual ply stresses in the conventional fiber direction are found from global stresses using classical laminate theory. The ply stresses are then

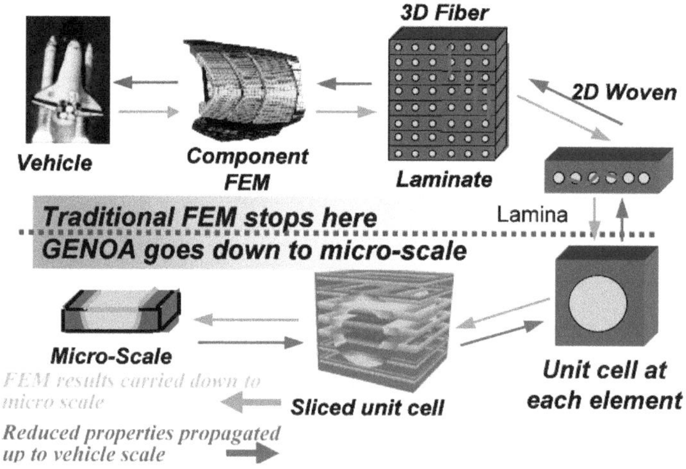

Figure 5.3 Micro- and macroscale interaction in progressive failure analysis [21].

checked against the strength limits. The hierarchal approach, shown in Fig. 5.3, is used.

Normally, 18 failure criteria are checked; 15 are listed in Table 5.1. The remaining three include failure criteria for honeycomb material and fracture toughness based criteria. These failure criteria focus on fiber, matrix, and ply failure. To ensure that only interlaminar shear strength properties are checked, in-plane shear, out-of-plane shear, and relative rotation between plies (strength-based failure criteria) were chosen for estimating the ILSS properties.

5.3 Results and Discussion

In this section, simulation results for multiscale composites are presented. The composite considered was made of conventional carbon fibers coated with multiwalled CNTs, MWCNTs oriented in random directions (Fig. 5.1a) with an EPON 862 matrix. The first step was to calculate the properties of anisotropic enhanced matrix containing only aligned MWCNTs via analytical Mori-Tanaka model using isotropic material properties (Table 5.2). Note that

Table 5.1 List of fiber/matrix or interactive predefined failure criteria available in damage and durability software, GENOA [21, 22]

Mode of failure	Description	Failure type
Longitudinal tensile	Fiber tensile strength	Fiber
Longitudinal compressive	Rule of mixture based delamination; Fiber micro-buckling; and fiber crushing	Fiber
Transverse tensile	Matrix tensile strength	Fiber
Transverse compressive	Matrix compressive strength	Matrix
Normal tensile	Delamination due to normal tension	Delamination
Normal compressive	Crushing of laminates	Delamination
In-plane shear	In-plane shear relative to laminate	Matrix
Transverse normal shear	Shear failure acting on transverse cross-section oriented in a normal direction of the ply	Delamination
Transverse longitudinal shear	Shear failure on transverse cross-section oriented in a normal direction of the ply	Delamination
Longitudinal normal shear	Shear failure on longitudinal cross-section oriented in a normal direction of the ply	Delamination
Modified distortion energy	Combined stress failure criteria used for anisotropic materials	Failure due to all strength
Relative rotation	Failure if adjacent plies rotate excessively with one another	Delamination
Strain invariant failure	Combined strain failure criteria used for composite material	
Ply strain limit	Matrix or ply strain cutout	Matrix/Ply
User defined	Combined strength such as Tsi-Wu	Failure due to all strength

the MWCNT isotropic material properties in Table 5.2 are average representation of literature data.

Figures 5.4 and 5.5 graphically show the output from the Mori–Tanaka closed-form analytical model using the matrix and nanoparticles listed in Table 5.2. Figures 5.4 and 5.5 show the effect of MWCNT volume fraction and aspect ratio, respectively, on

Table 5.2 Average representative isotropic material properties for (a) multiwalled carbon nanotubes [6] and (b) EPON 862 matrix

Property (MWCNT)	**Value**
Aspect ratio (L/D)	1000/30 [nm/nm]
Young's modulus (E_{NT})	450 [GPa]
Poisson's ratio (υ)	0.19
Tensile strength (σ_{ten})	63 [GPa]
Interfacial shear strength (ISS)	50–250 [MPa]
Property (EPON 862)	**Value**
Young's modulus (E_m)	3.5 [GPa]
Poisson's ratio (υ)	0.35
Tensile strength (σ_{ten})	139.7 [MPa]

the stiffness (normalized by matrix modulus) and strength of the enhanced matrix for ISS = 250 MPa. Note in Fig. 5.5 $m_f = 5$ wt%.

Figure 5.4 clearly shows that higher volume fraction of MWCNT in the matrix results in higher stiffness and strength. This is possible only if large amounts of MWCNTs can be uniformly distributed throughout the matrix. It also requires good MWCNT/matrix interface properties and better control on matrix viscosity during manufacturing processes such as VARTM. This is extremely difficult to realize in actual experimental work. Research indicates that functionalizing (chemical modification of the surfaces) the MWCNTs improves the bonding between the MWCNTs and the matrix. For stiffness, the Mori–Tanaka formulation does not consider variation in interfacial shear strength. While the stiffness increase is exponential in nature, the tensile strength shows a linear behavior.

Figures 5.4a–e show linear to exponential increase with MWCNT volume fraction. Figures 5.5a–e indicate nearly linear increment in longitudinal and transverse modulus and tensile strength as well as a linear decrement in the in-plane and out-of-plane shear moduli. All properties reach a plateau once the MWCNT aspect ratio reaches approximately 100 or reaches the dimensions of conventional fibers. Figures 5.4 and 5.5 indicate that MWCNT volume fraction has more significant effect on the material properties than the aspect ratio.

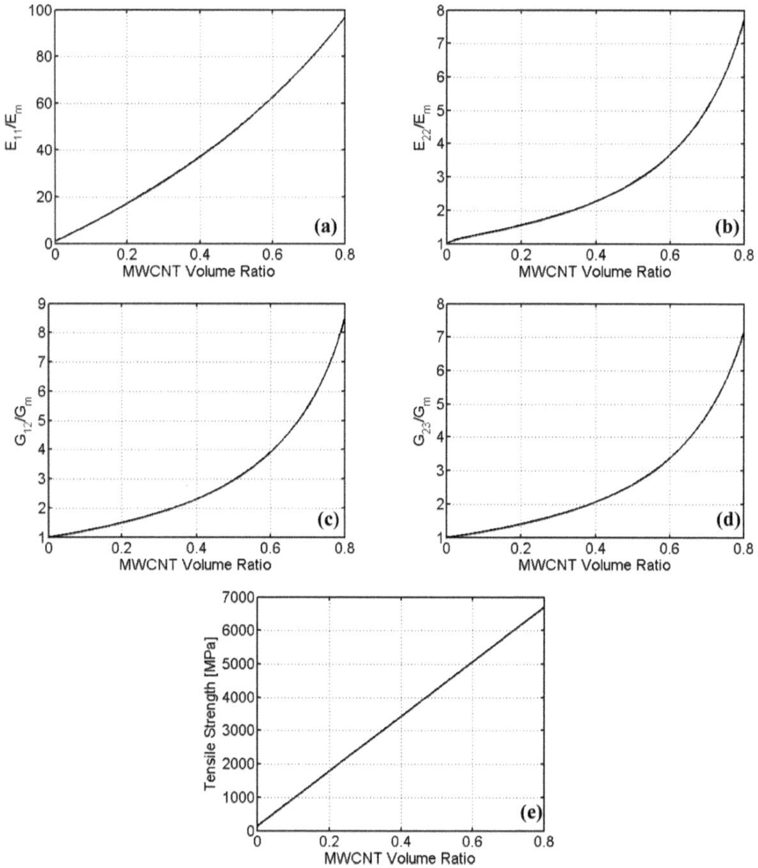

Figure 5.4 Stiffness E_{11}/E_{m} (a), E_{22}/E_{m} (b), G_{12}/E_{m} (c), G_{23}/E_{m} (d), and tensile strength (e) versus MWCNT volume ratio. ISS = 250 MPa. Output from closed-form Mori–Tanaka formulation [10] using input from Table 5.2.

Table 5.3 summarizes the calculated effective unidirectional ply properties (enhanced matrix containing aligned MWCNTs only) estimated using the analytical approach and varying the interfacial shear strength (ISS) and mass fraction (m_{f}) of MWCNTs in matrix:

(1) ISS = 250 MPa and $m_{\text{f}} = 1$ wt%;
(2) ISS = 250 MPa and $m_{\text{f}} = 5$ wt %;
(3) ISS = 50 MPa and $m_{\text{f}} = 5$ wt%.

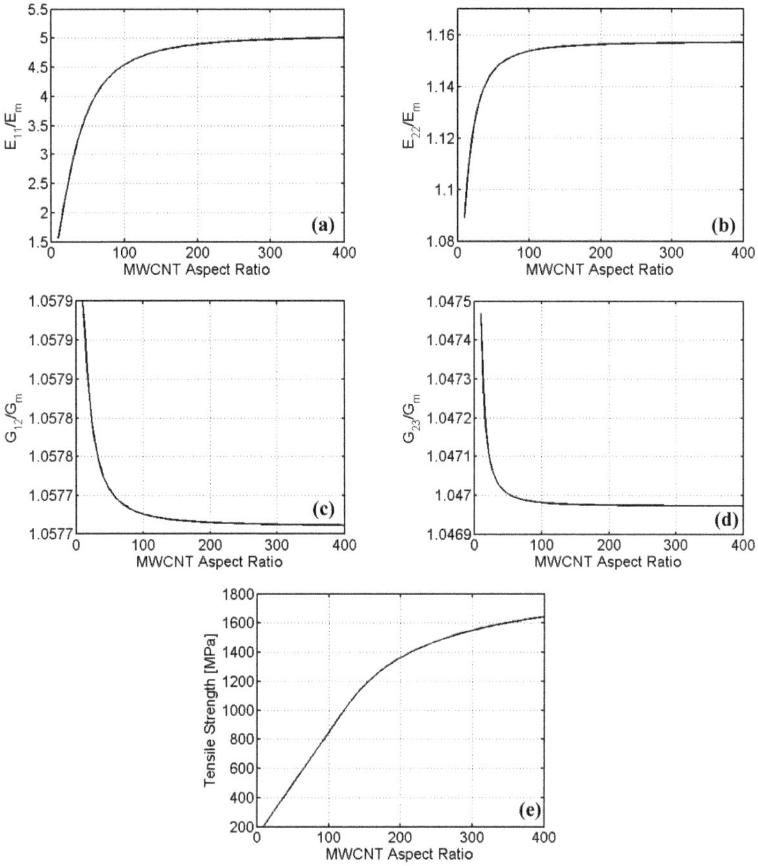

Figure 5.5 Stiffness E_{11}/E_m (a), E_{22}/E_m (b), G_{12}/E_m (c), G_{23}/E_m (d), and tensile strength (e) versus MWCNT aspect ratio; ISS = 250 MPa; mf = 5 wt%. Output from closed-form Mori–Tanaka formulation [10] using input from Table 5.2.

Table 5.3 also highlights the inability of the analytical models to capture the effect of ISS decrease on the aligned enhanced matrix stiffness. The ISS affects the tensile strength of the nanocomposite material, showing a direct relationship.

In the second step, the unidirectional ply properties were used with a quasi-isotropic laminate lay-up to estimate the isotropic enhanced matrix properties with GENOA progressive failure analysis.

Table 5.3 Calculated effective unidirectional ply properties estimated using Mori–Tanaka analytical approach. Each ply corresponds to the neat matrix that contain only aligned MWCNTs

Case No.	m_f [%]	ISS [MPa]	E_{11} [GPa]	$E_{22} = E_{33}$ [GPa]	$\nu_{12} = \nu_{13} = \nu_{23}$	G_{12} [GPa]	$G_{23} = G_{13}$ [GPa]	σ_{ten} [MPa]
1	**1.0**	250	**5.02**	**3.67**	0.35	**1.31**	**1.31**	188.86
2	**5.0**	**250**	**10.67**	**3.98**	0.35	**1.37**	**1.36**	**369.12**
3	5.0	**50**	10.67	3.98	0.35	1.37	1.36	**182.46**

m_f: mass fraction $\upsilon_{12} = \upsilon_{23} = \upsilon_{13}$: Poisson's ratio
ISS: Interfacial shear strength G_{12}: In-plane shear modulus
E_{11}: Longitudinal modulus $G_{23} = G_{13}$: Out-of-plane shear modulus
$E_{22} = E_{33}$: Young's modulus σ_{ten}: Tensile strength

Table 5.4 Calculated effective isotropic material properties for enhanced matrix containing MWCNTs in random direction in neat matrix and no conventional fibers using quasi-isotropic ply lay-up from progressive failure analysis for (a) $m_f = 1$wt% & ISS = 250 MPa and (b) $m_f = 5$ wt% & ISS = 250 MPa

(a) Property (Enhanced Matrix)	Value
Young's modulus (E_m)	4.03 [GPa]
Poisson's ratio (υ)	0.35
Tensile strength (σ_{ten})	201.1 [MPa]
(b) Property (Enhanced Matrix)	**Value**
Young's modulus (E_m)	6.016 [GPa]
Poisson's ratio (υ)	0.341
Tensile strength (σ_{ten})	333.43 [MPa]

ABAQUS was the FE solver. Tables 5.4a,b show an increment in strength and stiffness as a result of increased randomly oriented MWCNT volume fraction in isotropic enhanced matrix.

In the third step we determine the efficacy of randomly oriented MWCNT-coated conventional carbon fibers by simulating short-beam strength (SBS) tests to examine the interlaminar shear strength (ILSS). Four different material properties (cases) were considered: (a) conventional fibers in neat matrix, (b) coated (ISS = 250 MPa; $m_f = 1$ wt% MWCNTs) conventional fibers in neat matrix,

Table 5.5 Average representative carbon (RCCT) fiber material properties [23]

Property	Value
Diameter (D_f)	4.0 [μm]
Young's modulus ($E_{f11} = E_{f22}$)	14.48 [GPa]
Poisson's ratio ($\upsilon_{12} = \upsilon_{23}$)	0.23
Shear modulus (G_{f12})	6.895 [GPa]
Shear modulus (G_{f23})	1.586 [GPa]
Tensile strength (σ_{ten})	275.8 [MPa]
Compressive strength (σ_{comp})	413.7 [MPa]

(c) coated (ISS = 250 MPa; m_f = 5 wt% MWCNTs) conventional fibers in neat matrix, and (d) coated (ISS = 50 MPa; m_f = 5 wt% MWCNTs) conventional fibers in neat matrix. Table 5.5 lists the properties of the carbon (RCCT) fiber used for simulating SBS test.

Figure 5.6 provides snapshots of combined FE-PFA simulations at the final failure load. The simulations were performed using GENOA with ABAQUS FE solver. Figures 5.6a–c visually show the decrease in delamination damage area with increase in nanoparticle weight percentage (red elements) immediately after maximum load. Also shown are the damage summary panels, listing the specific failure criteria that trigger delamination. The increase in ILSS from the simulations is a direct result of improved interphase (strength and stiffness) in between the neat matrix and the conventional fibers. Figure 5.6d shows more delamination when the ISS is weaker. The lower support on the bottom surface results in slight damage also, which is visible in all cases in Fig. 5.6.

Figure 5.7 shows the corresponding load displacement results obtained at the top rigid loading pin. The rigid loading pin imposes the transverse and compressive loads in the short beam model. The drop in the load for the base material is due to matrix failure. Fan et al. (2008) observed similar drop in load in their tests. Figures 5.8 and 5.9 further support the reduction in delamination by showing the decrease in percent damage volume and strain energy with increase in MWCNT weight fraction and higher ISS, respectively. The percent damage volume corresponds to the ratio of volume of damaged elements to total initial volume.

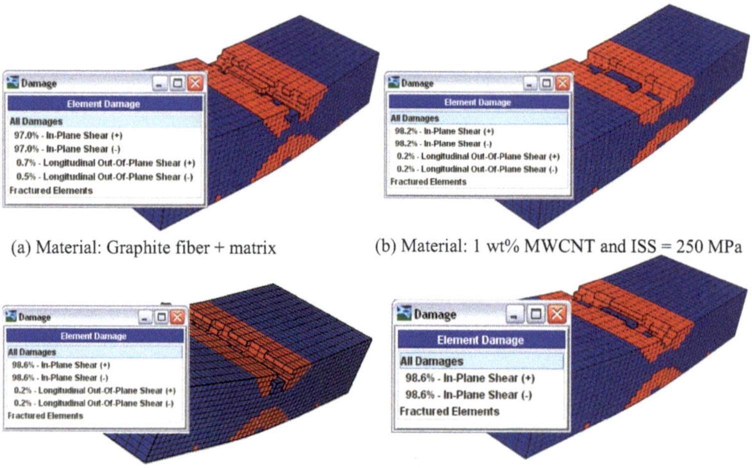

(a) Material: Graphite fiber + matrix (b) Material: 1 wt% MWCNT and ISS = 250 MPa

(c) Material: 5 wt% MWCNT and ISS = 250 MPa (d) Material: 5 wt% MWCNT and ISS = 50 MPa

Figure 5.6 Predicted final failure of short-beam three-point bending using progressive failure analysis in GENOA using calculated multiscale composite material properties: (a) ILSS = 34.40 MPa; (b) 41.81 MPa; (c) 44.28 MPa; (d) 40.56 MPa. Ply lay-up is $[0/45/-45]_s$; fiber volume ratio (FVR) = 0.6 and void volume ratio (VVR) = 0.0.

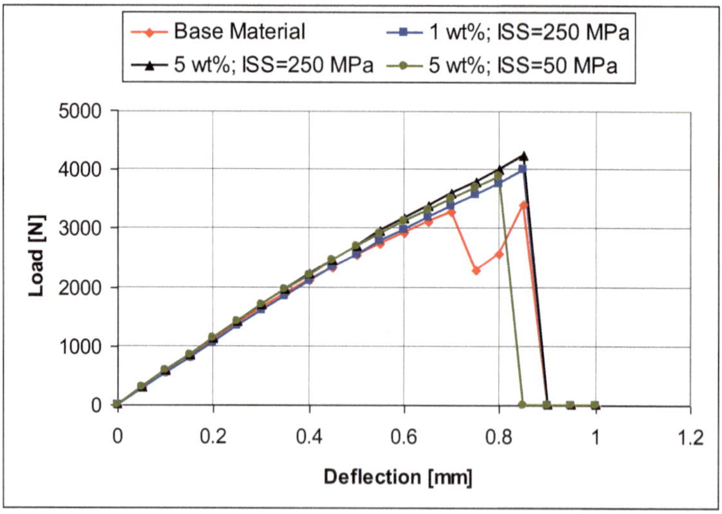

Figure 5.7 Load versus deflection from short-beam shear (SBS) simulations in GENOA using coated conventional fibers in the neat matrix. Ply lay-up is $[0/45/-45]_s$, fiber volume ratio (FVR) = 0.6, and void volume ratio (VVR) = 0.0.

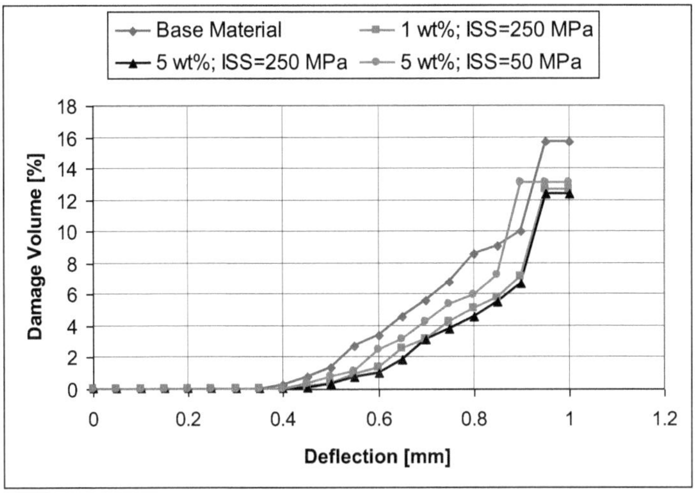

Figure 5.8 Percent damage volume versus deflection from short-beam shear (SBS) simulations in GENOA using coated conventional fibers in neat matrix; interfacial shear strength (ISS). Ply lay-up is $[0/45/-45]_s$; fiber volume ratio (FVR) = 0.6 and void volume ratio (VVR) = 0.0.

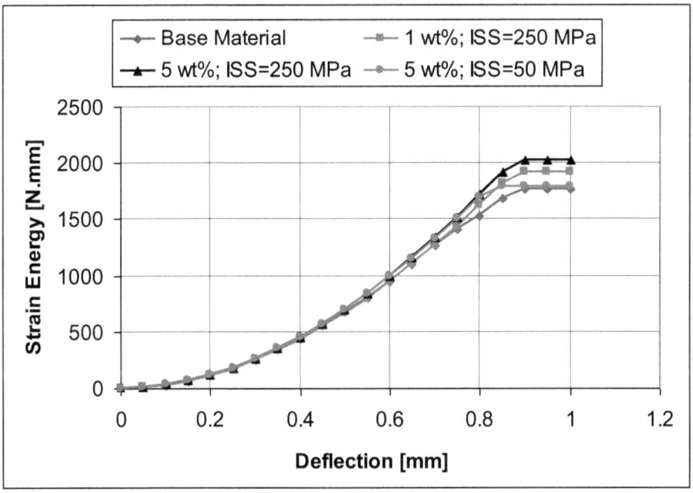

Figure 5.9 Strain energy versus deflection from short-beam shear (SBS) simulations in GENOA using coated conventional fibers in neat matrix; interfacial shear strength (ISS). Ply lay-up is $[0/45/-45]_s$; fiber volume ratio (FVR) = 0.6 and void volume ratio (VVR) = 0.0.

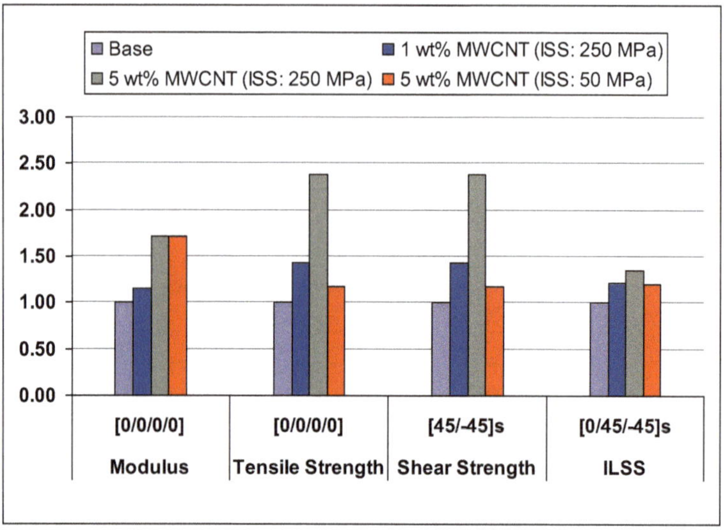

Figure 5.10 Comparison of results: normalized data for aligned enhanced matrix (modulus, tensile, and shear strength) and coated conventional fibers in neat matrix (ILSS).

Figure 5.10 also highlights the advantage of multiscale composites for improving strength and stiffness of FRP laminates. The values were normalized with the corresponding neat matrix and neat matrix plus conventional fiber properties. The ILSS values obtained for the simulations are listed in the caption for Fig. 5.6. The ILSS improved by 21.51% on addition of 1 wt% MWCNT in our simulation, while Fan et al. (2008) report 18.2% rise for similar amount of MWCNTs added in the matrix. Results validate the new suggested approach, showing a very close-to-average correlation with test data available in literature [24]. In order to run the ILSS simulations, the interphase failure theory [19], five additional parameters were required to define the interphase properties as input in GENOA and are listed below.

1. Ratio of tensile strength (interphase to matrix)
2. Ratio of shear strength (interphase to matrix)
3. Ratio of interphase thickness to fiber radius
4. Ratio of normal modulus (interphase to matrix)
5. Ratio of Poisson's ratio (interphase to matrix)

The values taken were based on experimental studies and calculated effective isotropic enhanced material properties. They were 1.44, 1.11, 0.13, 1.15, 0.99 and 2.39, 1.25, 0.13, 1.72, 0.97 for the 1 wt% and 5 wt% cases, respectively. The interphase thickness and glass fiber radius were assumed to be 500 nm [17] and 4 μm, respectively. The interphase shear strength ratio values were taken from the adhesion shear strength experimental data [24].

Future work will include refining and validating the above approach further and simulating the CNT FE models that consider van der Waal forces between the adjacent tubes in MWCNTs [25].

5.4 Conclusions

A unique approach that combines the analytical/numerical and progressive failure analysis was used to simulate multiscale hybrid composites. First, effective, transversely isotropic, enhanced matrix properties were estimated using Mori–Tanaka formulation. Then, isotropic (random) properties were estimated using classical laminate theory by generating quasi-isotropic ply lay-ups, where each ply has unidirectional properties determined from the first step. Lastly, the effective isotropic properties were used as interphase properties between the conventional fibers and the neat matrix. The advanced material was finally characterized by simulating standard ASTM tests, such as tension and short-beam strength. The simulation results obtained easily capture complete deformation behavior of these materials, including failure strength, while minimizing the dependency on complex MD simulations and tests. The short-beam strength, three-point bending simulation indicated that ILSS improved by 21.51% on addition of 1 wt% MWCNT—a significant improvement. The simulation results were comparable to the tests in literature and captured the effect of nanoparticle aspect ratio, material properties, and even ISS. More efforts are anticipated in the near future toward simulating static, fatigue, and impact along with validation with test data.

Acknowledgment

We would like to thank Dr. Chris Chamis from NASA Glenn (Cleveland, OH) and Prof. Levon Minnetyan from Clarkson University (Potsdam, NY) for taking the time for valuable discussions on the progressive failure analysis process.

References

1. Gojney, F. H., Wichmann, M. H. G., Fiedler, B., Bauhofer, W., and Schulte, K. (2005). Influence of nano-modification on the mechanical and electrical properties of conventional fibre-reinforced composites. *Composites: Part A* **36**, 1525–1535.
2. Garcia, E. J., Wardle, B. L., Hart, J., and Yamamoto, N. (2008). Fabrication and multifunctional properties of a hybrid laminate with aligned carbon nanotubes grown in situ. *Comp. Sci. Tech.* **68**, 2034–2041.
3. Njuguna, J., Pielichowski, K., and Alcock, J. R. (2007). Epoxy-based fibre reinforced nanocomposites. *Adv. Eng. Mater.* **9**, 835–847.
4. Veedu, V. P., Cao, A., Li, X., Ma, K., Soldano, K., S., Ajayan, P. M., et al. (2006). Multifunctional composites using reinforced laminae with carbon-nanotube forests. *Nat. Mater.* **5**(6), 457–462.
5. Yeh, M. K., Tai, N. H., and Lin, Y. J. (2007). Mechanical properties of phenolic-based nanocomposites reinforced by multi-walled carbon nanotubes and carbon fibers. *Composites, Part A* **39**, 677–684.
6. Ganguli, S., Aglan, H., Dennig, P., and Irvin, G. (2006). Effect of loading and surface modification of MWCNTs on the fracture behavior of epoxy nanocomposites. *J. Reinf. Plast. Compos.* **25**, 175–188.
7. Qui, J., Zhang, C., Wang, B., and Liang, R. (2007). Carbon nanotubes integrated multifunctional multiscale composites. *Nanotechnology* **18**, 11.
8. Fan, Z., Santare, M. H., and Advani, S. G. (2007). Interlaminar shear strength of glass fiber reinforced epoxy composites enhanced with multi-walled carbon nanotubes. *Composites: Part A* **39**, 540–554.
9. Mori, T., and Tanaka, K. (1973). Average stress in matrix and average elastic energy of materials with misfitting inclusions. *Acta Metall.* **21**, 571–574.

10. Tandon, G. P., and Weng, G. J. (1984). The effect of aspect ratio of inclusions on the elastic properties of unidirectionally aligned composites. *Polym. Compos.* **5**, 327–333.

11. Tucker, C. L., and Liang, E. (1999). Stiffness predictions for unidirectional short-fiber composites: review and evaluation. *Compos. Sci. Tech.* **59**, 655–671.

12. Eshelby, J. D. (1957). The determination of the elastic field of an ellipsoidal inclusion and related problems. *Proc Roy. Soc. A* **241**, 376–396.

13. Courtney, T. H. (2000). *Mechanical Behavior of Materials* (2nd ed.). New York, McGraw-Hill.

14. Barber, A. H., Cohen, S. R., Eitan, A., Schadler, L. S., and Wagner, H. D. (2006). Fracture transitions at a carbon-nanotube/polymer interface. *Adv. Mater.* **18**, 83–87.

15. Frankland, S. J. V., Caglar, A., Brenner, D. W., and Griebel, M. (2002). Molecular simulation of the influence of chemical cross-links on the shear strength of carbon nanotube-polymer interfaces. *J. Phys. Chem. B* **106**, 3046–3048.

16. Xiao, K. Q., and Zhang, L. C. (2004). The stress transfer efficiency of a single-walled carbon nanotubes in epoxy matrix. *J. Mater. Sci.* **39**, 4481–4486.

17. Thostenson, E. T., Li, W. Z., and Ren, Z. F., Chow, T. W. (2002). Carbon Nanotube/carbon fiber hybrid multi-scale composites. *J. Appl. Phys.* **91**(9), 6034–6037.

18. Qian, D., Dikey, E. C., Andrews, R., Rantell, T. (2000). Load transfer and deformation mechanisms in carbon nanotube-polystyrene composites. *Appl. Phys. Lett.* **76**, 2868.

19. Minnetyan, L., Su, X., and Abdi, F. (2008). Fiber-matrix interphase effects on damage progression in composite structures. *AIAA*, 2008–1732.

20. Garg, M. (2005). Mechanics of deformation of carbon nanotubes. Master's thesis. Massachusetts Institute of Technology, Cambridge, MA.

21. GENOA Durability and Damage Tolerance and Life Prediction Software, Alpha Star Corporation (2008). User Manual, http://www.ascgenoa.com.

22. Huang, D., Abdi, F., and Mossallam, A. (2003). Comparison of failure mechanisms in composite structure. *SAMPE* Conference Paper.

23. Sokolinsky, V. S., Housner, J., Surdenas, J., and Abdi, F. (2006). Progressive failure analysis of shuttle reinforced carbon-carbon plate specimens. *AIAA*, 2006-1789.

24. Hsiao, K.-T., Alms, J., and Advani, S. G. (2003). Use of epoxy/multiwalled carbon nanotubes as adhesives to join graphite fibre reinforced polymer composites. *Naonotechnology* **14**, 791–793.
25. Garg, M., Pantano, A., and Boyce, M. C. (2007). An equivalent orthotropic representation of the nonlinear elastic behavior of multiwalled carbon nanotbues. *J. Eng. Mat. and Tech.* **129**, 431–439.

Chapter 6

Validation for Multiscale Composites: Glass/Epoxy/Silica Nanoparticles

Mohit Garg[a] **and Parviz Yavari**[b]

[a]*AlphaStar Corporation, 5150 E. Pacific Coast Highway, Suite 650, Long Beach, CA 90804, USA*

[b]*California State University, 1250 Bellflower Blvd, Long Beach, CA 90840, USA*

parviz.yavari@csulb.edu

A multiscale micromechanics technique has been proposed to characterize an advanced multiscale composite system constituting continuous microscale diameter fibers and a matrix reinforced/infused with spherical-shaped nano-size particles of silica. The technique was used to predict effective constituent (fiber/matrix) strength and stiffness material properties using limited test data that is available from Uddin and Sun (2008). The effective constituent properties were then used to simulate longitudinal tension and compression and transverse tension unidirectional coupons using combined progressive failure analysis and finite element method. The predicted ply properties compared reasonably well with the test data within $\pm 4\%$ error.

Characterization of Nanocomposites: Technology and Industrial Applications
Edited by Frank Abdi and Mohit Garg

ISBN 978-981-4669-02-3 (Hardcover), 978-1-315-36489-6 (eBook)
www.panstanford.com

6.1 Introduction

Advanced fiber-reinforced plastics (FRP) composite materials are being used increasingly for aerospace applications because of their higher strength-to-weight ratio compared to metals. However, composite materials are weaker in transverse direction (transverse to the fibers) than in longitudinal direction. This makes the composite material prone to delamination due to matrix failure. In addition, the longitudinal composite compressive properties are also highly dependent on the stiffness and strength of matrix material (Uddin and Sun, 2008).

Research work has indicated that infusing matrix with nanoparticles can significantly enhance both strength and stiffness properties of the neat matrix. Using the stiffened matrix material improves the FRP material system as well. Manufacturing these advanced multiscale composite material systems is also quite challenging, and the challenges include uniformly dispersing the nanoparticles in the matrix, improving the interface between the matrix and the particles for proper load transfer from matrix to the nanoparticles, and removing voids, among others.

This chapter will characterize and predict the "effective" fiber and the matrix (with and without uniformly dispersed nanoparticles) strength and stiffness material properties. The limited test data are obtained from Uddin and Sun (2008). The effective fiber and matrix properties are different from the bulk fiber and matrix properties and lump the thermal residual stresses due to coefficient of expansion mismatch between the constituents during curing process of the composite plies.

6.1.1 *Experimental*

Uddin and Sun (2008) presented their experimental process to create a multiscale composite material system constituted of E-glass fibers reinforcing silica-nanoparticle-infused epoxy. The epoxy used in their experimental work was diglycidylether of bisphenol A (DGEBA). The nanoparticles consisted of spherical-shaped nano-size (~20 nm diameter) silica particles (Nanopox F400). The nanoparticles were uniformly distributed in the matrix

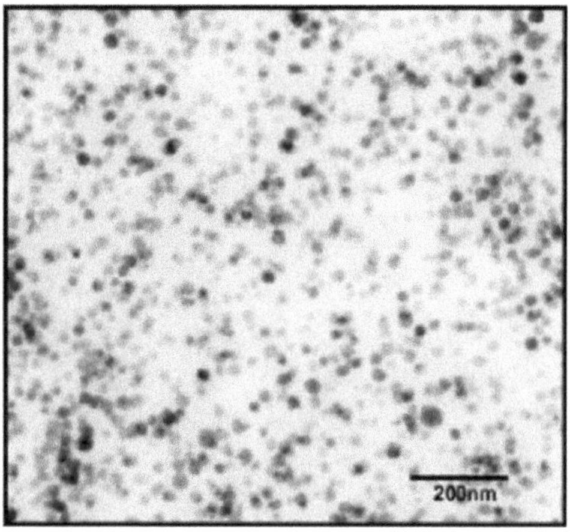

Figure 6.1 Transmission electron microscope image of 15 wt% silica/ epoxy resin (Uddin and Sun, 2008).

Table 6.1 Stiffness and compressive strength properties for DGEBA epoxy infused with and without 15 wt% silica nanoparticles

Material properties	Neat resin (DEGBA)	Nanocomposites (DEGBA + Nanopox F400)
Young's Modulus [MPa]	1660	2410.0
Compressive Strength [MPa]	92	108.0

and constituted 15 wt% in the end product (epoxy/silica). Figure 6.1 shows the transmission electron microscope image of Nanopox F400 particles uniformly dispersed in epoxy.

Table 6.1 lists the constituent and ply properties extracted from the experimental work from Uddin and Sun (2008).

6.2 Methodology

6.2.1 *Multiscale Micromechanics*

The constituent material properties can be estimated using limited lamina test data and multiscale micromechanics relations (Murthy

Table 6.2 Stiffness and strength properties for E-glass fiber reinforced DGEBA epoxy infused with and without 15 wt% silica nanoparticles. Tensile and compressive coupons consisted of 6 and 16 unidirectional plies, respectively

Material properties	Fiber volume ratio [%]	Baseline (glass/epoxy)	Multiscale composite (glass/epoxy/nanoparticles)
Long. Stiffness [MPa]	47	24000.0	24000.0
Trans. Stiffness [MPa]	47	4700.0	6627.0
Long. Tensile Strength [MPa]	47	600.0	666.0
Trans. Tensile Strength [MPa]	47	17.5	23.1
Long. Comp. Strength [MPa]	42	360.0	650.0
Long. Comp. Strength [MPa]	50	420.0	680.0

and Chamis, 1986). Commercial durability and damage-tolerant (DADT) software was used to predict the constituent properties. The module uses the micromechanics relations to predict the lamina/laminate properties. The formulation in the module also accounts for the manufacturing variables such as fiber and void volume ratio. The following process (Fig. 6.2) was adopted to predict

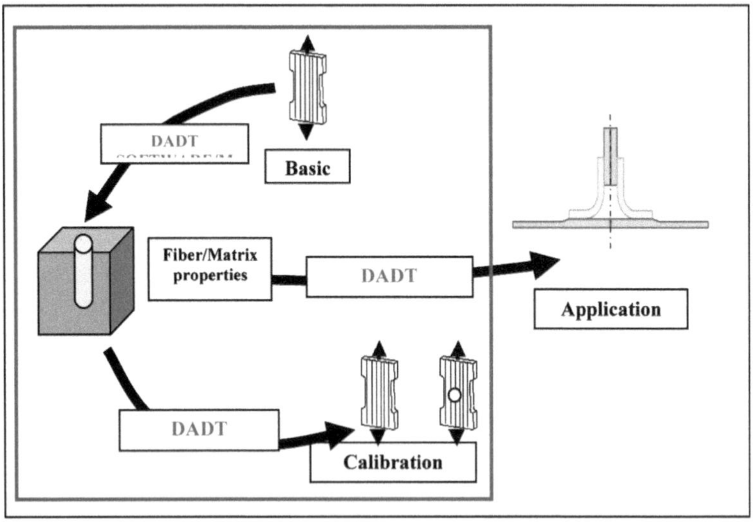

Figure 6.2 Material characterization process in DADT software using multi-scale micromechanics (Garg et al. 2008).

and validate the material properties using the limited test data (Uddin and Sun, 2008):

1. Obtain results for five tests: longitudinal tension and compression, transverse tension and compression, and in-plane shear.
2. Obtain the fiber and void volume ratio for the layup.
3. Obtain approximate bulk constituent (fiber/matrix) properties from literature or manufacturer.
4. Reverse-engineer/predict constitute material properties using micromechanics equations so that the ply properties match the five tests listed under step 1. Note that if five tests are not available, then the manufacturer properties are assumed to be the constituent properties.
5. Once the constituent material properties are known, progressive failure analysis is used in conjunction with finite element methods to predict the five tests for validity.

6.2.2 *Progressive Failure Analysis*

Progressive failure analysis is the approach widely used for failure analysis in composite materials. In this approach, failure in each ply constituting the laminate, under investigation, is assessed on the basis of the ply limit strength/strain properties, as shown in Fig. 6.3. Some failure criteria are maximum stress- or strain-based, and some interactive. The damaged plies are then either removed from the FE model or given a negligible stiffness to simulate a broken ply and study its effect on overall laminate. Later, once all the plies in an element are damaged, the element is removed. The stresses in individual plies are obtained using classical laminate theory, which are then compared against the limit stresses or strains parameters using default failure criteria (Garg et al., 2008).

Micromechanics is usually used to find the ply strength properties if fiber/matrix constituent properties are used as an input to the code. A commercial damage and durability tolerance software was used in conjunction with commercial FE solvers. The finite element solver was used to evaluate the stresses, strains, and displacement values at the element level and DADT software was used to evaluate damage at the ply level constituting the element in the FE model.

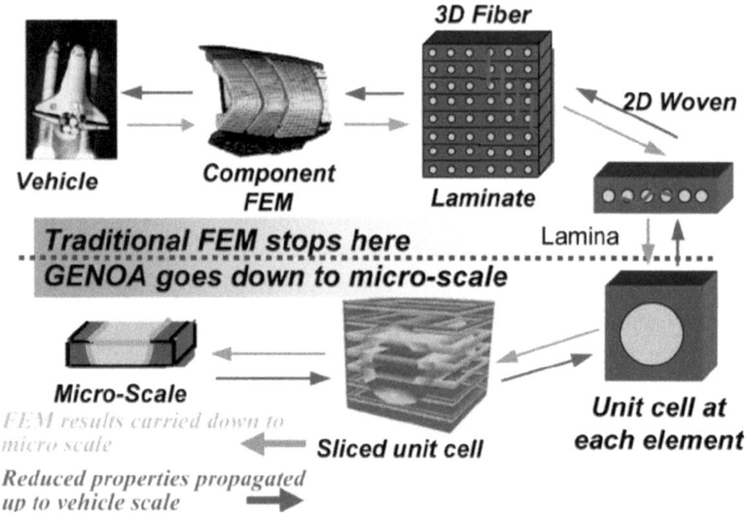

Figure 6.3 Progressive failure analysis: micro–macro scale interaction in DADT software (Garg et al., 2008).

The process was iterative and went back and forth between DADT software and the finite element solver. Once DADT software assessed that the element was damaged, it automatically controlled the material stiffness matrix, which was fed as an input to finite element solver for further stress calculations. The process was repeated until the structural component failed. The DADT software has about 25 strength- and strain-based failure criteria. It has the option to define the multiscale composite material system constituent properties as either fiber (transversely)/matrix (isotropic) or ply (anisotropic) properties. The PFA approach was tested elsewhere for various material system and different layups (Garg et al., 2008). The software is an extension to the DADT software originally developed at NASA (Minnetyan et al., 1990).

6.2.3 *Material Modeling Assumptions*

Owing to lack of the desired five tests listed earlier, some assumptions were made to characterize the fiber/matrix material properties:

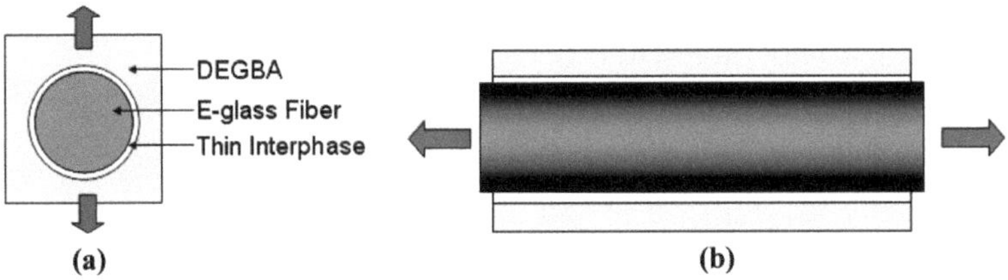

Figure 6.4 Modeling interphase effects for progressive failure analysis: (a) matrix-dominant failure (cross-section view); (b) fiber-dominant failure (side view).

1. linear elastic material properties for all constituents
2. void content assumed to increase linearly with increase in epoxy amount for both baseline (glass/epoxy) and multiscale composite (glass/epoxy/silica nanoparticles) material systems
3. the combined and effective enhanced matrix (epoxy/silica nanoparticles) material properties determined as enhanced
4. weak interface between E-glass fiber and epoxy when enhanced with silica nanoparticles interface (multiscale composite)
5. no interface between E-glass fiber and epoxy (baseline case)

Note that the interface mostly affects the transverse tensile properties as they are dependent on matrix properties and not the longitudinal properties, which are dependent on fiber properties, as shown in Fig. 6.4. Further details can be found in Minnetyan et al. (1990).

6.3 Results

Using the micromechanics approach in DADT software and the limited test data listed under Tables 6.1 and 6.2, the constituent (E-glass fiber and DGEBA epoxy) effective material properties were predicted and are listed in Tables 6.3–6.5. The effective properties were found using unidirectional tape with fiber volume ratio (FVR) of 47% and void volume ratio of 3.8%. As per Uddin and Sun (2008), it was difficult to create void free coupons. Only tensile strength of the interface was assumed to be lower than the enhanced matrix surrounding the E-glass fibers, and the rest of the properties were

Table 6.3 Predicted effective epoxy (DGEBA) matrix material properties

Material properties	Symbol	Before	After	Units
Young's Modulus	Em	3350	1657	[MPa]
Poisson's Ratio	vm	0.35	0.35	[–]
Tension Strength	SmT	80	33.8	[MPa]
Compression Strength	SmC	120	92	[MPa]
Shear Strength	SmS	68.95	59	[MPa]

Table 6.4 Predicted effective epoxy (DGEBA) matrix infused with 15 wt% silica nanoparticles material properties

Material properties	Symbol	Before	After	Units
Young's Modulus	Em	1657	2405	[MPa]
Poisson's Ratio	vm	0.35	0.35	[–]
Tension Strength	SmT	33.8	148	[MPa]
Compression Strength	SmC	92	108	[MPa]
Shear Strength	SmS	59	72.1	[MPa]

Table 6.5 Predicted effective E-glass fiber material properties

Material properties	Symbol	Before [7]	After	Units
Longitudinal Young's Modulus	Ef11	72390	51300	[MPa]
Transverse Young's Modulus	Ef22	72390	45700	[MPa]
Poisson's Ratio	vf12	0.2	0.25	[–]
Poisson's Ratio	vf23	0.2	0.25	[–]
Shear Modulus	Gf12	30130	20520	[MPa]
Shear Modulus	Gf23	30130	20520	[MPa]
Longitudinal Tension Strength	Sf11T	2482	1306	[MPa]
Longitudinal Compression Strength	Sf11C	2482	1480	[MPa]

*Kaddour et al. (2008).

taken to be the same as the enhanced matrix (DGEBA/Nanopox F400).

Once the constituent effective material properties were obtained, PFA approach was used to simulate the unidirectional coupon tests for longitudinal tension and compression and transverse tension

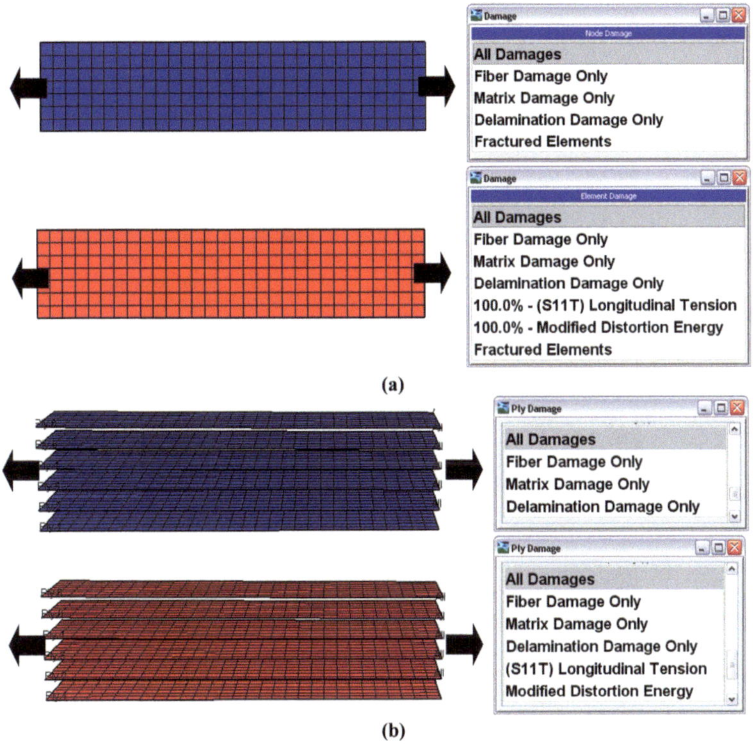

Figure 6.5 Finite element model used to simulate longitudinal tension with progressive failure analysis. (a) Laminate-level damage; (b) ply-level damage (blue and red color indicates no damage vs. damage, respectively).

cases. The FE model used to simulate the tension and compression cases consisted of a rectangular-shaped coupon made using 210 4-node shell elements. Figure 6.5 shows the laminate- and ply-level damage for the longitudinal tension case. Figures 6.6–6.8 show the comparison between the test and the predictions from PFA simulations for unidirectional baseline (E-glass/DGEBA) and multiscale (E-glass/DGEBA/Nanopox F400) composites. Figures 6.6 and 6.7 show the stress–strain curves obtained from simulation. Note that the test stress–strain curves were not available; therefore, strength and stiffness values from Uddin and Sun (2008) are shown in Figs. 6.6 and 6.7.

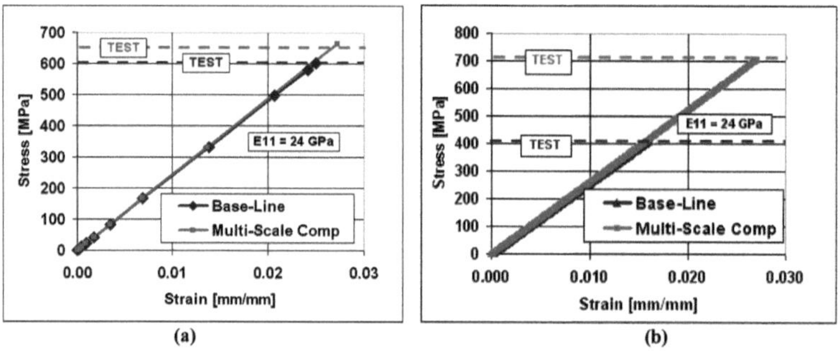

Figure 6.6 Progressive failure analysis simulated stress–strain curves for longitudinal unidirectional coupons; baseline (E-glass/DGEBA) and multiscale composite (E-glass/DGEBA/Nanopox F400); VVR = 3.8%: (a) longitudinal tension (FVR = 47%); (b) longitudinal compression (FVR = 50%) (E11 and E22 are the longitudinal [||to fiber] and transverse [to fiber] modulus).

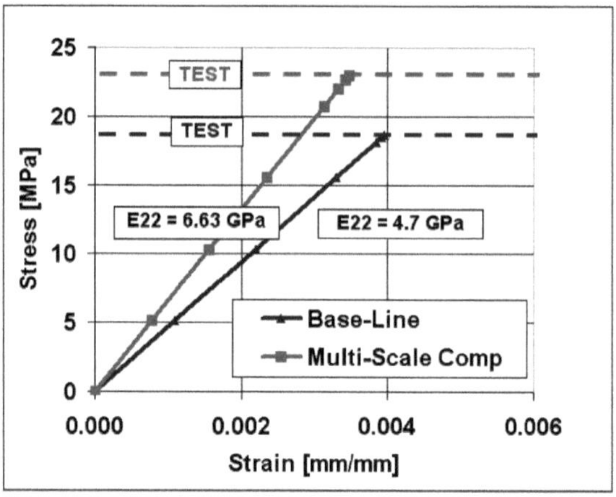

Figure 6.7 Progressive failure analysis simulated stress–strain curves for transverse unidirectional coupons; baseline (E-glass/DGEBA) and multiscale composite (E-glass/DGEBA/Nanopox F400); FVR = 47%, VVR = 3.8% (E11 and E22 are the longitudinal [||to fiber] and transverse [to fiber] modulus).

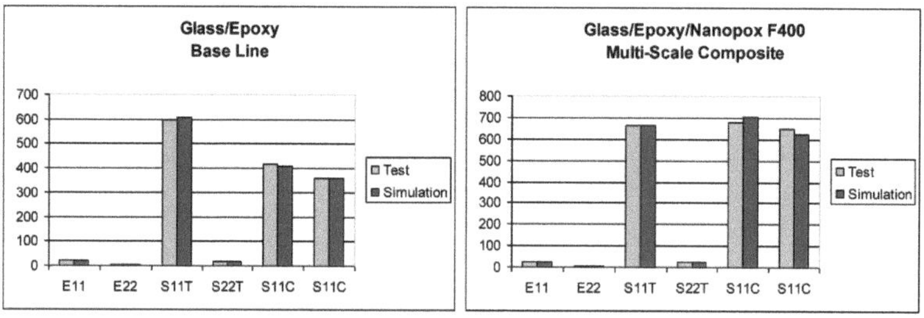

Figure 6.8 Comparison between test and simulation data: (a) baseline; (b) multiscale composites. Note that the first S11C is for FVR of 50% and the second for FVR = 42%. (E11, E22, S11T, S22T, and S11C are longitudinal and transverse modulus, longitudinal and transverse strength and longitudinal compressive strength, respectively.)

Table 6.6 Comparison between predicted and test data for unidirectional laminates made of baseline and multiscale composites

Material properties	FVR [%]	Baseline (glass/epoxy) Test	Simulation	% Error	Multiscale composite (glass/epoxy/nanoparticles) Test	Simulation	% Error
Long. Stiffness [MPa]	47	24000.0	24040.0	0.00	24.0	24.4	0.00
Trans. Stiffness [MPa]	47	4700.0	4708.0	0.00	6.6	6.6	0.00
Long. Tensile Strength [MPa]	47	600.0	606.6	1.10	666.0	666.0	0.00
Trans. Tensile Strength [MPa]	47	17.5	17.5	0.00	23.1	23.0	−0.43
Long. Comp. Strength [MPa]	42	360.0	360.0	0.00	650.0	624.0	−4.00
Long. Comp. Strength [MPa]	50	420.0	409.0	−2.62	680.0	704.0	3.53

*Uddin and Sun (2008).

Table 6.6 shows that, using the effective fiber and matrix properties, the PFA in conjunction with the FE method is able to predict the stiffness and strength properties of a complex material system with ease. The table also shows the maximum percent error difference between predictions and test data to be less than ±4%.

6.4 Conclusions

A multiscale micromechanics technique was used to characterize an advanced multiscale composite system constituting continuous microscale diameter fibers and matrix reinforced/infused with spherical-shaped nano-size particles of silica. The technique was used to predict constituent (fiber/matrix) strength and stiffness material properties using limited test data available in Uddin and Sun (2008). The predicted effective constituent properties were then used to predict the ply properties using progressive failure analysis (PFA) combined with finite element method (FEA). The effectiveness of the combined PFA and FEM was evident from the simulation results that predicted the unidirectional laminate properties, which compared with the test data within $\pm 4\%$ error. The calibrated properties can now be used to predict other higher-order tests such as impact, or any large complex parts used for automobile/aerospace and electronic applications.

References

1. Uddin, M. F. and Sun, C. T. (2008). Strength of unidirectional glass/epoxy composite with silica nanoparticle-enhanced matrix. *Composites Science and Technology* **68**, 1637–1643.
2. Murthy, P. L. N. and Chamis, C. C. (1986). Integrated composites analysis (ICAN). Technical paper 2515, NASA, Cleveland.
3. Garg, M., Abumeri, G. H., and Huang, D. (2008). Predicting failure design envelop for composite material system using finite element and progressive failure analysis approach. SAMPE, Long Beach, California, 18–22 May.
4. Minnetyan, L., Murthy, P. L. N., and Chamis, C. C. (1990). Composite structure global fracture toughness via computational simulation. *Computers & Structures* **37**, 175–180.
5. Kaddour, A. S., Hinton, M. J., Li, S., and Smith, P. A. (2007). Damage theories for fiber-reinforced polymer composites: the third world-wide failure exercise (WWFE-III). In Proceedings of the 16th International Conference on Composite Materials, Kyoto, Japan, 8–13 July.

Chapter 7

Influence of Nanoparticles and Effect of Defects on Mode I and II Fracture Toughness and Impact Resistance

Christos C. Chamis,[a,*] Frank Abdi,[b] Harsh Baid,[b] and Parviz Yavari[c]

[a]*Former Senior Aerospace Scientist, NASA Glenn Research Center, Cleveland, OH 44135, USA*
[b]*AlphaStar Corporation, 5150 E. Pacific Coast Highway, Suite 650, Long Beach, CA 90804, USA*
[c]*California State University at Long Beach, Long Beach, CA 90804, USA*
parviz.yavari@csulb.edu, christos.c.chamis@aol.com

Recent tests performed [1] on multiscale composites constituting of E-glass fiber with DEGBA epoxy enriched with 15 wt% silica nanoparticles showed about 27% and 55% improvement in Mode I and II fracture toughness over baseline (E-glass/DEGBA) material system. In addition, the multiscale composite system showed improvement in damage resistance capability when subjected to low-velocity impact tests. Analyzing such complex multiscale systems is a difficult task and usually requires lots of parameters to be considered that are difficult to parameterize because of the nanoscale involvement. In our previous paper [2], a multiscale

*Retired.

Characterization of Nanocomposites: Technology and Industrial Applications
Edited by Frank Abdi and Mohit Garg

ISBN 978-981-4669-02-3 (Hardcover), 978-1-315-36489-6 (eBook)
www.panstanford.com

micromechanics approach was used to predict the constituent material properties and later predict the laminate properties within ±4% error when compared to the test. Here we used the previously calibrated constituent material properties [2] for E-glass fiber and DEGBA epoxy enriched with and without the Nanopox F400 silica particles to simulate the double cantilever beam (DCB), end-notched flexure (ENF), and low-velocity impact tests. The simulated load–displacement curve for the DCB and ENF fracture toughness tests compared well with the test data [1] and the corresponding peak load was within 10% error. For low-velocity impact test, the simulation results for peak load were found to be within 33% error, while the maximum deflection and damage footprints were less than 10% error.

7.1 Introduction

The full potential of an advanced composite material system is not realized because of mostly brittle nature of the matrix (epoxy) used to bond the strong fibers together. The brittle behavior is the result of molecular cross-links in the epoxy. Although the cross-link density improves the stiffness, it adversely affects the toughness and delamination resistance properties of the epoxies. Uddin et al. (2008) recently showed experimentally that addition of 15 wt% silica nanoparticles (Nanopox F400) in DEGBA epoxy helped increase the Mode I and Mode II fracture toughness. In addition, reduction in damage footprint was observed for low-velocity impact (drop tower) tests. Similar experimental studies have been performed earlier that show improvement in fracture toughness and impact resistance by enriching the matrix with nanoparticles of different shapes and sizes.

In our earlier paper [2], we presented multiscale micromechanics approach to predict the constituent material properties and later used them to predict the laminate strength and stiffness properties using PFA approach. The results obtained were in close agreement with the test data (±4%). Here, the same constituent properties are used for simulating the fracture toughness and impact tests, as shown by the red box in Fig. 7.1.

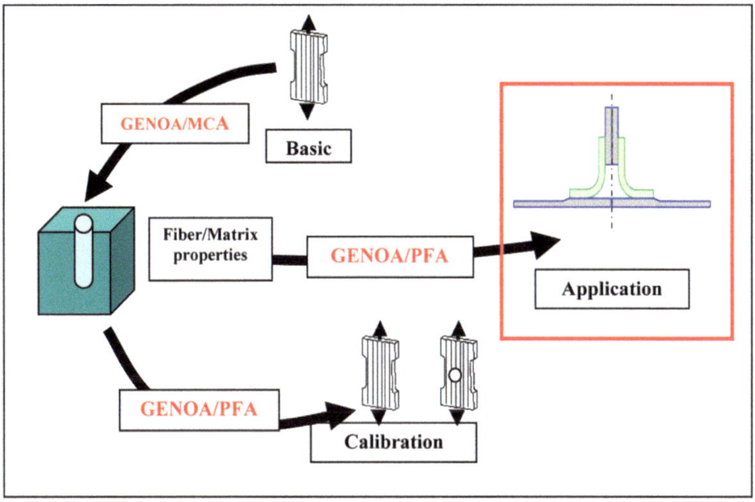

Figure 7.1 Material characterization process in GENOA using multiscale micromechanics [2].

In this chapter, we will simulate the double cantilever beam (DCB), end-notched flexure (ENF), and low-velocity impact tests. The DCB and ENF tests are simulated using linear elastic fracture mechanics (LEFM)-based methodology, virtual crack closure technique (VCCT), with finite element method (FEM), and low-velocity impact simulation using progressive failure analysis (PFA) in conjunction with FEM.

In the following sections, first the PFA- and LEFM-based VCCT approaches are explained briefly, followed by the FE model details, and finally the results and conclusions.

7.2 Methodology

7.2.1 *Virtual Crack Closure Technique (VCCT)*

A linear elastic fracture mechanics (LEFM)-based virtual crack closure technique (VCCT) was used to simulate Mode I and Mode II fracture toughness DCB and ENF tests, respectively. The VCCT is a widely used approach for simulating fracture initiation and subsequent growth for materials that exhibit linear elastic behavior

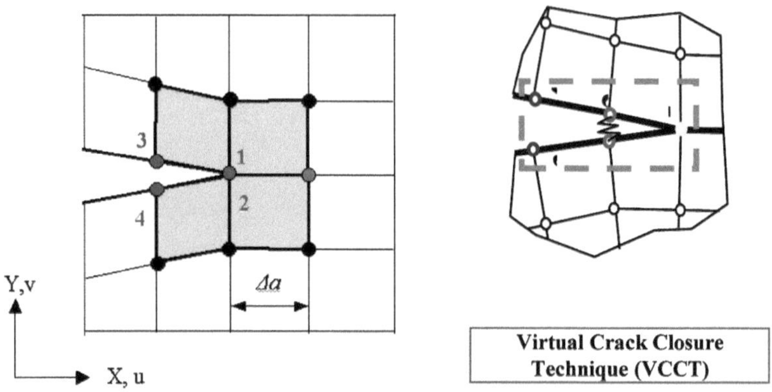

Figure 7.2 Material characterization process in GENOA using multiscale micromechanics [3].

until failure. A commercial durability and damage tolerant (DADT) software, GENOA, was used for this study. In GENOA, VCCT uses spring-type elements and applies the cohesive law as the spring internal force versus nodal separation.

In the chosen software, using VCCT technique with FE analysis explicitly determines the fracture mode separation. Using this technique, for crack propagation analysis, the crack advancement is controlled via sequentially releasing the nodes along the crack path (Fig. 7.2). The nodes in an FE model are initially tied together with highly stiff virtual spring elements and are released when Mode I, II, and III components of strain energy release rate exceed the mixed-mode fracture criteria, as shown in Eq. 7.1. The fracture criteria compare the instantaneous fracture energy release rate with the critical fracture energy release rate values of the material under consideration using the power law (Eq. 7.2). The instantaneous fracture energy release rates are calculated on the basis of the mesh size at the crack front, nodal forces at the crack tip, and relative nodal displacements behind the crack tip and crack edge. Further details of the approach can be found elsewhere [3].

$$G_{\mathrm{I}} \cong \frac{F_{y12}(v_3 - v_4)}{2B\Delta a}; \quad G_{\mathrm{II}} \cong \frac{F_{x12}(u_3 - u_4)}{2B\Delta a}; \quad G_{\mathrm{III}} \cong \frac{F_{x12}(w_3 - w_4)}{2B\Delta a} \tag{7.1}$$

$$\frac{G_{\mathrm{I}}}{G_{\mathrm{IC}}} + \frac{G_{\mathrm{II}}}{G_{\mathrm{IIC}}} + \frac{G_{\mathrm{III}}}{G_{\mathrm{IIIC}}} > 1 \tag{7.2}$$

Here G_{I}, G_{II}, G_{III}, v, u, w, B, Δa, and F are Mode I, II, and III fracture energy release rates, x, y, and z displacement components, thickness, crack length, and force components, respectively. The critical components of fracture energy release rates are shown with subscript "c."

7.2.2 *Progressive Failure Analysis (PFA) [2]*

Progressive failure analysis is the approach widely used for failure analysis in composite materials. In this approach, failure in each ply constituting the laminate, under investigation, is assessed on the basis of the ply limit strength/strain properties, as shown in Fig. 7.3. Some failure criteria are maximum stress- or strain-based and some interactive. The damaged plies are then either removed from the FE model or given a negligible stiffness to simulate a broken ply and study its effect on overall laminate. Later, once all the plies in an element are damaged, the element is removed. The stresses in individual plies are obtained using classical laminate theory, which are then compared against the limit stresses or strains parameters using default failure criteria. Micromechanics is usually used to find the ply strength properties if fiber/matrix constituent properties are used as an input to the code. Again GENOA is used in conjunction with another commercial FE solver, LS-DYNA, to simulate the impact analysis. The FE solver is used to evaluate the stress, strain, and

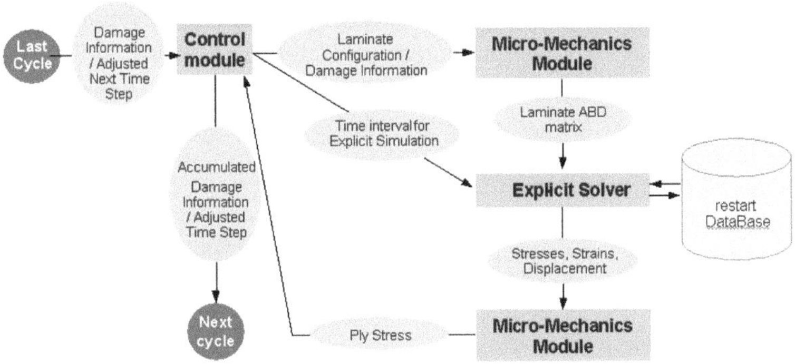

Figure 7.3 Flowchart for multiscale and hierarchical material/structural progressive failure dynamic analysis (PFDA) [4].

displacement values at the element level and GENOA is used to evaluate damage at the ply level constituting the element in the FE model. The process is iterative and goes back and forth between GENOA and the FE solver. Once GENOA assesses that the element is damaged, it automatically controls the material stiffness matrix which is fed as an input to FE solver for further stress calculations. The process repeats until the structural component fails. GENOA has about 25 strength- and strain-based failure criteria and has option to define ply properties as fiber/matrix or ply properties [2].

7.2.3 *Material Modeling Assumptions [2]*

Because of lack of the desired five tests listed earlier, some assumptions were made to characterize the fiber/matrix material properties:

1. Linear elastic material properties for all constituents
2. The void content assumed to increase linearly with increase in epoxy amount for both baseline (glass/epoxy) and multiscale composite (glass/epoxy/Nanopox F400) material systems
3. The combined and effective enhanced matrix (epoxy/Nanopox F400) material properties determined as enhanced
4. Weak interface between E-glass fiber and epoxy when enhanced with Nanopox F400 interface (multiscale composite)
5. No interface between E-glass fiber and epoxy (baseline case)

Note that the interface mostly affects the transverse tensile properties as they are dependent on matrix properties, and not the longitudinal properties, which are dependent on fiber properties, as shown in Fig. 7.4. Further details can be found elsewhere [6].

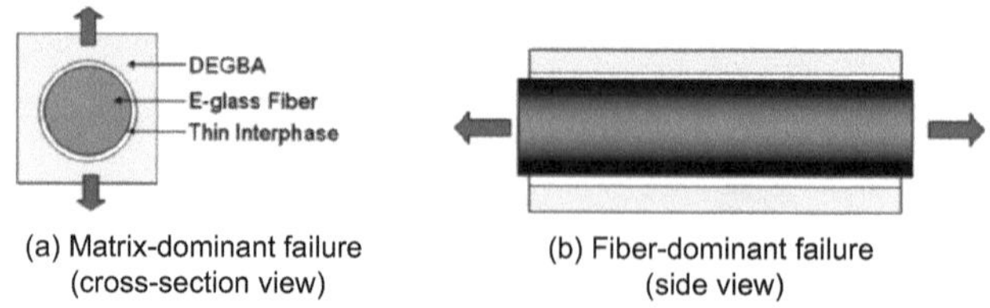

Figure 7.4 Modeling interphase effects for progressive failure analysis [2].

7.3 FE Model and Experimental Setup [1]

Uddin and Sun [5] presented their experimental process to create a multiscale composite material system constituted of E-glass fibers reinforcing silica nanoparticle infused epoxy. The epoxy used in their experimental work was diglycidylether of bisphenol A (DEGBA). The uniformly dispersed nanoparticles consisted of nano-sized spherical shape (~20 nm diameter) 15 wt% silica particles (Nanopox F400). A similar process and material system was later used for evaluating the fracture toughness and impact resistance tests [1]. Figures 7.5a and 7.6a show the experimental setup for DCB and ENF fracture toughness tests, while Figures 7.5b and 7.6b show the corresponding FE models. Figure 7.7 shows the FE model used for simulating the low-velocity impact analysis. The DCB and ENF tests

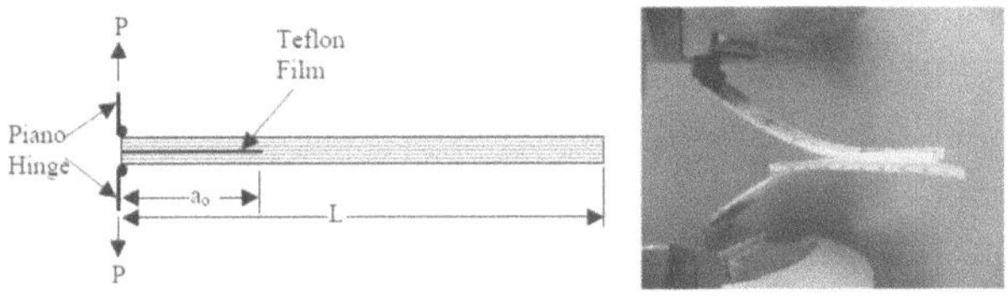

(a) Test Setup [1]

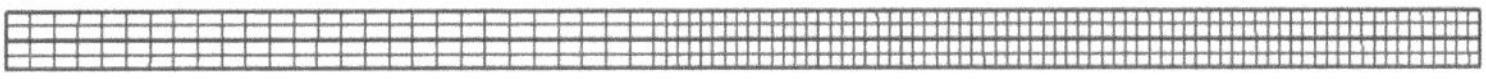

(b) FE model; 4-node shell elements = 480; 125.0 × 21.0 × 4.5 mm with a_0 = 55.0 mm

Figure 7.5 Double cantilever beam (DCB) test.

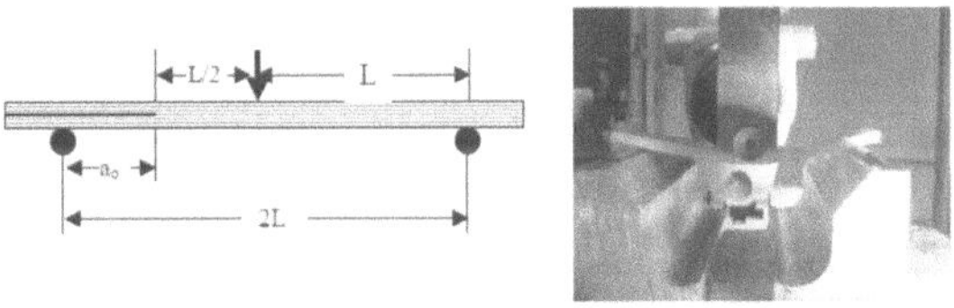

(a) Test Setup [1]

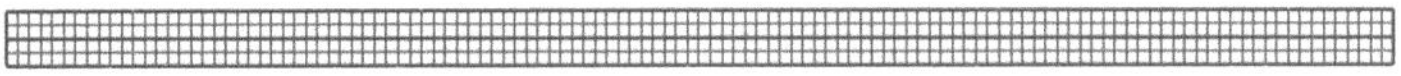

(b) FE model; 4-node shell elements = 480; L = 50 mm, b = 21.0 mm, t = 4.5 mm with a_0 = 25.0 mm

Figure 7.6 End-notch flexure (ENF) test.

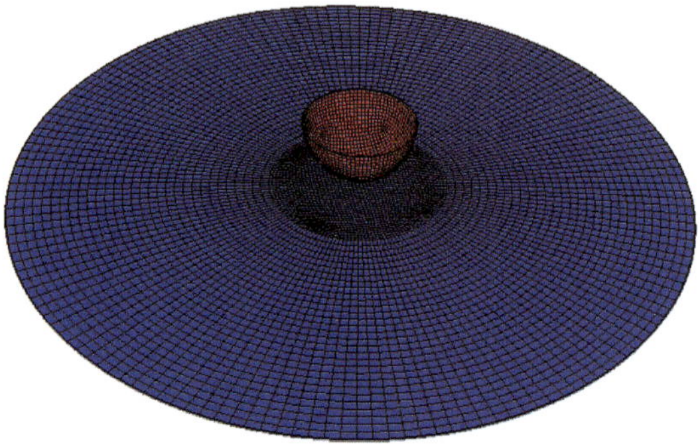

Figure 7.7 Finite element model for impact analysis; impactor (red) and panel (blue) diameters are 12.9 mm and 76.0 mm, respectively. The thickness of the panel is 3.5 mm; 8652 4-node shell elements.

were simulated with GENOA and MD NASTRAN, while low-velocity impact analysis was simulated with GENOA and LS-DYNA.

All DCB and ENF test coupons consisted of unidirectional plies with fiber volume ratio (FVR) of 50%, while the impact simulation specimen consisted of $[0_2/90_2]_s$ layup with the FVR of 43% and 44% for the neat and enriched epoxy cases, respectively. A void volume ratio (VVR) of 2.5% was assumed for fracture toughness simulations. The VVR was linearly varied as per previous work [2]. The impact simulation was performed for 5 J, 10 J, and 15 J impact energy cases with impactor mass of 5.77 kg.

Each ply was assigned the fiber matrix constituent material properties calibrated earlier [2]. Tables 7.1–7.3 show the effective material properties for DEGBA epoxy enriched with and without Nanopox F400 silica nanoparticles.

Table 7.1 Effective epoxy (DEGBA) matrix material properties [2]

Material properties	Symbol	Value	Units
Young's Modulus	Em	1.657	[GPa]
Poisson's Ratio	vm	0.35	[-]
Tension Strength	SmT	33.8	[MPa]
Compression Strength	SmC	92	[MPa]
Shear Strength	SmS	59	[MPa]

Table 7.2 Effective epoxy (DEGBA) matrix infused with Nanopox F400 (15 wt%) material properties [2]

Material properties	Symbol	Value	Units
Young's Modulus	Em	2.405	[GPa]
Poisson's Ratio	vm	0.35	[-]
Tension Strength	SmT	148	[MPa]
Compression Strength	SmC	108	[MPa]
Shear Strength	SmS	72.1	[MPa]

Table 7.3 Effective E-glass fiber material properties [2]

Material properties	Symbol	Value	Units
Longitudinal Young's Modulus	Ef11	51.3	[GPa]
Transverse Young's Modulus	Ef22	45.7	[GPa]
Poisson's Ratio	vf12	0.25	[-]
Poisson's Ratio	vf23	0.25	[-]
Shear Modulus	Gf12	20.52	[GPa]
Shear Modulus	Gf23	20.52	[GPa]
Longitudinal Tension Strength	Sf11T	1306	[MPa]
Longitudinal Compression Strength	Sf11C	1480	[MPa]

7.4 Results

7.4.1 *Mode I and Mode II Fracture Toughness Analysis*

The FE models shown in Figs. 7.5b and 7.6b are simulated using VCCT approach. The thickness of each shell element is set to 21.0 mm to account for the specimen width. The model contains a predefined crack path with duplicate nodes at the center of the FE model. The duplicate nodes are assigned with virtual spring elements leaving the initial crack opening. No micromechanics-based strength or strain-based failure criteria are used for such simulation. The failure is defined based on the power law (Eq. 7.2).

Figure 7.8 shows snapshots of the DCB and ENF test simulations, while Fig. 7.9 shows the corresponding load–displacement curve from simulations compared to the test data. For DCB test, the

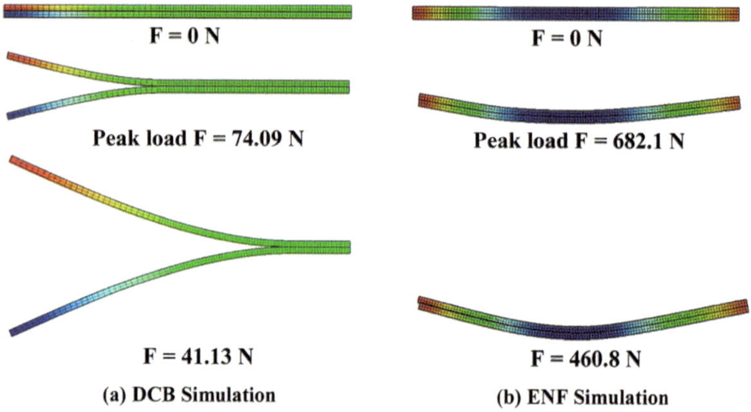

Figure 7.8 Snapshots of the DCB and ENF tests at initial, peak load, and after peak load. The color indicates the displacement (red positive, blue negative).

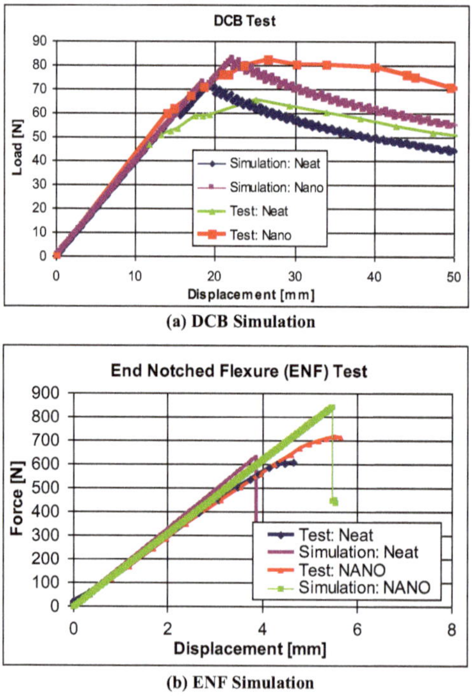

Figure 7.9 Comparison of simulated load–displacement curves for the DCB and ENF tests for base line and multiscale composite material system.

Table 7.4 Fracture energy release rate from test data [2]

Material system	GIC [J/(m.m)]	GIIC [J/(m.m)]
[E-glass/DEGBA]	1780	1300
[E-glass/DEGBA/Nanopox F400]	2540	2092

load–displacement test data were extrapolated from the compliance versus crack length and resistance curves provided in reference [1]. The simulated load–displacement curves for both baseline and multiscale composite material system cases agree well with the test data within acceptable errors. For both simulations, the critical values from reference [1] were taken as input to the simulation and specifically Eq. 7.2. Table 7.4 lists the critical energy release values used for the baseline and multiscale composite system.

7.4.2 *Low-Velocity Impact Analysis*

Unlike the fracture toughness DCB and ENF simulations in the previous sections, progressive failure dynamic analysis is performed using GENOA based user defined material card with strength-based failure criteria in LS-DYNA. Figure 7.10b shows the impact damage foot print predictions from simulations and are compared against the tested footprints in Fig. 7.10a. There is quite similarities between the tested and simulated footprints. The simulated data shows only delamination based footprints and is more detailed than what is available via tests. Figure 7.11 shows ply-by-ply damage from the simulation as well, for 10 J impact energy for base-line material system at the end.

Figure 7.12 and Table 7.5 below show the simulation results: load versus time and peak load and maximum deflection at peak load for the two material systems subjected to three impact energy levels.

Recall that these simulations were run using the effective fiber and matrix properties. Any variations in properties due to difference in batches were not considered here. The test impact analysis specimens were prepared from different batches according to

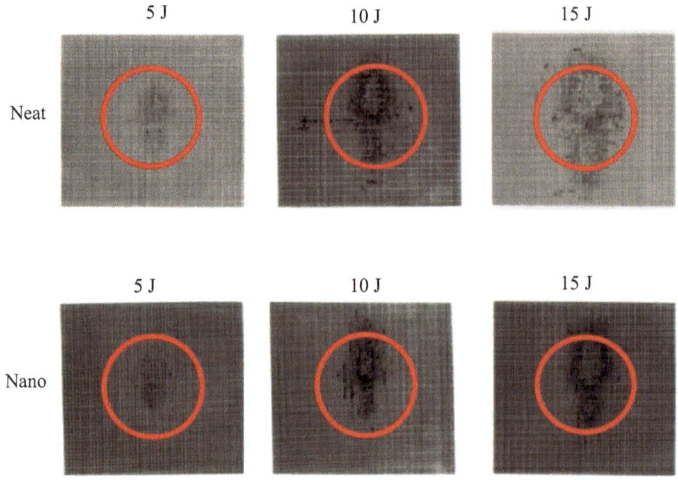

(a) Test Data [1]; panel dimensions of 97 × 97 × 3.5 mm; red circle diameter = 76 mm

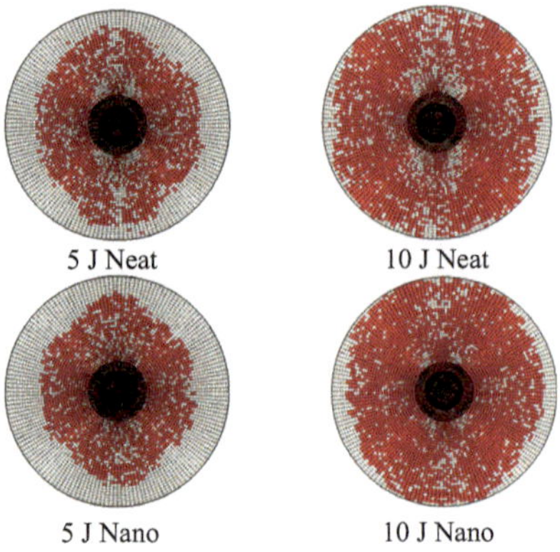

(b) Simulation Data; panel diameter = 76 mm

Figure 7.10 Comparison of tested and simulated damage (mode II delamination) footprints; red shows the damage accumulation at the end of the simulation. Notice slightly less damage pattern in nano case versus neat case.

reference [1]. However, Figs. 7.10–7.12 and Table 7.5 support the argument that the PFDA approach in conjunction with the FEM reasonably predicts the damage footprint, maximum deflection at peak load, and the load versus time curve.

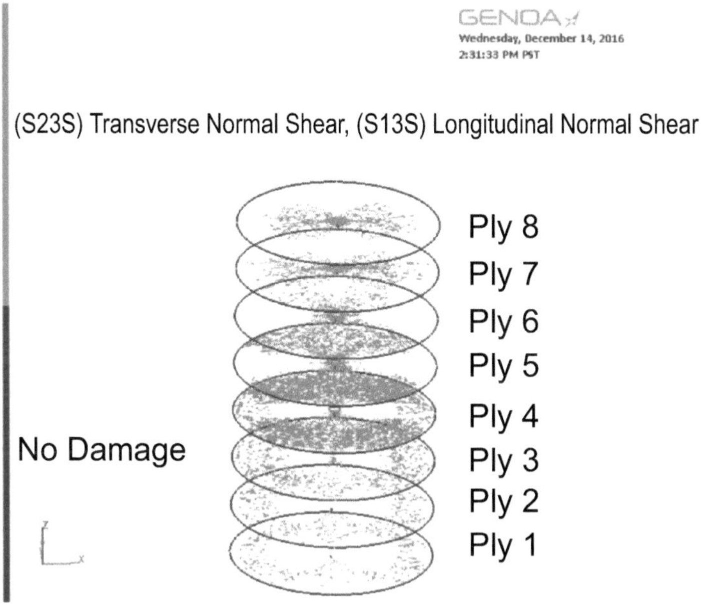

Figure 7.11 Simulated ply-by-ply damage for base-line material system at the end (impact energy of 10 J).

Table 7.5 Progressive failure dynamic analysis results for impact analysis; neat (E-glass/DEGBA); nano (E-glass/DEGBA/Nanopox F400)

Test results [Ref]				
	5 J		**10 J**	
Parameters	**Neat**	**Nano**	**Neat**	**Nano**
Peak Load [kN]	1.71	1.77	2.54	2.61
Deflection at Peak Load [mm]	4.5	4.13	6.1	6
Simulation results				
Peak Load [kN]	2.1	2.35	2.63	2.83
Deflection at Peak Load [mm]	4.27	3.72	5.574	5.91
% Error				
% Difference (Peak Load)	22.81	32.77	3.54	8.43
% Difference (Displacement)	−5.11	−9.93	−8.69	−1.50

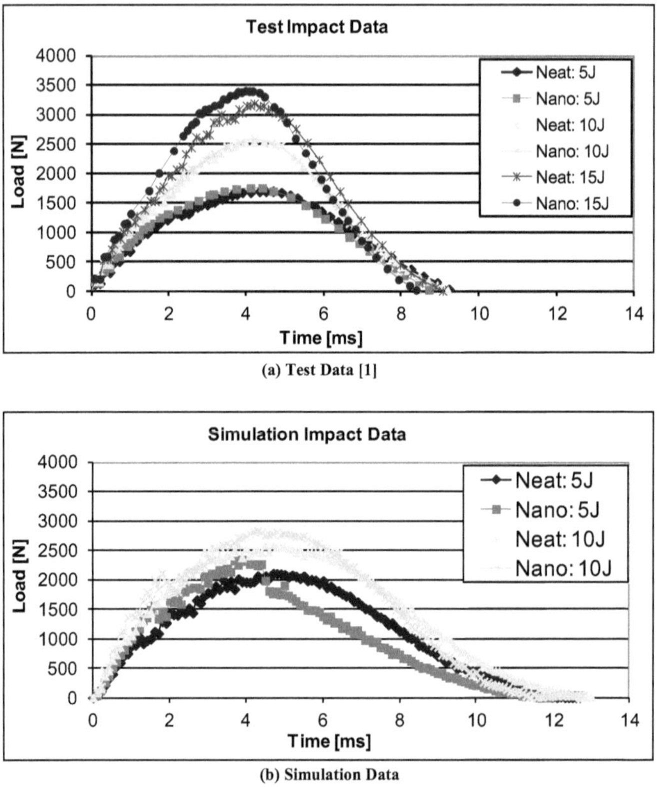

Figure 7.12 Progressive failure dynamic analysis simulated load versus time curves; neat (E-glass/DEGBA) and nano (E-glass/DEGBA/Nanopox F400).

7.5 Conclusions

Analyzing a complex multiscale composite material system is a very challenging problem. In addition, predicting failure is even more challenging. The material system usually requires several expensive tests because of the nanoscale particles in the composite system. In this chapter, we showed that using PFA and linear elastic fracture mechanics–based methods along with FE method can prove to be helpful in predicting the complete load–displacement curves, validating the fracture energy release rates, damage footprints at laminate and ply-by-ply levels, and the corresponding load and displacements during the impact simulations. For the selected material system, the simulated load–displacement curve for the DCB and ENF fracture toughness tests compared well with the test data [1] and the corresponding peak load was within 10% error. For low-velocity impact test, the simulation results for peak load were found to be within 33% error, while the maximum deflection and damage footprints were less than 10% error.

References

1. Uddin, M. F., and Sun, C. T. (2008). Impact resistance and toughness of composite laminates with nanoparticle-enhanced matrix. SAMPE 2008, Long Beach, CA, May 18–22.
2. Chamis, C. C., Garg, M., Morgan, D., Yavari, P., and Abdi, F. (2009). Material properties characterization for multi-scale composites: Glass/epoxy/silica nanoparticles. The Third International Conference on Advances and Trends in Engineering Materials and Their Applications, Canada.
3. Xie, D., Garg, M., Huang, D., and Abdi, F. (2008). Cohesive zone model for surface cracks using finite element analysis. May 2008 AIAA, Illinois.
4. Garg, M., Abdi, F., Zammit, A., Bayandor, J., and Song, S. (2009). Impact damage resistance and compression-after-impact strength of sandwich composites. SAMPE Baltimore, 2009.
5. Uddin, M. F., and Sun, C. T. (2008). Strength of unidirectional glass/epoxy composite with silica nanoparticle-enhanced matrix. *Composites Science and Technology* 68:1637–1643.

Chapter 8

Prediction/Verification of Composite Electrical Properties and Nano-Insertion Improvement*

Levon Minnetyan,[a] Frank Abdi,[b] Christos C. Chamis,[c,†] and Dade Huang[b]

[a]*Civil and Environmental Engineering, Mail Stop 5710, Clarkson University, Potsdam, NY 13699-5710, USA*
[b]*AlphaStar Corporation, 5150 E. Pacific Coast Highway, Suite 650, Long Beach, CA 90804, USA*
[c]*Former Senior Aerospace Scientist, NASA Glenn Research Center, Cleveland, OH 44135, USA*
lminnety@clarkson.edu, christos.c.chamis@aol.com

Composite equivalent electrical properties are computed on the basis of simplified micromechanics equations.[1] Dielectric constant, dielectric strength, and conductivity are computed utilizing the constituent material properties, effective medium, and percolation theories. 2D/3D composite electrical properties' simulation requires the algorithm development that combines the effective medium and percolation theories. A physically based percolation model is proposed to characterize the effective electrical conductivity of particle-filled composites by means of the combination of

*Mesomechanics 2009.
†Retired.

Characterization of Nanocomposites: Technology and Industrial Applications
Edited by Frank Abdi and Mohit Garg

ISBN 978-981-4669-02-3 (Hardcover), 978-1-315-36489-6 (eBook)
www.panstanford.com

effective medium (EM) and percolation equations with universal exponents. It is shown that the present model agrees well with the experimental data. Furthermore, the effect of particle shape on the percolation threshold and the effect of electrical conductivity of particle on the effective properties of composites are discussed.

8.1 Introduction

The algorithm development involves consideration of (a) electrical conductivity of the matrix considering the effect of voids, (b) resistivity, (c) longitudinal electrical conductivity of unidirectional ply (rule of mixtures), (d) transverse electrical conductivity of unidirectional ply, and (e) laminate electrical conductivity Ke and the inverse of conductivity or resistivity Ωe. This chapter validates against test data the prediction [1] of lamina and laminate dielectric strength and the assembly of laminate dielectric strength properties. In addition, the model considers the effect of nanomaterial insertion such as carbon nanotube (CNT) random fiber composites [2]. Nanotechnology test results suggest the presence of a CNT–matrix interface that lowers the composite conductivity compared to predictions without considering the presence of a CNT–matrix interface. The test results justify the lower than expected conductivity of the composite according to the percolation model that uses the CNT, matrix, and interface conductivities. Test results indicate that even at high concentrations, CNTs do not actually touch each other because they are covered by the interface zone. Therefore, the model must take into account the presence of the CNT–matrix interface zone and its electrical properties. The predictive method compares the conventional and nanocomposite electrical properties with CNT concentrations in the matrix, including a CNT–matrix interface. Simulation results are validated with test data with good agreement.

Nomenclature

Ke_{m} electrical conductivity of matrix
k_{v} fraction of voids of composite by volume
k_{m} matrix volume fraction of composite

k_{f} fiber volume fraction of composite
Ke_{v} electrical conductivity of voids
Ke_{l11} lamina longitudinal electrical conductivity
Ke_{l22} lamina transverse electrical conductivity
Ke_{l33} lamina normal electrical conductivity

8.1.1 *Composite Electrical Conductivity Equations Based on Simplified Micromechanics Theory*

When voids are present the electrical conductivity of the matrix is modified by [3, 4]

$$Ke_{\mathrm{m}} = \left(1 - \sqrt{k_{\mathrm{v}}}\right) Ke_{\mathrm{m}} + \frac{\sqrt{k_{\mathrm{v}}} Ke_{\mathrm{m}}}{1 - \sqrt{k_{\mathrm{f}}}\left(1 - \frac{Ke_{\mathrm{m}}}{Ke_{\mathrm{v}}}\right)} \tag{8.1}$$

Longitudinal electrical conductivity of unidirectional ply is computed by the rule of mixtures:

$$Ke_{l11} = k_{\mathrm{f}} Ke_{f11} + k_{\mathrm{m}} Ke_{\mathrm{m}} \tag{8.2}$$

Transverse electrical conductivity of unidirectional ply is given by

$$Ke_{l22} = \left(1 - \sqrt{k_{\mathrm{f}}}\right) Ke_{\mathrm{m}} + \frac{\sqrt{k_{\mathrm{f}}} Ke_{\mathrm{m}}}{1 - \sqrt{k_{\mathrm{f}}}\left(1 - \frac{Ke_{\mathrm{m}}}{Ke_{f22}}\right)} \tag{8.3}$$

$$Ke_{l33} = Ke_{l22} \tag{8.4}$$

Laminate electrical conductivity properties are computed in exactly the same manner as those of thermal conductivity.

Dielectric Constant Formulation: The dielectric constant γ is defined as the ratio of the permittivity of the material, ε, to the permittivity of vacuum, ε_{o}, or $\gamma = \varepsilon/\varepsilon_{\mathrm{o}}$. The permittivity of vacuum is $\varepsilon_{\mathrm{o}} = 8.85 \times 10^{-12}$ F/m $= 8.85 \times 10^{-12}$ Coulombs/(volt-meter) $= 0.225 \times 10^{-12}$ Coulombs/(Volt-inch). Similar to electrical conductivity, dielectric constant γ is a transport property as it is associated with the formation and transport of dipoles to accumulate charge, and therefore the composite dielectric constant can be computed similar to other transport properties such as electrical conductivity (cf. Agarwal and Broutman, p. 105; Zeng et al., p. 10972). Accordingly, longitudinal dielectric constant of unidirectional ply is computed by the rule of mixtures:

$$\gamma_{l11} = k_f \gamma_{f11} + k_m \gamma_m \tag{8.5}$$

Transverse dielectric constant of unidirectional ply is given by

$$\gamma_{l22} = \left(1 - \sqrt{k_f}\right) \gamma_m + \frac{\sqrt{k_f} \gamma_m}{1 - \sqrt{k_f}\left(1 - \frac{\gamma_m}{\gamma_{f22}}\right)} \tag{8.6}$$

$$\gamma_{l33} = \gamma_{l22} \tag{8.7}$$

8.2 Computation of Equivalent Electrical Properties of a Composite Material

GENOA composite mechanics module has been enhanced utilizing the effective medium and percolation theories to estimate (a) equivalent electric properties (MCA module): dielectric constant, dielectric strength, resistivity, and conductance for the composite.

A physically based percolation model was utilized to characterize the effective electrical conductivity of particle-filled composites by means of the combination of effective medium (EM) and percolation equations with universal exponents. It is shown that the present model agrees well with the experimental data. Furthermore, the effect of particle shape on the percolation threshold and the effect of electrical conductivity of particle on the effective properties of composites are discussed. The algorithm development involves consideration of (a) electrical conductivity of the matrix considering the effect of voids, (b) resistivity, (c) longitudinal electrical conductivity of the unidirectional ply (rule of mixtures), (d) transverse electrical conductivity of the unidirectional ply, and (e) laminate electrical conductivity Ke and the inverse of conductivity or resistivity Ωe. Figure 8.1 shows the software-generated ply electrical properties from a set of fiber and matrix and comparison with the test data.

8.3 Carbon Nanotube–Inserted Composites

It has been determined that insertion of a very small amount of carbon nanotubes (CNTs) in a composite improves the structural

Fiber Properties	**Matrix Properties**
Dielectric Strength in Longitudinal Direction =3.5E5 Dielectric Strength in Transverse Direction = 3.5E5 Dielectric Constant in Longitudinal Direction = 6.0 Dielectric Constant in Transverse Direction = 6.0 Conductivity in Longitudinal Direction = 0.01 Conductivity in Transverse Direction = 0.01	Dielectric Strength = 24000 Dielectric Constant = 4.0 Conductivity (1/Resistivity) =1.0E-13
Ply Properties Fiber Volume Fraction=0.4 Dielectric Strength in Longitudinal Direction = 3.33E4 Dielectric Strength in Transverse Direction = 4.64E4 Dielectric Constant in Longitudinal Direction = 4.46 Dielectric Constant in Transverse Direction =4.49 Conductivity in Longitudinal Direction = 3E-3 Conductivity in Transverse Direction = 1.66E-13	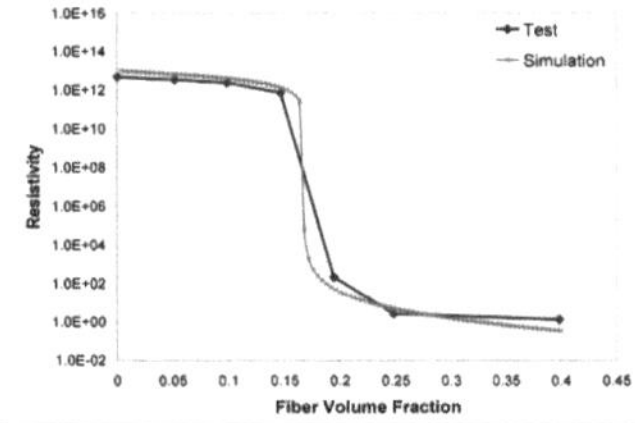

Figure 8.1 Electrical conductivity of a composite.

durability and also modifies the electrical conductivity of the matrix. A CNT in a composite matrix has been determined to form a thin interphase between the CNT and the matrix. The diameter of a CNT is 20 nm and the thickness of the interphase is approximately 2.5 nm. The average length of CNTs dispersed in a composite matrix is 25,000 nm. The conductivity of CNTs has been determined as 1850 S/cm, whereas the interphase conductivity is determined to be eight orders of magnitude lower at 1.85×10^{-5} S/cm. The conductivity of CNT/interphase combination is determined on the basis of the logarithmic average of the ratios of conductivity to distance of each phase. Therefore, we write the effective electrical conductivity of the CNT/interphase system as

$$\sigma_{\text{CNT/Interphase}} = (t + L) \times 10^{\left[0.5\times\left(\log\left(\frac{\sigma_{\text{INT}}}{t}\right)+\log\left(\frac{\sigma_{\text{CNT}}}{L}\right)\right)\right]} \qquad (8.8)$$

where t is the interphase thickness and L is the CNT length. For a typical CNT with the physical parameters given above, Eq. 8.8 gives the combined CNT/interphase conductivity as 18.50 S/cm. Simulation of the composite CNT/matrix conductivity using this value agrees well with test data, as shown in Fig. 8.2.

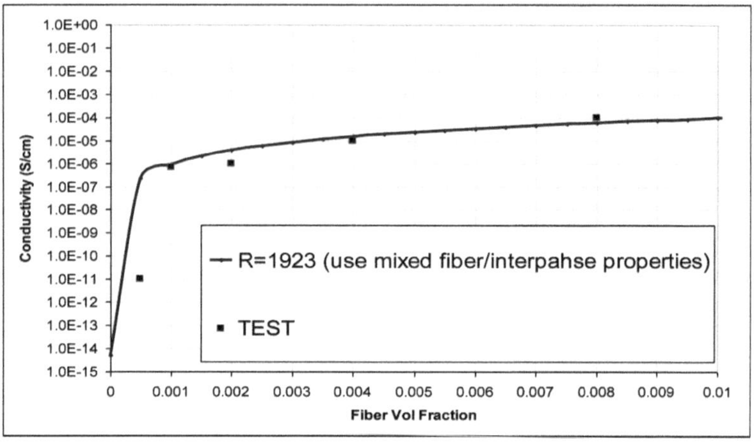

Figure 8.2 Comparison of simulation with test data for CNT-inserted composite conductivity.

References

1. Cai, W.-Z., and Tu, S. T. (2006). A physically based percolation model of the effective electrical conductivity of particle filled composites. *Journal of Composite Materials* 40:2131–2142.
2. Yan, K. Y., Xue, Q. Z., Zheng, Q. B., and Hao, L. Z. (2007). The interface effect of the effective electrical conductivity of carbon nanotube composites. *Nanotechnology* 18:255705.
3. Chamis, C. C. (1987). Simplified composite micromechanics for predicting microstresses. *Journal of Reinforced Plastics and Composites* 6:268–289.
4. Zeng, X. C., Bergman, D. J., Hui, P. M., and Stroud, D. (1988). Effective-medium theory for weakly nonlinear composites. *Physical Review B* 38(5):10970–10973.

Chapter 9

Polymer Nanocomposites as Ablative Materials: A Comprehensive Review

J. H. Koo,[a] M. Natali,[b] J. Tate,[c] and E. Allcorn[a]

[a] *The University of Texas at Austin, Dept. of Mechanical Engineering, Austin, TX 78739, USA*
[b] *University of Perugia, Dept. of Civil and Environmental Engineering, Terni, Italy*
[c] *Texas State University-San Marcos, Ingram School of Engineering, San Marcos, TX 78666, USA*
jkoo@mail.utexas.edu

A comprehensive review was conducted on the research and development of polymer nanocomposites used as thermal protection systems (TPS) for re-entry vehicles and as nozzle and internal insulation materials for propulsion systems. This chapter summarizes the significant R&D efforts on the studies of nanostructured ablative materials by scientists from China, India, Italy, Iran, and the US for the past decade. Thermoset, thermoplastics, elastomers, and thermoplastic elastomers were used as polymer matrices. Nanomaterials, such as montmorillonite organoclays, polyhedral oligomeric silssequioxanes (POSS), nanosilicas, nano-alumina, carbon black, carbon nanofibers (CNFs), carbon nanotubes (CNTs), flake graphite, and micro- and nanoscale ceramic particles were incorporated into the different resin matrices as fillers. Conventional fibers, such as asbestos, carbon, glass, graphitic felt,

Characterization of Nanocomposites: Technology and Industrial Applications
Edited by Frank Abdi and Mohit Garg

ISBN 978-981-4669-02-3 (Hardcover), 978-1-315-36489-6 (eBook)
www.panstanford.com

graphitic foam, Kelvar®, and silica were used as reinforcements. Ablation mechanisms were studied and proposed by the researchers on their nanostructured ablatives.

9.1 Introduction

Thermal protection materials are required to protect structural components of space vehicles during the re-entry stage, solid rocket motors (SRMs), liquid rocket engines, and missile launching systems. A thorough literature survey was conducted by Koo et al. (2006b, 2007b, 2007c) to review the development of these thermal protection materials for different military and aerospace applications. The series of literature surveys were grouped into (a) numerical modeling (Koo et al., 2006b), (b) materials thermophysical properties characterization (Koo et al., 2007b), and (c) experimental testing (Koo et al., 2007c).

The introduction of inorganic nanomaterials as additives into polymer matrix systems has resulted in polymer nanocomposites (PNCs) exhibiting multifunctional, high-performance polymer characteristics beyond what traditional polymer composites possess (Pinnavaia and Beall, 2000; Koo, 2006; Morgan and Wilkie, 2007; Mittal, 2010a, 2010b, 2012; Beall and Powell, 2011). Multifunctional features attributable to PNCs consist of improved thermal resistance, flame resistance, ablation resistance, moisture resistance, decreased permeability, chemical resistance, charge dissipation, and electrical/thermal conductivity (Pinnavaia and Beall, 2000; Koo, 2006; Morgan and Wilkie, 2007; Mittal, 2010a, 2010b, 2012; Beall and Powell, 2011). Through control/adjustment of the additive at the nanoscale level, one is able to maximize property enhancement of selected polymer systems to meet or exceed the requirements of current military, aerospace, and commercial applications.

One of the most important and pioneering research studies on the use of the montmorillonite (MMT) poly(caprolactam) (nylon 6) to a 2 wt% and 5 wt% loaded layered silicate/poly(caprolactam) nanocomposite was performed by Vaia et al. (1999). These materials were synthesized by in situ polymerization of ε-caprolactam in the presence of dispersed organically modified MMT nanoclay. This concept was further developed by researchers worldwide using different types of polymers such as nylon, phenolic, elastomer, and thermoplastic elastomer with different types of nanomaterials such as nanoclay, POSS, carbon nanofiber, multi-walled carbon nanotube, nano-alumina, nanosilica, and others. Traditional reinforcements such as asbestos, carbon, glass, and silica were also used. This chapter reviews the open literature on how researchers processed, characterized, and tested their polymer nanocomposites as thermal protection materials for low and high energy hyperthermal environments.

9.2 Behavior of Thermal Protection Materials

Polymeric composites have been used as ablative thermal protection systems (TPS) for a variety of military and aerospace applications. Thermal protection materials, such as carbon phenolics and carbon–carbon composites are used for spacecraft heat shields during the re-entry stage, thermal protection of missile launching system, and in SRMs as internal insulation and nozzle materials. Thermosetting polymer matrix composites are constructed of layers of woven carbon, glass, quartz, or silica fibers impregnated with a resin matrix. Phenolic resin is the most commonly used polymer matrix for ablatives. These thermal protection systems are exposed to a thermochemical flow and subjected to high temperatures in excess of 3000°C with very high heating rates.

Thermochemical ablation refers to the phenomenon of surface recession of an ablative due to severe thermal attack by an external heat flux. The initial heat transfer into the ablative occurs by pure conduction, and the resulting temperature rise causes material expansion (swelling) which may be attributed to pure thermal expansion as well as vaporization of any traces of moisture. When the material reaches a sufficiently high temperature, thermochemical degradation or pyrolysis of the polymer matrix begins. The pyrolysis reactions result in the production of decomposition gases and solid carbonaceous char residue. The thermal expansion and the disappearance of solid material due to the decomposition result in an increase in porosity and permeability

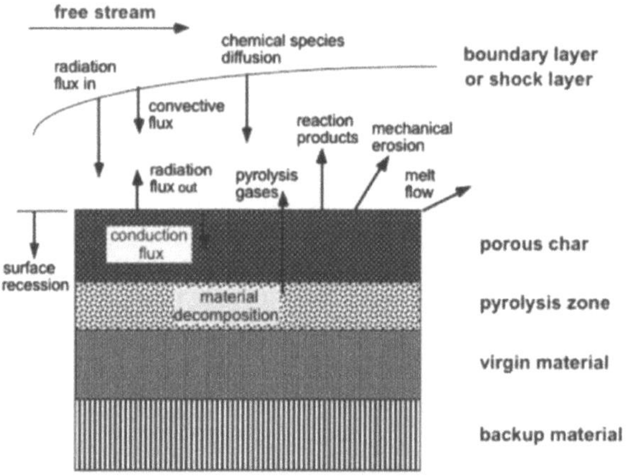

Figure 9.1 Energy accommodation mechanisms of ablative TPS materials.

of the polymeric material. Thus pyrolysis gases begin to escape through the polymeric material. For the thermal protection to be effective all these reactions must be strongly endothermic. The gases which flow through the char structure remove energy by convention, thus attenuating the conduction of heat to the reaction zone. This phenomenon is depicted in Fig. 9.1 (Laub et al., 2003).

9.3 Polymer Nanocomposites Review

A comprehensive literature review was conducted on open literature and relevant references are included in the chapter. There are only a small number of research groups in the world investigating polymer nanocomposites as advanced thermal protection materials that included the works of

- Air Force Research Laboratory (AFRL) Studies
- China Studies: Harbin Institute of Technology
- India Studies: Defence Research and Development Organisation (DRDO)
- Iran Studies: Tarbiat Modares University
- Italy Studies: University of Perugia, Politecnico di Torino, and University of Rome

- Mississippi State University Studies
- Texas State University–San Marcos Studies
- The University of Texas at Austin Studies

It is more meaningful to group all these studies in terms of the polymer matrix used in the investigators' studies. Table 9.1 shows a summary of the different polymer systems, types of nanocomaterials, and conventional fiber reinforcements used in their studies.

A. *Thermoplastic Nanocomposite Studies*

A.1 Nylon–Clay Nanocomposite Studies by AFRL, USA—Vaia et al. (1999)

The ablative performance of poly(caprolactam) (nylon 6) nancomposites was examined by Vaia et al. of Air Force Research Laboratory (ARFL) (Vaia et al., 1999). A relatively tough, inorganic char forms during the ablation of these nylon–clay nanocomposites resulting in at least an order-of-magnitude decrease in mass loss (erosion) rate relative to the neat polymer. This occurs for a small amount of 2 wt% ($\sim$0.8 vol%) exfoliated mica-type layered silicate. The nanoscopic distribution of silicate layers leads to a uniform char layer that enhances the ablative performance. The effect of concentration and size of the silicate layers as well as specific bonds between the polymer and the silicate on degradation kinetics and on the microstructure of the char layer are determined.

Poly(caprolactam) (nylon 6, N, Ube) was compared to a 2 wt% (NCH2) and 5 wt% (NCH5) layered silicate/poly(caprolactam) nanocomposites (Ube). These materials were synthesized by in situ polymerization of ε-caprolactam in the presence of dispersed organically modified montmorillonite. Additionally, a nominally 5 wt% layered silicate/poly(caprolactam) nanocomposite (NLS5) formed during melt processing of poly(caprolactam) (Capron B125WP, Allied Signal) and Cloisite 30A (organically modified MMT, Southern Clay Products) was examined. This exfoliated nanocomposite was fabricated using a co-rotating twin-screw extruder (MPC/V-30, Haake). In contrast to in situ polymerization of NCH2 and NCH5, melt processing of NLS5 is not expected to result in a large fraction

Table 9.1 Summary of the different polymer matrix used by the various research groups studying ablation using polymer nanocomposites

	Polymer matrix/institution		Nanoadditive/reinforcement	Remarks
A.1	Nylon 6	AFRL, USA	MMT organoclay	Improvement in ablative performance of nanocompoistes is associated with the char-forming characteristics of the polymer nanocomposites.
A.2	Poly(phenylene oxide)/Polystyrene (PPO/PS)	Politecnico di Torino, Italy	Modified clays (MMT, hydrotalcite, and sepiolite) and equiaxial, oxide and non-oxide powders	Sepiolite-added and alumina-based PPO composites showed the most promising behaviors. Only TGA, DTGA, cone calorimeter, and microstructural characterization were conducted in this study.
B.1	Different Polymers	AFRL, USA	MMT organoclay	Focused on polymer/layered silicate materials for use in space applications.
C.1	EPDM	AFRL, USA	POSS	Detailed the strategy employment of using sacrificial nanocompoistes along with applied research in the areas of SRM insulation and space-survival materials.
C.2	EPDM	Harbin Institute of Technology, China	Organo-MMT (OMMT) with Kevlar fibers	Up to 11 phr OMMT in EPDM erosion rate decreased, beyond it ablation resistance decreased.
D.1	NR	Tarbiat Modares Univ., Iran	MMT organoclay	Improvement in aging hardness, curing time, mechanical, flammability properties by the addition of 3 wt% of MMT clay.
E.1	Thermoplastic polyurethane (TPU)	The University of Texas at Austin, USA	MMT organoclay, CNF, or MWNT	MMT organoclay-based TPU nanocomposite (TPUN) has the best ablation performance, followed by MWNT-based TPUN.

F.1	Phenolic	Mississippi State Univ., USA	CNF with carbon fibers	The nano-scale dimensions of the VGCF causes major changes in the heat transfer rates and affects the resulting combustion/decomposition chemistry.
F.2	Phenolic	The Univ. of Texas at Austin, USA	MMT organoclay, POSS, or CNF with and without carbon fibers	High loading of MMT organoclay in phenolic, low loading of POSS in phenolic impregnated with carbon fibers, and high loading of CNF in phenolic without carbon fibers all gave better ablation performance than carbon phenolic baseline.
F.3	Phenolic	Harbin Institute of Technology and National Univ. of Defense Technology, China	POSS with carbon fibers (CFs); flake graphite reinforced barium-phenolic resin composite	Phenolic-POSS-CF produced the best charred surface, hence has a best ablation performance. Composite with flake graphite diameter of about 0.5 mm has the lowest ablation rate.
F.4	Phenolic	DRDO, India	Nanosilica with carbon fibers	Ablation resistance increased with the nanosilica content up to 2.0 wt% beyond it ablation resistance decreased.
F.5	Phenolic	Tarbiat Modares Univ., Iran	MMT organoclay with asbestos fibers	The 6 wt% MMT with asbestos fibers has the best ablation performance.
F.6	Phenolic	Univ. of Perugia, Italy	Nanosilica, carbon black, or MWNT with E-glass fibers	Exploit the synergy between traditional and nano-sized fillers were investigated.
F.7	Phenolic	Texas State Univ.-San Marcos, USA	MWNT with carbon fibers	Mass loss, recession rate, and in-depth temperature all reduced with the increased of MWNT wt% (up to 2 wt%).
F.8	Phenolic	Univ. of Rome	Graphitic felt and graphitic foam	All felt-based samples passed undamped the longest oxy-acetylene exposure tests.

of poly(caprolactam) chains tethered through covalent bonds to the organic modifiers on the silicate surface.

The ablation rate (erosion rate) of the materials was determined from exposure to combustion gases in a mock SRM firing rig. Fifty grams of pelletized resin were compression molded into 7'' (180 mm) by 3'' (75 mm) triangular sections approximately 0.2'' (5 mm) thick in an autoclave. To enhance retention of material during the test, a second series of plaques were produced by consolidating the resins into a 1/8'' (0.32 mm) polyimide mat. In addition to the poly(caprolactam) nanocomposites, ethylene-propylene diene (EPDM) with 30% silica filler, EPDM with 15% chopped Kevlar fiber (state-of-the-art (SOTA) internal insulation for SRMs) and phenolic impregnated canvas were also examined. Table 9.2 summarizes the samples. Ablation tests were performed in a 4'' (10.2 cm) diameter char motor. Two triangular plaques were arranged along the inside of a tapered, rectangular chamber. The propellant consisted of 88% ammonium perchlorate with hydroxylated polybutadiene binder. Temperature of the combustion gases was approximately 3000°C and the test duration is 8 s. The thickness of the remaining virgin

Table 9.2 Summary of char motor ablation rate results

	Erosion rate (mil/s),[a] ± 15%	
	Low mass flux (Mach 0.013, 12.8 m/s)	**High mass flux (Mach 0.041, 39.6 m/s)**
	(0.10 lb/in^{2}s) (0.69 kPa/s)	**(0.40 lb/in^{2}s) (2.75 kPa/s)**
N(poly(caprolactam))	nvmr	nvmr
NCH2	5	nvmr
NCH2, mat	4	9
NCH5	4	9[b]
NCH5, mat	3	10
NLS5	5	14
NLS5, mat	5	11
Phenolic canvas	7	8
EPDM/Kevlar	6	10
EPDM/30% silica	6	11

[a] mil/s = 0.0254 mm/s
[b] At 0.35 lb/in^{2}s
nvmr: no virgin material remaining

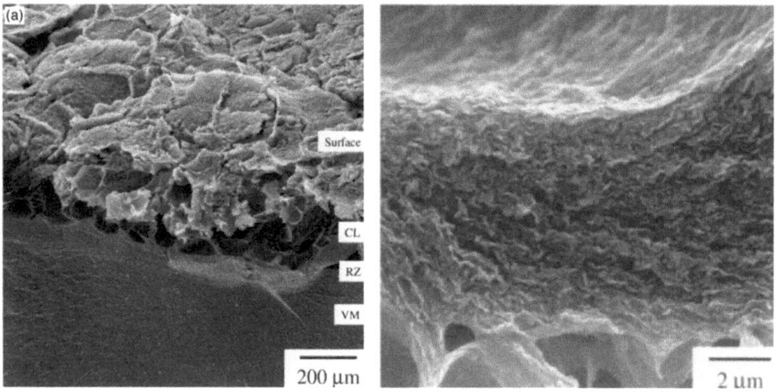

Figure 9.2 Cross-section (left) SEM image of the burnt layered silicate/poly (caprolactam) nanocomposite. In the left image, the char layer (CL), the reaction zone (RZ), and the virgin material (VM) are clearly visible. In the right image, the nanostructured nature of the char is displayed. Reprinted from Vaia et al., (1999) with permission from Elsevier.

material (after removal of the degraded char layer) was determined from the average of at least four measurements.

Since the presence of the layered silicate does not alter primary polymer decomposition in the reaction zone, the initial nanostructure of the dispersed silicate must influence the microstructure and composition of the char layer, thereby enhancing the ablation performance. Figure 9.2 shows the scanning electron micrographs of the surface char region of NCH5 ablative sample. The characteristic ablation regions (virgin material, porous reaction zone, and dense char layer) are apparent in all samples. The microstructural features of these char imply that increasing silicate content results in a tougher char. In these chars, nanoscopic (10–20 nm thick) densely packed plate-like morphology is observed. Increased silicate content results qualitatively in a spatially more uniform and thicker char. Specific characteristics of the silicate such as layer size, uniformity of layer distribution, and strength of the interphase result in a finer surface texture for NCH5 char than NLS5 char. In general, the spatially uniform arrangement of the silicate layers on an ultra-fine (nanometer) level facilitates the formation of a spatially uniform inorganic char. At 5 wt% the layers are on average only 20–25 nm apart. Thus, the efficient distribution of the inorganic precursors necessary to form a uniform char layer is the key structural characteristic of the nanocomposites leading to the enhanced ablative performance at remarkable low additions of a reinforcing phase.

A.2 Poly(phenylene oxide)/Polystrene (PPO/PS) Studies by Politecnico de Torino, Italy—Lombardi el al. (2012) and Fino et al. (2012)

Lombardi and co-workers (Lombardi et al., 2012; Fino et al, 2012) investigated PPO-based composites for potential ablative TPS application. The first part of their studies focused on PPO composites containing layered modified clay, such as MMT, hydrotalcite, and sepiolite (Lombardi et al., 2012). A commercial PPO/PS blend, Noryl® 914 (GE Plastics) was used as the polymer matrix. It has a density of 1.09 g/cm^3 and a melting point of 290°C. Three layered silicates used: Cloisite® 15A (C, Southern Clay Products) is MMT clay; Perkalite® A100 (H, AkzoNobel Polymer Chemicals) is an organically modified hydrotalcite; and Sepiolite (S, Tolsa) is a hydrated magnesium silicate were used in the authors' studies.

Brabender internal mixer was used to melt compound these nanomaterials into the PPO/PS matrix blends. Degradation of these polymer composites were studied using thermal analyses (TGA) and cone calorimeter tests. Thermal analyses were conducted in Ar heating up to 1500°C with heating rate of 20°C/min, DTG analyses performed up to 550°C with a heating rate of 10°C/min. Cone calorimeter tests were conducted with a heat flux of 35 kW/m^2. SEM analyses revealed all the composites still presented micron-sized aggregates platelets or needles. On the basis of the DTA results, PPO+S and PPO+C presented the best behavior for ablative materials, since they showed a more endothermic decomposition in comparison in the same temperature range of the neat polymer. Based on cone calorimeter results, PPO added with C and S presented a longer combustion time and a higher charring ability. Overall investigation showed the Sepiolite-added PPO composites can be selected as promising materials for innovative ablative shields.

In a second study (Fino et al., 2012), equiaxial, oxide, and non-oxide powders, such as magnesium aluminate spinel (SP), aluminum hydroxide (boehmite) (B), α- and α-alumina, silicon carbide (SiC), silicon nitride (N), and molybdenum disilicide (M) were used as fillers incorporated into the PPO/PS matrix polymer. From DTA data, it is evident that PPO decomposition was unaffected by oxide fillers, except in PPO+SP material. Among the non-oxide fillers, SiC decreased the material ability to absorb heat during decomposition. Thermal behaviors of each filler were summarized in this study (Fino et al., 2012).

B. Polymer-Clay Nanocomposite Studies

B.1 Polymer-Clay Nanocomposite Studies by AFRL, USA—Lincoln et al. (2000)

Lincoln et al.'s study (Lincoln et al., 2000) focuses on polymer/layered silicate materials to highlight the potential of nanocomposites for use in space applications. They also focus on fabrication methods for nanocomposites by controlled dispersion of preformed constituents, such as passivated nanoparticles or organically modified mica-type silicates. Filler particles on the nanoscale can alter the fundamental structure of the matrix material, especially in the case of semi-crystalline polymers, which can have a profound effect on the physical properties. The filler can also affect the materials' interaction with electromagnetic radiation. The improvement of nanocomposites over the neat polymer has the potential to offer space system designers a choice of a new, less expensive material for use in their systems. These nanocomposites are polymer-containing materials. Lincoln et al. (2000) discussed the benefits of using polymer nanocomposites for space applications, such as environmental stability, launch systems (solid rocket motors and liquid fuel tanks), and satellite bus (platform, deployable membranes, photovoltaic arrays, and solar concentrators). Nanocomposites are novel materials that potentially offer drastic cost savings for the space industry. They are lightweight, which is most desirable in the age where it cost thousands of dollars to place spacecraft into orbit. They are also easy to process and manufacture into useable parts, making them inherently less expensive and/or more reliable. The most attractive nanocomposite property is their potential to be space survivable. Nanocomposites can offer protection to vital systems to keep spacecraft operational for the duration of their lifetimes.

C. EPDM Nanocomposite Studies

C.1 EPDM-POSS Studies by AFRL, USA—Philips et al. (2002)

Air Force Research Laboratory has studied how the incorporation of POSS can be used for protective coatings in oxygen rich environments. Philips et al. (2002) detailed the strategy employment for using sacrificial nanocomposites along with applied research in the areas of SRM insulation and space-survival materials. Proprietary POSS monomer (5–50%) was blended into the EPDM pre-polymer using a standard braebender, followed by extended curing in a standard oven. The flexible POSS-EPDM polymer was then cut and fitted into conical phenolic sections for testing. POSS-dianiline and POSS-polyimide have been synthesis and characterization. A film was cast from the resulting POSS-poly(amic acid) on a clean glass plate and place in a clean oven under nitrogen to cure in specific conditions and the films were dried at room temperature for 24 hours.

The testing of POSS-EPDM samples for ablation performance was achieved using standard tactical and boost SRM propellants. Both the insulation standard and the POSS-EPDM sample were present in the same SRM firing. Post-firing analyses of the samples involved the use of industrial standard scraping of the formed char layer followed by measurements of the remaining polymer at various stages in the sample, which represent different mass fluxes the sample was exposed. At higher mass flux, the POSS-EPDM sample performs significantly better than the standard, with ablation rates of up to 50% less than the industry standard. These results are very significant, since a 50% reduction in the ablation rate of current SOTA materials results up to a 7.4% increase in payload to orbit. For SRM tests, analysis of post-fired samples showed the formation of a glass SiO_2 layer. Significant reduction in ablation rates was reproducibly demonstrated to the SOTA SRM insulation.

C.2 EPDM-Clay Studies by Harbin Institute of Technology, China—Gao et al. (2010)

Gao et al. (2010) of China focused their attention to nanomodify the SOTA SRM insulation material, EPDM/Kevlar by adding an organo-montmorillonite (OMMT) to the EPDM matrix and the synergy of these nanofiller with the traditional components of the formulation was studied. The content of OMMT was increased from 1 phr to 15 phr. After all the additives—namely the nanosilica, the short fibers, the flame retardants, the high char retention resin and the peroxide—were well dispersed in the matrix with a two-roll mill for about 20 min, the materials were vulcanized (160°C and 15 MPa for 40 min). XRD analysis of the EPDM/OMMT was performed on the cured materials and TEM was used to further evaluate the morphology of nanocomposites. The mechanical properties of the produced materials, mainly the elongation to break, were also measured and discussed. This property is very important since when it placed in the case of a SRM, the ablative must absorb the mechanical stresses induced to the SRM during propellant casting, storage, transportation, and flight. It also helps in anchoring the propellant grain to the inner wall of the SRM case. As a result, the insulator should have high elongation to break (in excess of 200%). Moreover, the ablative properties were evaluated using an oxyacetylene torch according to ASTM E-285-80.

The results can be summarized as follows. The XRD patterns revealed that the interlayer spacing between the nanoclay layers of the OMMT was increased due to the intercalation of EPDM molecules in the filler tactoids. The TEM images also showed that the platelets were homogeneously dispersed in the matrix by the strong shear forces of rollers. In terms of mechanical properties, the elongation to break of the EPDM/Kevlar composites was analyzed as a function of the fiber content and these results were compared with the corresponding data of the EPDM/Kevlar/OMMT system with respect to the OMMT percentage. For the baseline EPDM/Kevlar it appeared that above a fiber content equal to 6 phr, the elongation to break strongly reduced with the increasing content of the fibers: over 8 phr of fibers, the elongation to break of the EPDM/Kevlar (about 150%) became less effective to meet the requirements of SRM liners. The 12 phr EPDM/Kevlar formulation was chosen for the next step of the research in which the OMMT were introduced in the heat shielding material. Even at this high fiber content, due to the presence of the nanoclay, the material exhibited an impressive increase of the maximum strain. This result was confirmed also at a high nanoclay contents: at 13 phr OMMT the elongation to break displayed by the EPDM/Kevlar/OMMT system showed a value equal to approximately 560%.

However, over an OMMT content of 7 phr, the elongation to break tended to decrease. From 1 phr to 7 phr OMMT, the enhancement in the elongation at break can be directly attributed to the good dispersion of OMMT in EPDM matrix. The SEM images of the EPDM/Kevlar/OMMT fracture surfaces evidenced that the nanoclay platelets reduced the interfacial adhesion of the aramid fibers with matrix and thus the OMMT acted as a lubricating agent. At the higher filler, content the reduction of the elongation to break can be attributed to the aggregation of the silicate layers. In terms of ablative properties, up to 11 phr OMMT, with the increase of the nanoclay content, the loss of mass as well as the erosion rate decreased. Quantitatively, the erosion rate passed from 0.18 mm/s at 0 OMMT phr to about 0.07 mm/s at an OMMT content of 11 phr. Above this threshold, the erosion rate and the loss of mass slightly increased. The result is particularly interesting since the OMMT based formulations showed the potentiality to be very good candidates as a replacement of the traditional EPDM/Kevlar shielding elastomers.

D. Natural Rubber (NR) Nanocompostie Studies

D.1 NR-Clay Studies by Tarbiat Modares University, Iran—Khanlari et al. (2010)

The effect of organically modified MMT clay in natural rubber on the thermal, flammability, hardness, and mechanical properties using melt compounding were investigated by Khanlari and Kokabi

(2010). The authors shown by the addition of 3 wt% of organically modified MMT (Cloisite 15A) to NR, decreases the curing time by 33%, its aging hardness rise was decreased more than 55%, and elastic modulus increased by 63%. The ignition time was delayed by 150% and a reducation in peak heat release rate of 54% when compared to pristine NR. The thermal stability for the NR-clay nancomposite was also enhanced.

E. Thermoplastic Polyurethane Nanocomposties (TPUN) Studies

E.1 TPUN Studies by the University of Texas at Austin, USA—Koo et al. (2003–2012)

For SRM internal insulation, Koo and co-workers [Koo et al. (2003, 2005, 2006, 2006a, 2007a, 2009, 2010, 2011c); Ho et al. (2009, 2010); Bruns et al. (2009); Lee et al. (2009a, 2009b, 2009c, 2010, 2011); and Allcorn et al. (2011, 2012)] used two thermoplastic polyurethane elastomers (TPUs) as the base polymer and incorporated them with different amounts of MMT clay, CNF, POSS®, and multi-walled carbon nanotubes (MWNT). Selective properties improvement was observed with these TPU nanocomposites (TPUNs) as compared to industry standard Kevlar®-filled EPDM insulation material. The effects of weight loadings of layered clay, CNF, and MWNT on the thermal and flammability performance of this novel class of materials have been explored using a variety of test protocols and methods, such as thermogravimetric analysis (TGA), radiant panel experimental apparatus, UL 94, and cone calorimeter. Selective TPUNs were then tested using scaled SRM, oxyacetylene torch, as well as scaled hybrid rocket motors with the intent to develop them for SRM insulation application. Two commercially available TPUs were melt-blended with various loadings of these nanofillers using twin-screw extrusion. The morphological, physical, thermal, flammability, thermophysical properties, and kinetic parameters of these three families of TPUNs were characterized. The *processing-structure-property* relationships of this class of novel TPUNs were established.

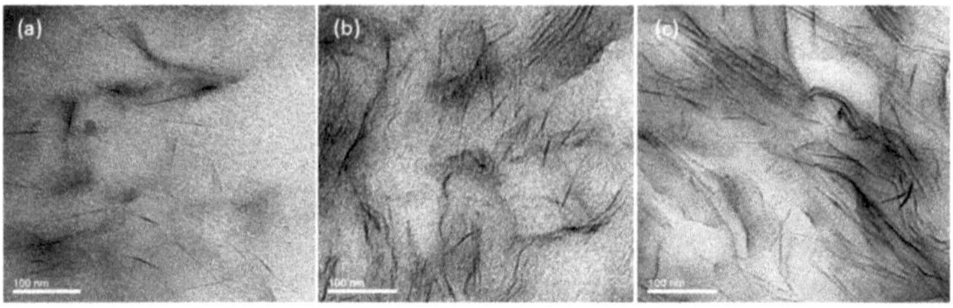

Figure 9.3 TEM images of (a) 2.5%, (b) 5%, and (c) 10% Cloisite 30B in TPUN.

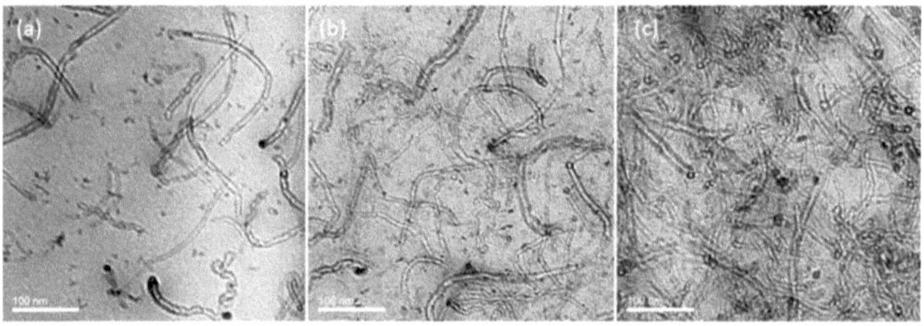

Figure 9.4 TEM images of (a) 2.5%, (b) 5%, and (c) 10% MWNT in TPUN.

Morphological results. TEM images of both the MWNT and nanoclay TPUs were performed in order to determine the visual degree of dispersion (Figs. 9.3 and 9.4). In the 2.5 wt% nanoclay sample, the individual platelets are easily identifiable and good dispersion is observed. The 5 and 10 wt% images show some remaining stacks, however the majority of the images show individual nanoclay platelets. It should also be noted that it does not appear to have any directional orientation from the injection molding process. The MWNT agglomerates are clearly debundled as shown in the two lower weight percentages. In the 10 wt% MWNT TEM image, there is a high density of MWNTs. These are not in bundle form. The bulk material dispersion is assessed by observing TEM images at different locations and view similar images. Again, there does not appear to be any preferred orientation.

Oxyacetylene test bed results. High heat flux experiments have been conducted. The material is exposed to the hottest portion of the flame, with a 7 mm flame diameter, for 25 s. Samples of 12.7 × 12.7

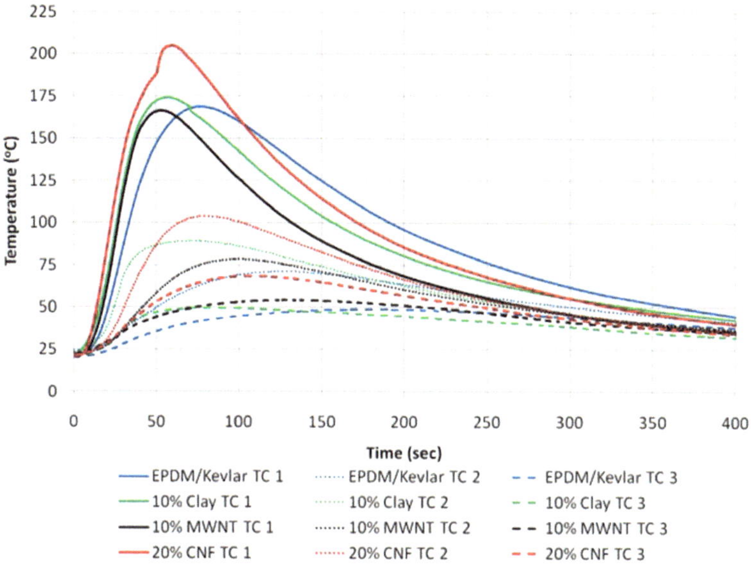

Figure 9.5 Representative oxyacetylene torch.

× 50 mm are cut out of a 12.7 mm × 10.16 cm × 10.16 cm (1/2'' × 4'' × 4'') compression pressed sheet. The torch is directed towards the square face (12.7 × 12.7 mm). In this orientation, the sample may be tested twice. At 5 mm increments three thermocouples are embedded into the center of the specimen to measure the in-depth heat soak temperature during testing. Thermocouples are drilled into the specimen at both ends for the possibility of testing a single specimen twice depending on the ablation rates.

Thermal results. A representative temperature profile of each specimen is shown in Fig. 9.5. The specific materials are color coded and the thermocouples (TC) are distinguished as: TC 1 solid line, TC 2 dotted line, and TC 3 dashed line. After the 25 s flame exposure is removed the heat is transferred conductively through the specimen. In this set of tests, the peak temperature of TC 1 in the CNF sample is 205°C and occurs at 60 s. The CNF has the highest heat soak temperature. The MWNT and nanoclay TC 1 peaks at 166°C and 174°C and occur at 50 and 64 s, respectively. The EPDM/Kevlar TC 1 peak, 168°C, shows the most delay, 75 s. Figure 9.6 shows

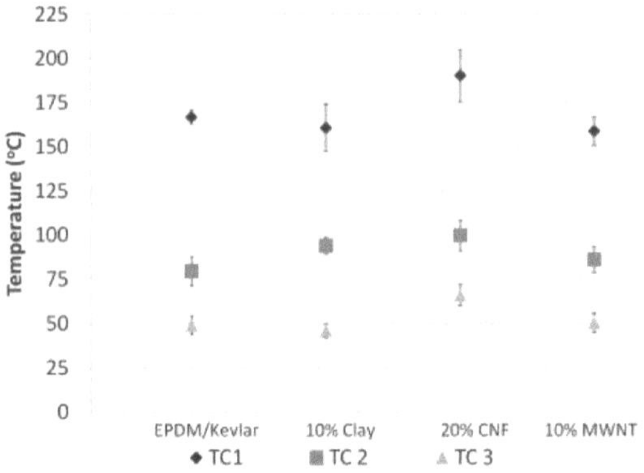

Figure 9.6 Peak in-depth temperatures of temperature. EPDM/Kevlar profile and PNCs.

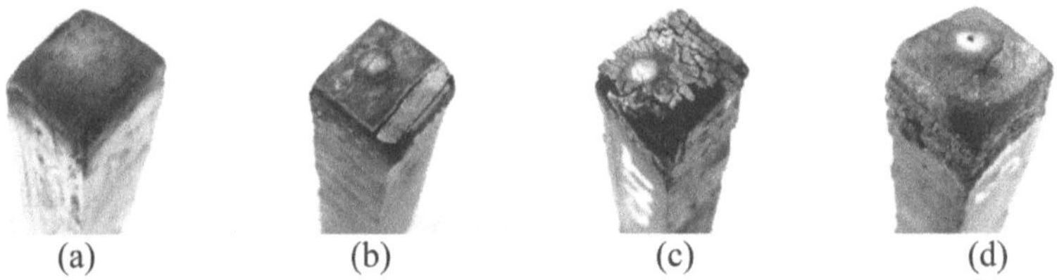

Figure 9.7 (a) Kevlar-EPDM, (b) 10% clay, (b) 10% MWNT, and (c) 20% CNF PNC post oxyacetylene torch burn.

the peak in-depth temperatures of EPDM/Kevlar and PNCs between thermocouples. The nanoclay, MWNT, and CNF PNCs lose 13.5%, 33.7%, and 46.5% more mass, respectively, compared to the mass loss of Kevlar/EPDM.

Post firing images. The samples after firing are shown in Fig. 9.7. A 1 mm compact char layer is left at the end of the firing of Kevlar/EPDM. Although compact, the virgin and char material does not have good adhesion. The nanoclay PNC post test material has about twice the char thickness of the Kevlar/EPDM. Although the char has a crack, the char was maintained well within the virgin material. Similar to the results found in the vertical UL 94 test, many cracks are observed in the MWNT char surface and result in a weak 2 mm thick char. The CNF PNC char is very thick, greater than 5 mm,

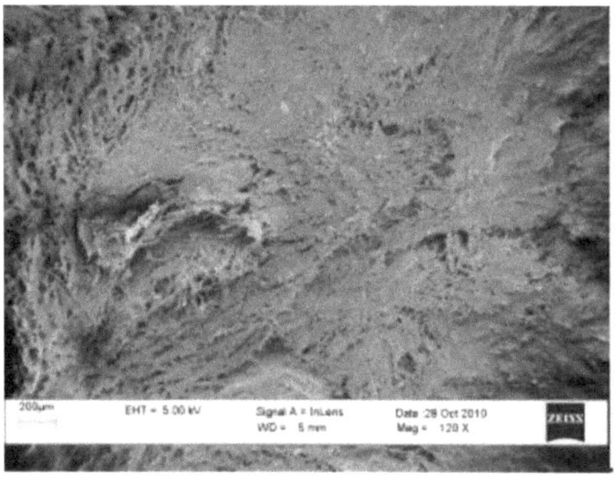

Figure 9.8 SEM of Kevlar/EPDM char post oxyacetylene torch burn (unit bar of left is 200 μm and right is 2 μm).

Figure 9.9 SEM of 10% nanoclay PNC char post oxyacetylene torch burn in progressive magnification (unit bar of left is 100 μm and right is 10 μm).

as the first thermocouple was in the char layer. The CNF char surface is flat without cracks; however, it is still very weak. The results from this set of experiments are consistent with the flammability and hybrid rocket tests (Lee et al., 2010, 2011).

SEM images of the post-test samples also were taken. The Kevlar/EPDM char material is compact and shows that the individual Kevlar fibers hold the material together as shown in Fig. 9.8 (Lee et al., 2010, 2011). The nanoclay char is also a compact composition. A granular morphology is also observed (Fig. 9.9). Bubble formations observed in SEM images found in the UL 94 burn tests are observed on the MWNT char surface as well as crack

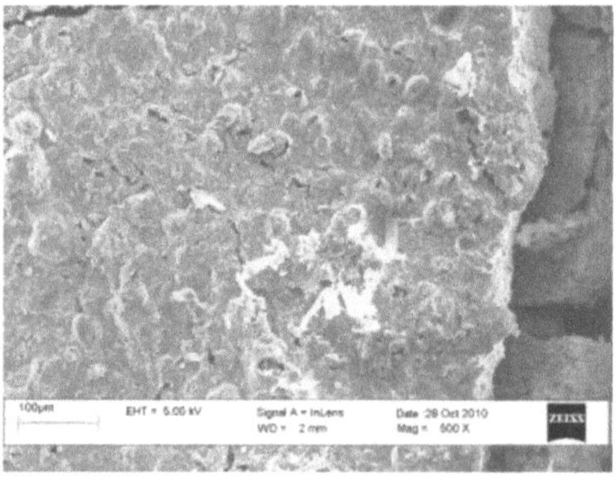

Figure 9.10 SEM of 10% MWNT PNC char post oxyacetylene torch burn in progressive magnification (unit bar of left is 100 μm and right is 200 nm).

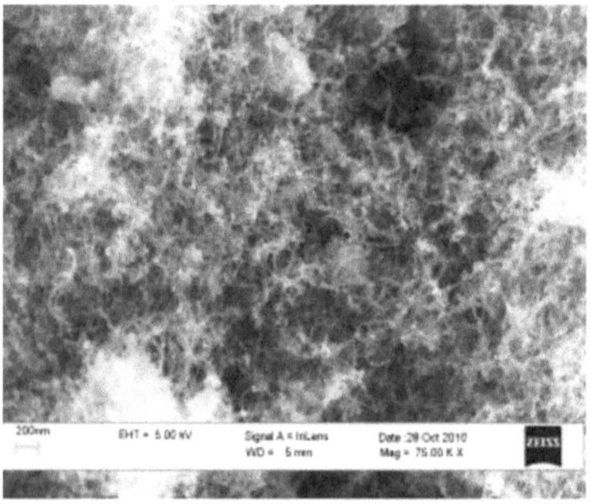

Figure 9.11 SEM of 20% CNF PNC char post oxyacetylene torch burn in progressive magnification (unit bar of left is 20 μm and right is 2 μm).

formations (Fig. 9.10). An in-depth SEM image shows the bundle form of MWNT that is also observed in the UL 94 burn test. The weak CNF char is shown to have a very porous structure (Fig. 9.11). Individual CNFs are also able to be observed.

In summary, the results from the oxyacetylene torch test show that both the nanoclay PNC and MWNT char layers are about twice the thickness of the Kevlar/EPDM char. However, the MWNT char layer is weak compared to the nanoclay PNC char. The CNF char is also very weak and was over five times the thickness of the Kevlar/EPDM char. However, the mass loss from the CNF sample is almost twice that lost from the Kevlar/EPDM sample. In-depth thermal measurements also show that the thermal protection of the PNCs is comparable to that provided by the Kevlar/EPDM, except for the CNF sample which had a much higher peak temperature due to the fact that the first embedded thermocouple ended up in the char region. A conventional advanced fiber is therefore needed to boost up the PNC ablation performance. More detailed ablation test results of these TPUNs can be found in Allcorn et al. (2011, 2012).

F. Phenolic Nanocomposites Studies

F.1 Phenolic-CNF Nanocomposite Studies by Mississippi State University, USA—Patton et al. (1999a, 1999b)

The ablation, mechanical, and thermal properties of vapor grown carbon fiber (VGCF) and phenolic resin composites were evaluated by Patton et al. (1999a, 1999b). The objective of Patton et al.'s study is to determine the potential of using this material in SRM nozzles. Composites with varying VGCF loadings (30–50 wt%) including one sample with ex-rayon carbon fiber plies were prepared and exposed to a plasma torch with a heat flux of 16.5 MW/m^2 at approximate 1650°C for 20 s. Low erosion rates and little char formation were observed, confirming that these materials were promising for SRM nozzle materials. When fiber loadings increased, mechanical properties and ablative properties improved. The VGCF composites had low thermal conductivities (approximately 0.56 W/m-K), indicating they were good insulating materials.

VGCF (Pyrograf III^{TM}) manufactured by Applied Sciences, Inc., Cedarville, OH, USA, and phenolic resin, SC-1008 manufactured by Borden Chemical, Inc. (now own by Momentive) were used in this study. VGCF/phenolic resin composite specimens were prepared by high shear mixing followed by thermal curing. Nineteen composite specimens were made by different mixing methods. Four types of samples were made, they were expressed in wt/wt as follows: 30/70 VGCF/Phenolic, 40/60 VGCF/Phenolic, and 40/60 VGCF/Phenolic (where ball milled as-received VGCF was used). Two types of fibers were used: as-received and compacted fibers. All specimens used in ablation testing were made as-received VGCF.

The flexural strengths and moduli were lower for these VGCF/phenolic samples compared to the authors' previous data of VGCF/epoxy using identical mixing techniques (Patton, 1999b). The flexural strength of 95.9 MPa and a modulus of 6.24 GPa for VGCF/epoxy composite (Patton et al., 1999b), when the epoxy resin was replaced by phenolic resin, a flexural strength of 25.6 MPa, and a modulus of 1.07 GPa (Patton et al, 1999a). This reduction in mechanical properties may be due to poor fiber-to-matrix adhesion for the VGCF/phenolic composites. Samples made with compacted fiber had high flexural moduli and strengths. The compacted fiber was a different batch of fiber and this may have affected the properties of the composite samples made from it.

All the composite samples exhibited similar thermal conductivities (0.54–0.62 W/m-K) as compared to 0.28 W/m-K of cured phenolic resin without fiber. There did not seem to be any trends in the data to indicate that increased carbon loading or different manufacturing techniques significantly affect the conductivity. The thermal conductivities in three dimensions were obtained on a 40/60 VGCF/phenolic ablation test composite sample. They indicate that the material is slightly anisotropic, different by less than ±9% from the average in all three directions. The specimen has a thermal conductivity which is closed to the average of all three specimens.

Ablation was characterized by erosion rates, specimen weight loss, and the increase in load. NASA's standard nozzle material, MX-4926 was used as the baseline. The MX-4926 composite is composed of 50 wt% woven ex-rayon carbon fiber, 15 wt% carbon black filler, and 35 wt% SC-1008 phenolic resin. The total carbon loading is 65 wt% in MX-4926. The VGCF/phenolic tested in this study contained only 30–45 wt% VGCF (only 46–69% as much as carbon as MX-4926 composite). As compared to the MX-4926 baseline, the VGCF composites experienced: (1) higher (58–114%) erosion rates, (2) lower weight loss (10–26%), and (3) lower load changes (60.5–68.5%). Therefore, more weight was being lost or near the surface of the specimen due to higher erosion rates vs. weight loss from subsurface thermal decomposition. The load changes for the VGCF composites are significantly lower than that of the MX-4926 composite. Load change is an indirect measure of neat penetration and char depth; increases in loading change indicate an increase in heat penetration and char depth. Thus, the VGCF composites appear to be far better insulators than the MX-4926. The higher erosion rates of the VGCF samples probably reflect their lower carbon content. The erosion rates/wt% of carbon present in the composites was very similar between MX-4926 and the VGCF composites.

The set of VGCF/phenolic composites exhibited extremely good erosion resistance (considering their low carbon loadings) while exhibiting less weight loss and load change than current SOTA MX-4926. Higher carbon contents are needed to match the erosion resistance of MX-4926. The carbon loadings used in this study have been much lower (30–45 wt%) that the MX-4916 composite (45 wt%). The mechanical properties of the VGCF composites need improvement. Lower void volume, better surface treatments for fiber/matrix adhesion and mixed VGCF/carbon fiber weave/carbon filler combinations should be investigated. VGCF/phenolic composites, should be made with higher carbon loadings (50–65 wt%) and subjected to ablation testing. The nanoscale dimensions of the VGCF causes major changes in the heat transfer rates and affects the resulting combustion/decomposition chemistry. The nano-sized fiber reduces the rate of heat transfer perpendicular to the surface into the composite compared with the rate of heat transfer in the MX-4926 material, which uses continuous carbon fibers.

F.2 Phenolic Nanocomposites Studies by the University of Texas at Austin, USA—Koo et al. (2002–2004)

The Univeristy of Texas at Autin group has been involved in the traditional R&D of ablative materials since early 1990 (Koo et al., 1992a, 1992b, 1993; Cheung et al., 1993; Wilson et al., 1994; Shih et al., 2003; Koo et al., 2011a, 2011b) and began the development of polymer nanocomposite (PNC) ablatives since 2002. Koo and co-workers [Koo et al. (2003, 2005, 2006, 2006a, 2007a, 2009, 2010, 2011c); Ho et al. (2009, 2010); Bruns et al. (2009); Lee et al. (2009a, 2009b, 2010, 2011); Allcorn et al. (2011, 2012)] reported two new classes of PNCs (SRM insulation and nozzle materials) that are lighter and exhibit better ablation performance and insulation characteristics than current SOTA ablative materials. These PNC materials exploit the ablation resistance of both resin and nanoparticles. For SRM nozzle ablatives, MMT nanoclays, CNF, and POSS® nanomaterials were incorporated into a resole phenolic resin, impregnated into rayon carbon fibers, and fabricated into phenolic-based polymer matrix composites (Koo et al., 2003, 2005, 2006, 2007a). These PNCs demonstrated significantly improved ablation and insulative performance versus the SOTA MX-4926 ablative as demonstrated by Koo's group. For SRM internal insulation, detailed results were presented in Section E.1. Selective properties improvement was observed with these thermoplastic polyurethane elastomer nanocomposites as compared to industry standard Kevlar-filled EPDM insulation material. In this section, the phenolic nanocomposites studies will be presented.

The NRAM (nanocomposite rocket ablative material) program is aimed at developing materials for SRM nozzles (Koo et al., 2003, 2005, 2006, 2007a). Both WAXD and TEM methods are used to determine dispersion uniformity prior to full ablation testing which is expensive. SC-1008 resole phenolic resin in isopropanol alcohol (IPA) is the baseline resin system. The commercial material, MX-4926, is a rayon carbon fabric impregnated with SC-1008 containing carbon black (CB) particles. On dispersing nanoclay into the resole phenolic resin with CB, followed by cure and examination by WAXD and TEM, it was concluded that CB particles in the baseline MX-4926 caused interference during dispersion of the nanoparticles in the resole phenolic resin. As a result, CB was eliminated in subsequent blending experiments. Use of nanoclay, Cloisite® 30B from Southern Clay Products, in loadings of 5, 10, and 15 wt% indicated that the nanoparticles dispersed satisfactorily into the SC-1008. Wide-angle x-ray diffraction (WAXD) of the three nanoclay loadings in SC-1008 and indicated undesirable dispersibility of the nanoclay into SC-1008. TEM images of nanoclay with SC-1008 indicate intercalation and not exfoliation of the nanoclay in the resin system. The TEM analyses were very convenient in determining the degree of dispersion/exfoliation and were a cost-effective and efficient technique for screening different formulations prior to committing to a large quantity of material.

The SC-1008 with 5, 10, and 15 wt% Cloisite® 30B replaced the 15 wt% CB in the original MX-4926 formulation. Table 9.3 illustrates the different loadings of Cloisite® 30B [$(HE)_2MT$], and trisilanolphenyl-POSS® [SO-1458] in the rayon fabric reinforced SC-1008 system. When a small amount of CNF (<1.5 wt%) was added in the SC-1008 phenolic resin, the viscosity of the resin increased significantly. It was impossible to process carbon prepregs under these conditions. It was decided that a molding compound will be fabricated to include large loadings of CNF dispersed in the SC-1008 resin. As a result, three loadings of PR-24-PS CNF at 20, 24, and 28 wt% were dispersed into SC-1008 without the rayon fabric reinforcement. Compositions and densities of the NRAMs are shown in Table 9.3. These nanomodified composites were manufactured and fabricated by Cytec Engineered Materials. The nanomodified materials are lower in density than the baseline material, MX-4926 (Table 9.3).

The SSRM is a small-scale supersonic liquid-fueled rocket motor burning a combination of kerosene and oxygen. It was used to study the ablation and insulative characteristics of the ablatives. The SSRM has been demonstrated to be a very cost-effective laboratory device to evaluate ablatives under identical conditions for initial screening and development (Koo et al., 1992a, 1992b, 1993, 2003, 2006, 2007a, 2011a, 2011b; Cheung et al., 1993; Wilson et al., 1994; Shih et al., 2003). The ablation rates (Fig. 9.12a) of MX-4926 and all three nanoparticles indicate that only 7.5% nanoclay composition shows a lower ablation rate than MX-4926 which is about 0.4 mm/s. All three CNF compositions are lower than MX-4926 with the 28% CNF-NRAM being the lowest. All three POSS® compositions have lower ablation rate with the 5% POSS®-NRAM being the lowest. Residual mass (Fig. 9.12b) of MX-4926 is about 92%. The POSS®-NRAMs exhibited the highest of all the nanoparticles in residual mass or about 93%. The clay-NRAMs were comparable to MX-4926 while the CNF-NRAMs were lower. A summary of backside temperature rise is shown in Fig. 9.12c with all the nanoparticles compositions exhibiting lower maximum backside heat-soaked temperature rise than MX-4926. The value of MX-4926 is 106°C with CNF-NRAMs being the lowest, from 54° to 72°C. The POSS®-NRAMs were intermediate, from 75° to 86°C and clay-NRAMs being the highest

Table 9.3 Specimen configuration for SSRM laminates fabrication

Material ID	Density (g/cc)	Rayon carbon fiber reinforcement (wt%)	Resin SC-1008 phenolic (wt%)	Filler (wt%)
MX-4926 (Control)	1.44	50	35	15 carbon black (CB)
MX-4926 ALT Clay 5%	1.42	50	47.5	2.5 Cloisite® 30B [(HE)$_2$MT]
MX-4926 ALT Clay 10%	1.43	50	45	5 Cloisite® 30B [(HE)$_2$MT]
MX-4926 ALT Clay 15%	1.43	50	42.5	7.5 Cloisite® 30B [(HE)$_2$MT]
PR-24-PS 20%/SC-1008	1.35	None	80	20 PR-24-PS CNF
PR-24-PS 24%/SC-1008	1.38	None	76	24 PR-24-PS CNF
PR-24-PS 28%/SC-1008	1.41	None	72	28 PR-24-PS CNF
MX4926 ALT SO-1458 2%	1.41	50	49	1 Trisilanolphenyl-POSS® [SO-1458]
MX4926 ALT SO-1458 6%	1.38	50	47	3 Trisilanolphenyl-POSS® [SO-1458]
MX4926 ALT SO-1458 10%	1.40	50	45	5 Trisilanolphenyl-POSS® [SO-1458]

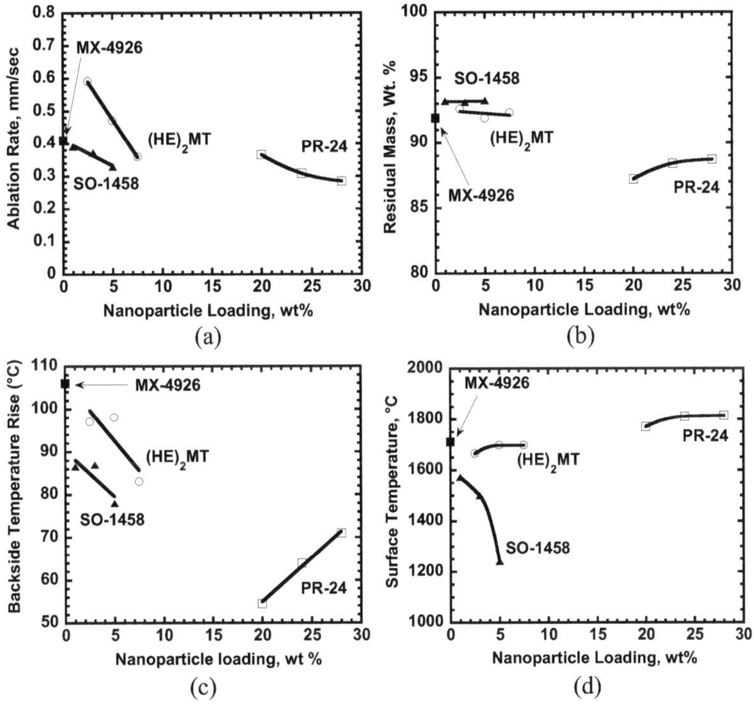

Figure 9.12 (a) Ablation rate, (b) residual mass, (c) backside temperature, and (d) surface temperature of MX-4926 and NRAMs with different types of nanoparticles [$(HE)_2$ MT-Cloisite® 30B, PR-24-PS CNF, and SO-1458 POSS®] at various loading levels.

of the nanoparticles, from 82° to 98°C, but still below the MX-4926 value of 106°C.

An IR pyrometer was used to measure the surface temperatures of all materials during SSRM firings. Figure 9.12d shows the surface temperatures of MX-4926 (1700°C) and the NRAMs. Surface temperatures of the CNF-NRAMs samples were higher than those of MX-4926, the clay-NRAMs, and the POSS-NRAMs. This finding suggests that better radial heat transfer is occurring rather than axial heat transfer. This is supported by the glowing heat of the material surface observed during and after material testing. This phenomenon was observed by other researchers (Patton et al., 1999a) and needs further study. The surface temperatures of clay-

NRAMs and POSS-NRAMs were lower than that of MX-4926. The amount of nanoclay in the clay-NRAMs had essentially no effect. The amount of POSS in POSS-NRAMs had a significant effect on the surface temperature of the POSS-NRAMs.

The feasibility of using NRAMs in rocket nozzle assemblies was clearly demonstrated using SSRM subscale ablation testing. MMT nanoclays, CNF, and POSS can be implemented into the existing semi-production line at Cytec Engineered Materials to manufacture fiber-reinforced prepregs and compression molded into laminates. Higher loadings of MMT, POSS, and CNF improve erosion resistance, and 28% CNF has the lowest erosion rate in the absence of carbon fiber reinforcements. Backside temperatures of all NRAMs were lower than that of baseline MX-4926, and CNF-NRAM as a group had lower temperatures than the MMT-NRAM and POSS-NRAM groups. Peak erosion of POSS-NRAM is 20% lower than MX-4926 at very low loading (5% POSS). Peak erosion of CNF-NRAM (28% CNF) without rayon fabric is 42% lower than MX-4926. Backside temperatures of CNF-NRAM, MMT-NRAM, and POSS-NRAM are 68%, 28%, and 26% lower than MX4926, respectively.

F.3 Phenolic Nanocomposite Studies by China—Liu et al. (2009), Hu et al. (2005), and Yu et al. (2012)

It clearly appears that the Chinese scientific community is spending significant efforts on the study of nanostructured ablative materials. Some of the most representative examples of their scientific outcomes will be summarized. Liu et al. (2009) focused their attention on the study of carbon/phenolic composites (PR/CF) in the presence of POSS. Two types of functionalized POSS were employed as nanomodifiers of the phenolic resin: a commercial octamethyl-POSS (Me-POSS) and a phenyl-POSS (Ph-POSS). The Ph-POSS was produced starting from a commercial OctaTrimethylsiloxy-POSS (TMS-POSS) and using styrene according to Laine (2005). For comparison purposes, a silica precursor tetraethoxysilicate (TEOS) was also selected to produce (Hu et al., 2005) another nanomodified resin used to impregnate the carbon/phenolic composite (T-SiO_2/PR/CF). The exact compositions of the different formulations in terms of filler percentages were not reported. The ablation performances of the produced composites are: T-SiO_2/PR/CF, Me-POSS/PR/CF, and Ph-POSS/PR/CF were studied by means of an oxy-acetylene test bed. The following conditions were selected: the distance from torch tip to sample was 20 mm and the exposure time was 20 s.

Experimental results both in terms of linear ablation rate as well as of loss of mass, the hierarchy displayed by the tested materials was the following: T-SiO_2/PR/CF > Me-POSS/PR/CF > Ph-POSS/PR/CF. With respect to the type of functionalization attached to the POSS, the Ph-POSS displayed a significantly lower erosion and loss of mass than the Me-POSS. This result can be directly related to the higher carbon content of the phenyl groups over methyl structures. The better thermal stability of Ph-POSS over Me-POSS embedded in a polymeric matrix was also confirmed by other studies (Tanaka et al., 2009; Franchini et al., 2009). The SEM analysis on the burnt samples also confirmed that the Ph-POSS/PR/CFs produced the best charred surface: a homogeneous hybrid carbonaceous/silica skin on the top region of the Ph-POSS/PR/CF samples protected the inner layers of the composites, whereas the other two formulations displayed a porous and irregularly eroded char, leaving the carbon fibers exposed to the flame.

A novel flake graphite reinforced barium-phenolic resin composite was made by roller coating technology and its thermal ablation was studied by Yu et al. (2012) under long pulse laser irradiation. Results show that ablation rate of the flake graphite reinforced barium-phenolic composite is 32.8 μg/J at 1700 W/cm^2 irradiation power density, which is lower than that of the barium resin and the barium-phenolic carbon fiber-reinforced composites. When the composites were exposed to the laser radiation, the flake graphite plays as a mirror and reflects part of the laser energy. As a result, the laser radiation deposited on the composites is reduced. The size of the flake graphite also affects the ablation rate. The composite with flake graphite diameter of about 0.5 mm has the lowest ablation rate.

F.4 Phenolic-Nanosilica Nanocomposite Studies by India—Srikanth et al. (2010)

The most significant research of the Indian scientific community on nanostructured polymeric ablatives was probably carrier out at the Defence Research and Development Organisation (DRDO). Srikanth et al. (2010) studied the effect of nanosilica on the ablation resistance and on the interlaminar shear strength (ILSS) of carbon/phenolic (C–Ph) composites. The through-thickness thermal conductivity was also measured at 300°C under steady-state conditions (ASTM E–1225-09). Moreover, the microstructure of the ablated materials was investigated by SEM. Rayon-based carbon fabric was used as a reinforcing material and a resole was used as a matrix. The nanosilica had an average diameter of 40–50 nm. Four different formulations were prepared by adding 0.0 (control), 0.5, 2.0, and 4.0 wt% of nanosilica in the phenolic resin. The pristine resin and the produced nanocomposites were used to impregnate the carbon fabric. Three samples from each of the laminate were machined to a size of 10 mm diameter by 20 mm (the length of the sample was perpendicular to the direction of the fabric). The specimens were exposed to a plasma arc jet at a flame velocity of about Mach 1 and at a stagnant flux of 2.5 MW/m^2 for 20 s. The ablation performances were determined by dividing the weight loss during the test by the exposure time.

Due to the presence of nanosilica, the ILSS increased with the nanosilica content up to a filler percentage equal to 2.0 wt%: above this threshold the ILSS decreased. This result indicated that nanosilica particles embedded in the matrix acted as an anti-slippage agent. It was also speculated that above 2.0 wt% the nanosilica decreased the adhesion between the fabric layers. An increased ILSS makes the composites more resistant to the high aerodynamic shear forces and is expected to result in an enhanced ablation resistance. Plasma arc jet test results were in good agreement with the ILSS trend: in fact, the ablation resistance increased with the nanosilica content up to 2.0 wt%, beyond which it was lower than the control recipe. The addition of nanosilica also brought down the thermal conductivity of the C–Ph composites: the thermal conductivity depends on the lattice vibrations and the microstructure (micro cracks, grain boundaries, porosity, etc.). The introduction of nanosilica increased the number of grain boundaries, resulting in lower thermal conductivity: similar observations for the C–SiC composites have been reported by Kumar et al. (2009). As a result of a reduced thermal conductivity, the ablation resistance resulted improved.

Since the formation of the char is an endothermic process, the amount of heat dissipated during the ablation depends on the efficiency of this process. Moreover, the char dissipates a large fraction of incident heat through surface re-radiation. Accordingly, a complete charring of the resin combined with high char retention gives the better heat dissipation during ablation. The SEM microstructure studies showed that the damage to the carbon fibers in the blank C–Ph is considerably higher compared to the damage occurring in the presence of nanosilica. The ablated surfaces of nanostructured C–Ph composites showed that SiC formed on carbon fibers during ablation. This result was also confirmed by XRD analysis. The authors proposed two main mechanisms for the ablation results and the related microstructural and compositional observations. First, under plasma arc jet testing, the temperature can reach 4000°C. In the case of C–Ph composites, the matrix cannot effectively hold the carbon layers together, leading to severe erosive losses. In the case of nanosilica loaded C–Ph composites, when the temperature goes beyond 1600°C the silica melts (Knacke et al., 1991) and forms a viscous layer which acts like a high-temperature binder able to hold the underlying phenolic matrix and the char thus ensuring a higher radiant emission and leading to a complete charring of the resin, thereby absorbing more heat. Moreover, the molten silica reacts with the carbon, forming a hard SiC phase (Xiao et al., 2007) (second process). The resulting carbon–silicon carbide (C–SiC) can withstand the aerodynamic shear forces under high-temperature erosive environments, thereby enhancing the ablation resistance (Lee et al., 2004).

F.5 Phenolic-Clay Nanocomposite Studies by Tarbiat Modares University, Iran—Bahramian et al. (2009)

The main contribution of the Iranian scientific community on the topic of nanostructured ablatives has been produced by the researchers at Tarbiat Modares University. They focused mostly on the improvement of asbestos/phenolic ablatives. Particularly, Bahramian et al. (2006, 2008, 2009, 2011) studied the ablative performance, the thermal decomposition, and temperature distribution through the thickness of asbestos/phenolic composites impregnated with layered silicate nanocomposite: these data were compared with the results for traditional asbestos/phenolic ablators.

In their work, the combination of solution and in situ intercalation methods was used to fabricate the nanocomposite. Ethyl alcohol was used to disperse the layered silicates (Cloisite 15A) and dissolve the phenolic resin. The phenolic resin in an excess of ethyl alcohol could be adsorbed into the nanoclay tactoids. However, once the ethyl alcohol was removed, the phenolic chains remained trapped in the layered silicate sandwiched structure. The asbestos cloth was impregnated with the nanocomposite. Three composites having different nanoclay percentages were studied: 3 wt% (NMA1), 4 wt% (NMA2), and 6 wt% (NMA3). Flat panel samples were used for the oxy-acetylene flame test (heat flux 900 W/cm^2) and K thermocouples were placed through the thickness of the sample at 2, 4, and 6 mm from the surface: the surface erosion measurement was carried out on cylindrical shaped samples (10 diameter and 25 mm height). XRD analysis was used to evaluate the dispersion of the nanoclay in the composites. The specific heat capacity measurements were performed according to ASTM D-1269. Thermo-gravimetric and DSC analysis were also performed on the produced materials.

The XRD patterns of nanocomposites exhibited a behavior typical of an exfoliated morphology, confirming the penetration of the resol molecules between the clay sheets. According to the TGA data, the presence of clay improved the thermal stability of the composites. Many factors are responsible of this improvement. Among them, the dispersed large aspect ratio platelets hindered the diffusion of the decomposed volatile products, as a direct result of the decreased permeability. For the thermal oxidation process, the mass transfer of oxygen into the gallery may also be slowed down, decreasing the degradation of the matrix. The ablative re-assembly of nanoclay platelets and the sintering of them at elevated heat fluxes led to the formation of a homogeneous silicate-carbonaceous char layer that works as a heat shield during burning. This is widely accepted as the main flame retardancy mechanism common in this class of nanocomposites (Bartholmai et al., 2004; Zanetti et al., 2001; Duquesne et al., 2003; Gilman et al., 2000). It resulted that in the range of the considered nanoclay content, the higher the nanoclay content, the better the thermal stability. TGA results were also used to acquire the parameters used to produce a model of the ablation process. The heat capacity measurements were carried out up a temperature of 500 K: the values of the specific heat of the asbestos/phenolic composite and of the NMA3 nanocomposite are in the range of 1000–1500 and 3300–5800 J/kgK, respectively. The higher value of the NMA3 nanocomposite is one of the key factors to cause the better insulation performance of the NMA3 over the asbestos/phenolic composite. The dynamic DSC scans also showed that the heat of ablation of the NMA3 composite was quite higher (about 950 kJ/kg) than the control material (45 kJ/kg).

The experimental temperature of the top surface and through the thickness of asbestos/phenolic and NMA3 samples were also acquired. At the same experimental conditions, the steady-state surface temperature of the asbestos/phenolic composite and of the NMA3 were about 1500°C and 2100°C, respectively. This difference clearly shown that the nature of top surface of the traditional composite and of the nanocomposite impregnated asbestos fabric is completely different. The higher value of surface temperature of the NMA3 is another key factor to cause the higher insulation performance than the traditional asbestos/phenolic composite. Since the re-radiated energy follows a T^4 law, for the NMA3 a decreased heat flux is transferred into the material thus increasing the lifetime of the material. The authors also evaluated the thermal diffusivity of the material as a function of the depth. It resulted that in the section one where the char layer is created due to high temperature ablation, the effective thermal diffusivity of the asbestos/phenolic composite is twice than the NMA3 composite. This phenomenon drives a clear difference in the char layer formation and in the ablation mechanism of the composite and of the nanocomposite based system.

F.6 Phenolic Nanocomposites Studies by University of Perugia, Italy—Natali et al. (2009–2011)

For over 15 years, the University of Perugia—STM group (Terni)—extensively worked in the field of ablative materials. At the end of 1990, in collaboration with ALENIA Spazio S.p.A, the group developed and tested a medium-density silicone-based ablative (ALS-051) for the heat shield of the Italian Space Agency recoverable capsule (CARINA). An accurate ablation model of the studied material was also developed by Torre el al. (1998a, 1998b, 2000). Since the beginning of 2000, the researchers were also focused on the study of nanostructured ablatives. Nanosilica (Natali et al., 2011a, 2011b) and MWNTs (Natali, 2012) were the main filler candidates. The possibility to exploit the synergy between traditional and nano-sized fillers was also investigated.

Glass and silica fiber phenolic composite ablators are a very important class of ablatives. As an example, they can be effectively used in the production of passively cooled rocket combustion chambers. To reduce the erosion phenomena of these ablators, the studies were focused on the selection of the best types of fiber reinforcement, in terms of short or long fibers. Research shown that in many hyperthermal conditions, liners produced with chopped reinforcements—fibers or fabrics cut in small squares [e.g., 1.27 by 1.27 cm (1/2 by 1/2 inch)]—could match the performance of configurations based on the use of more expensive long fibers (Natali et al., 2011b): accordingly, small parts as well as significant portions of a rocket combustion chamber can be effectively produced using ablative Bulk Molding Compounds (BMCs) (Pavli, 1968; Warga, 1979). A wide range of commercial ablators are sold as molding compounds (LR1504, Cytec). Research on glass or silica/phenolic liners also showed that silica in powder play a crucial role in the improvement of the ablative performance of composite liners (D'Alelio et al., 1971; Sutton et al., 2000; Peterson et al., 1969). In fact, when exposed to high temperature, micron-sized silica particles form a highly viscous melt layer which covers the charred surface, working as an anti-oxidizing protective barrier and without inhibiting radiative cooling. The melt layer also continues to adsorb a significant amount of heat by virtue of endothermic processes and due to the phase transitions of silica.

Starting from these results, Natali et al. (2011a) investigated the possibility to produce glass phenolic composites with nanosized silica as an alternative to micron-sized silicon dioxide. Typically, in glass/phenolic or silica/phenolic liners loaded with micron-sized silica, the amount of SiO_2 particles is of about 8–10 wt% and the fiber loading is of about 60 wt% (Peterson et al., 1969; LR1504). For example, Fiberite MX2600 consists of 31 wt% resin, 61 wt% chopped strands of silica, and 8 wt% of micron-sized silica particles. The authors first studied two nanosilica loaded nanocomposites (Natali et al., 2011a): the considered filler percentages were 5 wt% and 20 wt%. Then the pristine resin and the nanocomposites were used as matrices for the impregnation of E-glass chopped fibers. In all the cases, the fiber percentage was kept constant at a value equal to 60 wt%. The amount of nanosilica on the whole weight was 8 wt% (BMC3) when the 20 wt% loaded nanocomposite was used to impregnate the short fibers: this recipe strongly resembles the formulation of the MX2600 with the difference that the micron sized silica was replaced by the nanosized version. To further investigate the effects of nanosilica, another BMC recipe containing a nanofiller percentage equal to 2 wt% was studied (BMC2): this paste was produced using the 5 wt% loaded nanocomposite. A BMC recipe prepared with no nanosilica was used as a control material (BMC1). Oxy-acetylene torch testing (ASTM-E-285-80) was used to study the produced materials (heat flux equal to 850 W/cm^2 and the exposure time at 60 s). The flame was applied to one of the bases of the cylindrical shaped samples. In depth temperatures were acquired at 5 mm (T1) and 10 mm (T2) from the surface. The morphological structure of burnt materials was investigated by SEM analysis.

Figure 9.13a shows the burnt surface of the control recipe (BMC1) and Fig. 9.13b summarizes the ablation mechanism for this material. In absence of nanosilica, the glass fibers melted into low viscosity under exposure to the flame, leading to a high erosion rate. Unbounded spheres of glass easily flowed away from the zone touched by the torch plume leaving the charred surface unprotected: the glass droplets moved to a well-defined corona on the external edge of the flame. In terms of the contact angle with the charred surface, the marked spherical shape of the glass droplets clearly evidenced the poor compatibility of the melt with the substrate. The results of the BMC2 recipe are shown in Fig. 9.14a. Under exposure to the flame the glass fibers and nanosilica melted together to form a blend. Even this low nanosilica percentage was able to increase the viscosity of the blend. The droplets of melted glass fibers exhibited

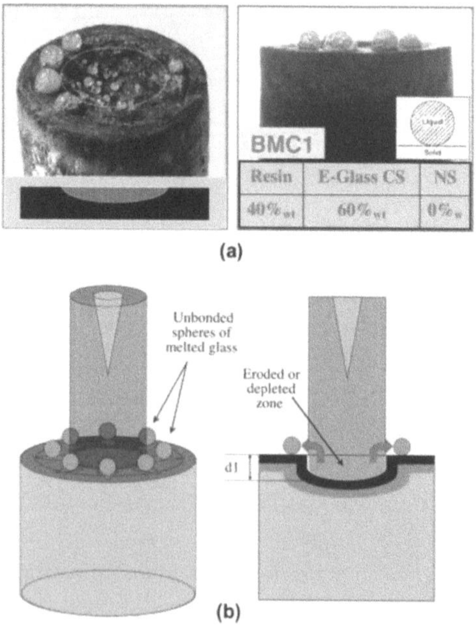

Figure 9.13 Burnt surfaced of control recipe (BMC1) (a). Ablation mechanism (b). Reprinted from Natali et al. (2011b) with permission from Elsevier.

a better adherence on the burnt surface. However, this percentage of nanofiller did not completely freeze the glass droplets under the zone hit by the flame as summarized in Fig. 9.14b: the shear forces produced of the flame gases were strong enough to remove the melt from the zone hit by the flame.

However, in comparison to the BMC1 recipe, glass droplets moved to a corona with a smaller diameter, confirming the enhanced stickiness of the charred surface. The contact angle of the melted droplets with the charred surface indicated that the burnt substrate increased its adhesive capability and compatibility to the melt. Accordingly, the erosion rate experienced by the BMC2 material resulted decreased. As a matter of fact, the most interesting result was obtained from the BMC3 recipe (Fig. 9.15a). In this case, the higher nanosilica content was able to effectively block the high viscosity, silica rich, and droplets of melted glass exactly under

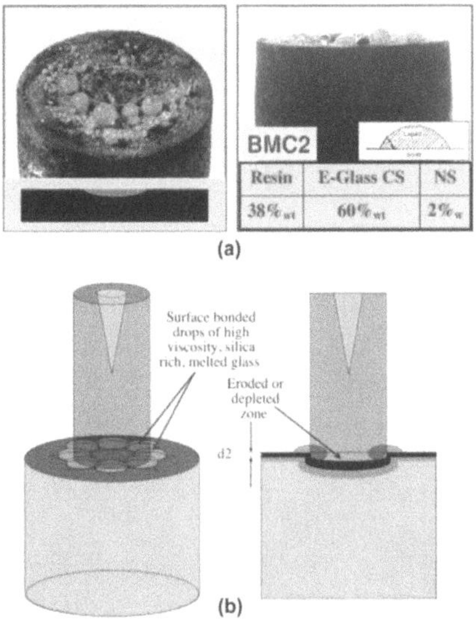

Figure 9.14 Burnt surfaced of control recipe (BMC2) (a). Ablation mechanism (b). Reprinted from Natali et al. (2011b) with permission from Elsevier.

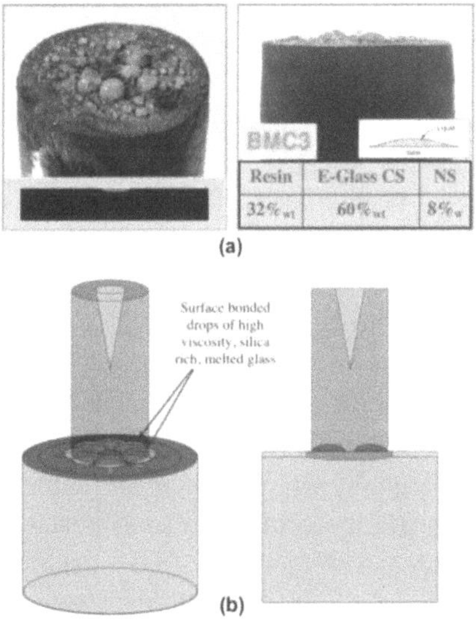

Figure 9.15 Burnt surfaced of control recipe (BMC3) (a). Ablation mechanism (b). Reprinted from Natali et al. (2011b) with permission from Elsevier.

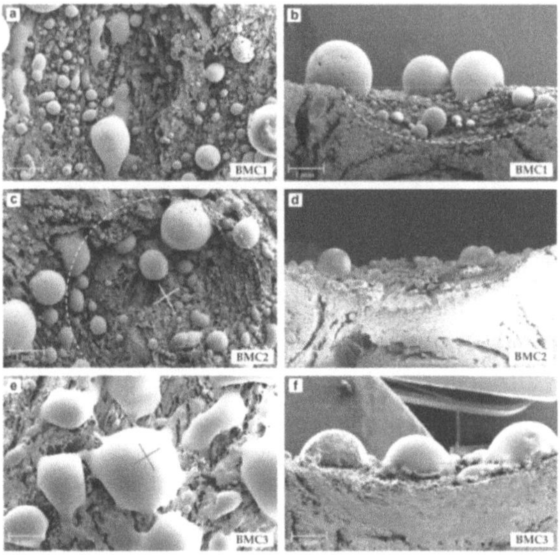

Figure 9.16 SEM micrographs of the burnt samples, (a) and (b) refer to BMC1; (c) and (d) refer to BMC2; (e) and (f) refer to BMC3. (For interpretation of the references to colour in this figure legend, the reader is referred to the web version of this article.) The crosses on the images (a), (c), and (e), mark the point where the core of the torch impinged the sample. In all the pictures, the dashed lines delimit the eroded areas. Reprinted from Natali et al. (2011b) with permission from Elsevier.

the zone impinged by the flame (Fig. 9.15b). The protection of the charred substrate and the preservation of the surface noticeably increased, thus leading to significantly lower erosion. Observing the contact angle of the melted droplets, it was possible to check the high compatibility between the charred surface and the droplets of melt. The adherence capability of the charred surface was very high, confirming the effectiveness of the use of silica nanoparticles dispersed in an organic matrix. Since the charred substrate was held above the virgin material for a long period, the re-radiation efficiency was increased. Figure 9.16 shows the results of the SEM characterization carried out on the burnt samples confirming the indication obtained with the traditional microscopy.

To further point out the role of the nanosilica, Fig. 9.17 shows a detail of the charred surface of the BMC3 recipe: at high

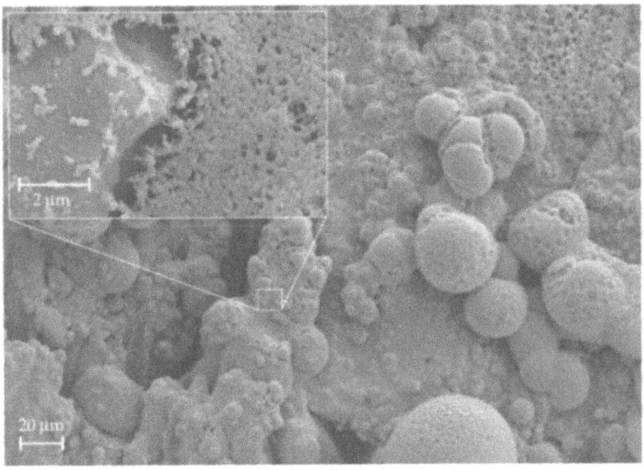

Figure 9.17 Morphology of charred surface of BMC3. Reprinted from Natali et al. (2011b) with permission from Elsevier.

magnification it is possible to identify sintered nanosilica particles. Moreover, the inset picture also well shows how nanosilica particles adhere on the glass filaments, acting as a binder. The sintering mechanism of silica nanoparticles was more deeply investigated by the same authors in a previous work on silica/phenolic nanocomposites (Natali et al., 2011b). The in depth temperatures showed that as the nanosilica content increased, the BMC paste exhibited an improved insulation. The higher the amount of silica nanoparticles, the higher the amount of heat absorbed by the filler. Moreover, increasing the nanosilica content, increased the thickness of charred region above the thermocouples, reducing the transferred heat.

For silica rich chars exposed to a not oxidizing hyperthermal environment, a reaction producing silicon carbide is allowed: the presence of SiC can enhance the ablation performance of the char (Wu et al., 2002; Wang et al., 2000; Zheng et al., 1995). The production of SiC, is possible only when a carbothermal reduction between silica and carbon takes place (Peterson et al., 1969). During oxy-acetylene tests, at the core of the plume, in the presence of a neutral flame, the burnt surface is relatively protected against the oxidation and SiC can be produced. However, since the temperature of the core exceeds about 2700°C, the SiC phase undergoes decomposition.

Natali et al. (2012) also investigated the ablative properties of two carbon nanofillers. In particular, carbon black (CB) and MWNTs were used to produce highly loaded (50 wt%) phenolic composites. This study is important because it allowed pointing out the main differences in terms of ablative performance among traditional carbonaceous filler like CB and relatively new filler, such as MWNTs. In fact, well before the appearance of the term nanocomposite, carbon black was extensively used in the production of polymeric ablators. For example, Cytec Engineered Materials MX-4926, a standard carbon/phenolic nozzle material, is composed by 50 wt% of carbon fibers, 15 wt% of carbon black, and 35 wt% of phenolic resin. Recently, in the class of nanosized carbon fillers, nano-filaments like nanofibers (CNFs) have been studied as filler for polymeric TPSs (see Section F.3, of the University of Texas at Austin). The aim and novelty of the Italian research was to evaluate carbon nanotubes (CNTs) as potential fillers for ablatives. In fact, MWNTs proved to be very effective flame retardant additives for polymers (Monti et al., 2011): the improved flame resistance was attributed to the formation of a continuous protective nanotube felt working as a heat shield (Kashiwagi et al., 2005a, 2005b; Cipiriano et al., 2007). In analogy with the mechanism of layered silicates, such a protective barrier reduces the mass loss rate and the flammability of the material. In agreement with the above mentioned mechanism, the flame retardancy resulted improved at higher CNT loads. Moreover, the CNTs embedded in the charred surface re-emit a large amount of the incident radiation into the gas phase, decreasing the heat transferred into the inner layers and thus reducing the polymer pyrolysis rate (Zhao et al., 2009; Rajatelara et al., 2010). As a drawback, carbon-based reinforcements are easily destroyed in an oxidized hyperthermal environment. Nonetheless, to date, also considering the flame retardant properties of MWNTs, there was a lack of research on CNTs used as nanofillers for ablatives. The research of Natali was important not only because the ablative response of MWNTs embedded in a polymeric matrix were investigated, but also because a direct comparison with CB certainly represents a valuable outcome for the scientific community and the industry. In this study two different mixtures were produced, both consisting of 50 wt% of the phenolic matrix and 50 wt% of the nanofiller: one with CB (Vulcan 7H) referred to as PR-CB, constituted by 50%-PR and 50%-CB, the other with MWNTs, constituted by 50%-PR and 50%-MWNTs (Arkema Graphistrength C100) referred to as PR-MWNT. The results concerning the thermogravimetric analysis, the evaluation of the heat capacity, the oxy-acetylene test and the post burning morphology will be here summarized.

In terms of thermal stability, at a heating rate of 20°C/min the TGA results in air and nitrogen showed that PR-CB and PR-MWNT exhibited a very close behavior. The study of the heat capacity as a function of the temperature can be very important for the comprehension of the ablation mechanism and it can also be used for the modeling of the process. It appeared that in the whole range of temperatures, both composites exhibited a higher heat capacity than the pristine matrix, with the PR-MWNT system having the highest values. The post burning images of the tested materials showed that in terms of the erosion rate, the PR-MWNT composite exhibited higher erosion than the PR-CB system. The PR-MWNT samples suffered a marked removal of material and the flame produced a crater exactly in a zone confined under the plume of the torch. During the tests, small flakes of the material were peeled off the surface of the PR-MWNT samples: this behaviour was also found by other authors for CNF based NRAMs (Patton et al., 1999a). The PR-MWNT samples also showed a very thin charred layer above the virgin material. The PR-MWNT samples showed a very limited number of in depth cracks propagating from the mantel surface to the center of the sample. For the PR-CB composites, the flame produced a broad and deep charred region having relatively good integrity: however, the flame did not produce any crater. Also, during the test, this material did not show any macroscopic ejection of burning material. The PR-CB burnt samples exhibited a remarkable number of macrocracks going from the mantel of the samples up to the lower limit of the charred region. It can be concluded that the PR-CB system suffered a char penetration extended on a higher and deeper region of the sample.

The morphology of the burnt surfaces was also investigated by SEM. At low magnification the PR-CB revealed the cracking phenomena also identified with the optical analysis. The top surface of the PR-CB showed a morphology characterized by micrometer and sub-micrometer sized open and closed bubble-like structures: the carbon black particles were not distinguishable from the charred layer. The morphology of the zones beneath the surface was very close to that one of the surface but, in some cases, a uniform and compact medium with no voids or bubble was found. If compared to the PR-CB system, the surface of the PR-MWNT samples showed a homogeneous morphology free of any macro cracks: the ablation appeared to be a surface phenomenon with a very thin transition zone—the char—between the virgin and burnt material. The char was constituted by many irregular aggregates mainly composed of entangled MWNTs. Beyond the undamaged and unburnt MWNT filaments, SEM also evidenced many bright spots clearly related to the metal precursors used to produce the nanotubes. The temperature profiles showed that the PR-MWNT system exhibited higher thermal diffusivity than the PR-CB recipe. The authors also concluded that if compared to the PR-CB system, the PR-MWNT composite possessed a higher thermal conductivity in the direction of the flame plume. Loss of mass data indicated that both materials experienced a similar remotion of mass. As a consequence, since the surfaces of PR-MWNT samples experienced a more severe erosion than the PR-CB specimens and considering that the amount of loss of mass is of the same order of magnitude for both materials, it can be concluded that the ablation mechanisms of the two materials is deeply different. For the PR-MWNT, the loss of mass was due to the physical erosion of material whilst, since the PR-CB did not erode appreciably, the loss of mass can be attributed to the higher degree charring of the organic matrix. As a conclusion CB and MWNTs exhibit a very different behavior.

F.7 Phenolic-MWNT Nanocomposites Studies by Texas State University-San Marcos, USA—Tate (2010–2011)

This research was collaborative work with Cytec Engineered Materials (CEM) as an attempt to enhance properties of the SOTA ablative, MX-4926. Therefore, test panels were manufactured using processes and materials similar to those used by CEM. MX-4926 MC

(Molding Compound) is rayon precursor-based carbon fabric with a MIL-R-9299 phenolic resin. Composite panels were manufactured using rayon-based carbon fabric supplied by CEM, which is 8-harness satin weave. The fabric has a weight of 261 g/m^2, a specific gravity of 1.84, and a thickness of 0.48 mm. Break strength in the warp direction is 0.496 MPa and 0.599 MPa in the fill direction, with 1.96 picks/mm warp and 2 picks/mm fill. SC-1008 phenolic resin meets the MIL-R-9299, was supplied by Hexion Specialty Chemicals. The viscosity is roughly 180–300 cps depending on storage conditions. SC-1008 contains roughly 20%–25% IPA as a solvent. Graphistrength™ C100 MWCNTs were supplied by Arkema, Inc. Typical diameter of these MWCNTs is 10–15 nm and a length between 1 and 10 μm. Graphistrength™ C100 MWCNTs tend to be agglomerated with dimensions of roughly 50–900 μm. High shear mixing and sonication have been used in previous studies to provide good dispersion and exfoliation of nanoparticles (Thorstenson et al., 2005; Cheng, 2006; Safadi, 2002). Tate et al. (2011) investigated the dispersion characteristics using sonication, high shear mixer, and combination of sonication and high shear mixer in previous studies. TGA results and SEM analysis suggested that dispersion of MWCNTs within IPA using sonication followed by high shear mixing produces optimal dispersion characteristics for use in ablative panels. This method was adapted to produce composites panels with MWCNT loading of 0.5, 1, and 2 wt%.

Oxyacetylene test bed (OTB) was developed to perform a small scale testing of newly developed ablatives. Samples of size 12.7 × 12.7 × 10 mm were prepared and two holes of approximately 1.59 mm (1/16 inch) diameter drilled at depth of 2.7 mm and 7.7 mm for thermocouple placement. Samples were exposed to a heat flux of approximately 1000 W/cm^2 (6 mm distance from the OTB) for 45 s. Six samples were tested in each category. Ablative performance of the material was analyzed based on: percentage mass loss, recession, peak temperatures at different depth in ablative samples and SEM to analyze dispersion of MWCNTs in phenolic resin. There was gradual decrease in percentage mass loss as wt% of MWCNTs increased. For control samples average mass loss was 26% and for 2 wt% MWCNT average mass loss was 23%. While for 0.5 wt% and 1 wt% MWCNT average mass losses were 25.5% and 25%, respectively. There is considerable reduction in mass loss as wt% of MWCNTs increased. Typically mass loss is because of pyrolysis of material and through erosion of char. Similarly, recession was also high for low wt% formulations and decreased gradually as wt% of MWCNTs increased. Average recession was 0.83 mm for control samples, while it was 0.38 mm for 2wt% samples. Average peak temperatures were reduced considerably with the addition of MWCNT to the composite (Fig. 9.17).

F.8 Phenolic-Graphitic Felt and Phenolic–Graphitic Foam Studies by University of Rome, Italy—Pulci et al. (2010) and Marra et al. (2010)

Sapienza University of Rome group developed, manufactured, and tested a family of carbon-phenolic ablative TPS materials with the aim of fulfilling the thermal and mechanical requirements corresponding to the actual loads experienced by re-entry vehicles (Pulci et al., 2010; Marra et al., 2012; Valente, 2011). Experimental investigations were conducted on two different composite systems: a resole resin coupled with a graphitic felt and a graphitic foam. Their aim is to optimize the manufactured procedure and to characterize the mechanical behavior and the insulation performance of the fabricated composites.

A commercially available resole phenolic resin was employed and two carbon-based reinforcements were selected for impregnation with the phenolic resin. Graphitic foam (Grafoam® manufactured by Graftech International, Clarksburg, WV, USA) and a rigid graphitic felt (Sigratherm® manufactured by SGL Carbon SE, Wiesbaden, Germany) were used. Grafoam carbon foam is an open-cell, rigid, carbon-based material that can be produced in densities ranging from 0.03 to 0.56 g/cm^3 and in a variety of sizes. Sigratherm felt have high thermal stability in vacuum up to 2500°C and under inert atmospheres up to 3000°C. It is well known that above 2000°C physical properties of the ablative can change. Figure 9.18 shows the microstructures of carbon-based reinforcements.

Infiltration process was performed by total immersion procedure: both reinforcements were totally soaked for 30 min in a phenolic solution with an isopropanol/resin ratio tailored in order

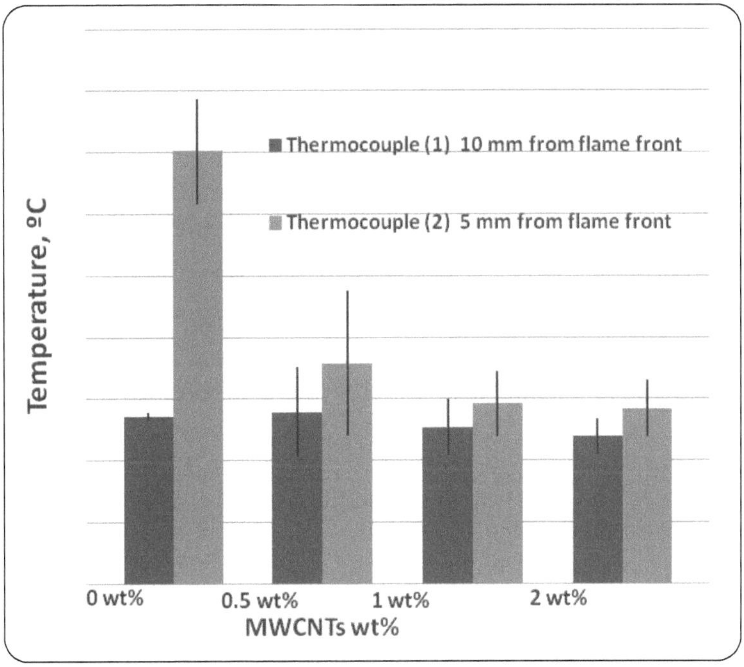

Figure 9.18 Average peak temperatures at two in-depth locations of the control sample and different phenolic-MWCNT nancomposites.

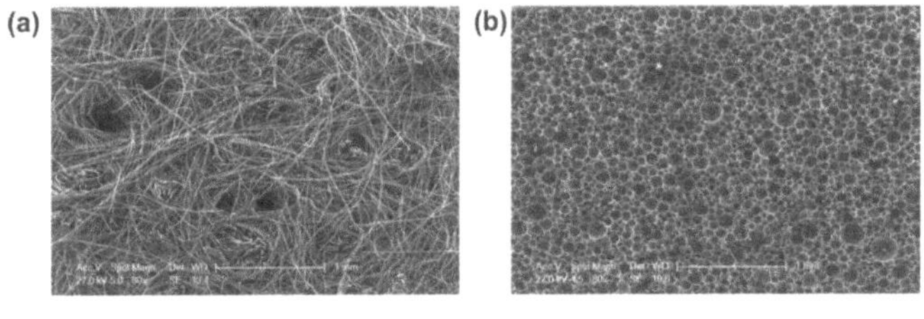

Figure 9.19 Microstructure of carbon-based reinforcements: (a) Sigratherm felt and (b) Grafoam foam.

to obtain the specific density (0.5 g/cm^3) after the cure process. After resin polymerization, a resin/composite volume ratio of 0.61 and 0.59 was calculated by Sigratherm and Grafoam composites, respectively. Resin pre-heating (50°C) and vacuum infiltration (3 min at 10^{-3} bar) after soaking were mandatory for removing excess solvent and efficient impregnation. All tests were carried out in these conditions.

Preliminary ablation tests were performed on the two fabricated systems according to ASTM E285-80 standard. The size of samples tested with oxy-acetylene torch was 100 × 100 × 20 mm or 50 × 50 × 20 mm; the distance between nozzle and samples was 20 mm and the temperature of the front and the back surface of the tiles were monitored by a two-color digital fiber-optic non-contact pyrometer (Impac Infrared, Frankfurt, Germany) and by a type-K thermocouple, respectively. Heat flux, as measured by a calorimeter (HFM 1000-Vatell Corp., Christiansburg, VA, USA) during experimental tests, was 1.5 MW/m^2. Four exposure times were investigated: 1, 5, 10, and 20 min. Exposure tests were performed on 5 impregnated and on 2 un-impregnated samples. Maximum temperature of about 4000 K was registered on the front face of exposed samples.

Higher hardness values on the surface than in the cross sectional areas were observed. Impregnated samples showed a homogeneous resin distribution throughout the thickness where the cross section of a 3 cm thick Sigratherm composite is observed (Pulci et al., 2010). Magnified views of impregnated samples cross sections, showing suitable degree of resin impregnation were achieved for both reinforcements (Pulci et al., 2010). All specimens exhibited a linear elastic behavior, Grafoam and Sigratherm composites show similar values of elastic modulus (about 2 GPa) but the felt composite showed a much higher (+144%) flexural strength than the foam composite. Pure resin has an elastic modulus of 4 GPa, whereas un-impregnated supports have a modulus of 0.25 GPa and 0.5 GPa for the felt and the foam, respectively. It is evident that the resin is almost completely responsible for the composite stiffness. Results of four-point bending tests evidenced a higher flexural strength of impregnated Sigratherm with respect to Grafoam.

Thermophysical properties (thermal diffusivity, thermal conductivity, and specific heat) of the composite samples are shown by Pulci et al. (2010). Similar behaviors were evidenced for the two types of materials. Foam composites exhibited a slightly lower thermal conductivity at high temperature (1500–1900°C), while impregnated felts exhibit lower thermal conductivity at temperatures lower than 800°C. Un-impregnated Grafoam and Sigratherm reinforcements showed a room temperature thermal conductivity of 0.12 and 0.22 W/(m-K), respectively. The results of a typical oxy-acetylene exposure test on Sigratherm and Grafoam composite samples, in terms of front and back surface temperature are reported by Pulci et at. (2010). The two materials show the same temperature evolution with time, evidencing comparable insulation properties. However, the analysis of samples after tests showed that marcroporous tiles were cracked during the flame exposure, whereas all felt-based samples passed undamped the longest oxy-acetylene exposure tests.

There is a clear indication that a carbon/phenolic composite material made of rigid non-woven graphitic felt impregnated with a resole phenolic resin can act as a very effective insulator in a very wide range of temperatures (from room temperature to about 2000°C), making it suitable for forming/machining operations. A very accurate optimization of the main steps of the manufacturing procedures is required to achieve the above. The research multi-step curing cycle of 12 days at different temperatures is required to achieve optimal microstructure and highest mechanical properties of the composite. The outstanding performance of the manufactured composite system exposed to the intense heat flux of the oxy-acetylene flame was achieved. Sigratherm-based ablators were selected for arc jet testing in a Plasma Wind Tunnel (Scirocco, CIRA ScoA, Italy) to simulate re-entry conditions.

In order to support the material/component design and to offer new and differential tools for the screening of TPS materials, this research group has developed a CFD (computational fluid-dynamics) model of the oxy-acetylene test apparatus per ASTM E285-80 standard. The CFD model simulates the combustion of the oxygen-acetylene mixture occurring in a conventional welding torch and calculates the heat flux incident on the material sample and the related temperature fields. The validation of the CFD model was obtained by performing suitable instrumented tests with direct measurements of both incident heat flux and temperatures of the front and back surfaces of an ablative sample. Two different models using Fluent® software have been developed for the simulation: the first model is intended for the evaluation of the heat flux actually generated by oxy-acetylene torch; and the second model simulates a real test on an ablative material. Two different experimental set-ups were prepared in order to validate the results obtained by the numerical simulations. Good agreement was found between numerical results and experimental measurements (Pucli et al., 2010; Marra et al., 2010).

9.4 Summary and Conclusions

A comprehensive review of polymer nanocomposites designed for advanced thermal protection materials was conducted. It has been demonstrated that the improvement in ablative performance of these polymer nanocomposites relative to traditional filled system is associated with the char-forming characteristics of the polymer nanocomposites. The spatially uniform distribution of nanoscale inorganic filler results in the formation of a uniform inorganic char at a relatively low fraction of inorganic. The nanostructured ablative materials, will provide multifunctional properties, such as ablation resistant, thermal stability, insulative, and weight saving. Ablation mechanisms were also studied and proposed by the researchers on their nanostructured ablatives.

References

Allcorn, E., Natali, M., and Koo, J.H., (2011) Ablation performance and characterization of thermoplastic elastomer nanocomposites, *Proc. of SAMPE 2011 ISTC*, Fort Worth, TX, Oct. 17–20, 2011.

Allcorn, E., Natali, M., and Koo, J.H., (2013) Ablation performance and characterization of thermoplastic elastomer nanocomposites, *Composites Part A: Applied Science and Manufacturing*, 45, 109–118.

ASTM E-1225-09, (2008) Standard Test Method for Thermal Conductivity of Solids by Means of the Guarded-Comparative-Longitudinal Heat Flow Technique, Annual Book of ASTM Standards.

ASTM E-285-80, (2008) Standard test method for oxyacetylene ablation testing of thermal insulation materials, Annual Book of ASTM Standards.

ASTM Standard E-457-08, (2008) Standard Test Method for Measuring Heat-Transfer Rate Using a Thermal Capacitance (Slug) Calorimeter, Annual Book of ASTM Standards.

Bahramian, A.R., Kokabi, M., Famili, M.H.N., and Beheshty, M.H., (2006) Ablation and thermal degradation behaviour of a composite based on resol type phenolic resin: Process modeling and experimental, *Polymer*, 47, 3661–3673.

Bahramian, A.R., Kokabi, M., Famili, M.H.N., and Beheshty, M.H., (2008) High temperature ablation of kaolinite layered silicate/phenolic resin/asbestos cloth nanocomposite, *Journal of Hazardous Materials*, 150, 136–145.

Bahramian, A.R., and Kokabi, M., (2009) Ablation mechanism of polymer layered silicate nanocomposite heat shield, *Journal of Hazardous Materials*, 166, 445–454.

Bahramian, A.R., and Kokabi, M., (2011) Numerical and experimental evaluation of the flammability and pyrolysis of a resole-based nanocomposite by cone calorimeter, *Iranian Polymer Journal*, 20 (5), 399–411.

Bartholmai, M., and Schartel, B., (2004) Layered silicate polymer nanocomposites: new approach or illusion for fire retardancy? Investigations of the potentials and the tasks using a model system, *Polymers for Advanced Technologies* 15, 355–364.

Beall, G.W., and Powell, C.B., (2011) *Polymer-Clay Nanocomposites*, Cambridge, UK: Combridge University Press.

Bruns, M.C., Koo, J.H., and Ezekoye, O.A., (2009) Population-based models of thermoplastic degradation: Using optimization to determine model parameters,*Polymer Degradation and Stability*, 94, 1013–1022.

Cheng, J., (2006) *Polysyanate ester/small diameter carbo.n nanotubes nanocomposite*, Master's thesis, The University of Texas at Austin.

Cheung, F.B., Koo, J.H., et al., (1993) Modeling of one-dimensional thermomechanical erosion of high-temperature ablatives, *Journal of Applied Mechanics*, 60, 1027–1032.

Cipiriano, B.H., Kashiwagi, T., Raghavan, S.R., Yang, Y., Grulke, E.A., Yamamoto, K., et al., (2007) Effects of aspect ratio of MWNT on the flammability properties of polymer nanocomposites, *Polymer*, 48, 6086–6096.

Cytec Engineered Materials. http://www.cytec.com/engineered-materials/selectorguide.htm.

D'Aelio, G.F., and Parker, J.A., (1971) *Ablative Plastics*, New York: Marcel Dekker.

Duquesne, S., Jama, C., Bras, M.L., and Delobel, R., Recourt, P., and Gloaguen, J.M., (2003) Elaboration of EVA-nanoclay systems-characterization, thermal behaviour and fire performance, *Composites Science and Technology*, 63, 1141–1148.

Fino, P., Lombardi, M., Antonini, A., Malucelli, G., and Montanaro, L. (2012) Exploring composites based on PPO blend as ablative thermal protection systems – Part II: The role of equiaxial fillers, *Composite Structures*, 94, 1060–1066.

Franchini, E., Galy, J., Gèrard, J.F., Tabuani, D., and Medici, A., (2009) Influence of POSS structure on the fire retardant properties of epoxy hybrid networks, *Polymer Degradation and Stability*, 94, 1728–1736.

Gao, G., Zhang, Z., Li, X., Meng, X.Q., Zheng, Y., and Jin, Z., (2010) Study on Mechanical and Ablative Properties of EPDM/OMMT Thermal Insulating Nanocomposites, *Journal of Nanoscience and Nanotechnology*, 10, 7031–7035.

Gilman, J.W., Jackson, C.L., Morgan, A.B., Harris, R., Manias, E., Giannelis, E.P., Wuthenow, M., Hilton, D., and, Phillips, S.H., (2000) Flammability properties of polymer-layered silicate nanocomposites: Polypropylene and polystyrene nanocomposites, *Chemistry of Materials,* 12, 1866–1873.

Gupta, R.A., Kennel, E., and Kim, K.J., Eds., (2010) *Polymer Nanocomposites Handbook*, Boca Raton, FL: CRC Press.

Ho, W.K., Koo, J.H., and Ezekoye, O.A., (2009) Kinetics and thermophysical properties of polymer nanocomposites for solid rocket motor insulation, *Journal of Spacecraft and Rockets*, 46(3), 526–544.

Ho, W.K., Koo, J.H., and Ezekoye, O.A., (2010) Thermoplastic polyurethane elastomer nanocomposites: Morphology, thermophysical, and flammability properties, *Journal of Nanomaterials*, Article ID 583224.

Hu, L., Zhang, X., and Sun, Y., (2005) Hardness and elastic modulus profiles of hybrid coating, *Journal of Sol-Gel Science and Technology*, 34, 41–46.

Kashiwagi, T., Du, F., Winey, K.I., Groth, K.M., Shields, J.R., Bellayer, S.P., *et al.,* (2005a) Flammability properties of polymer nanocomposites with single-walled carbon nanotubes: effects of nanotube dispersion and concentration, *Polymer*, 46, 471–81.

Kashiwagi, T., Du, F., Douglas, J.F., Winey, K.I., Harris, R.H., and Shields, J.R., (2005b) Nanoparticle networks reduced the flammability of polymer nanocomposites, *Nature Materials*, 4, 928–33.

Khanlart, S., and Kokabi, M., (2010) Thermal stability, aging properties, and flame resistance of NR-based nanocomposites, *Journal of Applied Polymer Science*, 119, 855–862.

Knacke, O., Kubaschewski, O., and Hesselman, K., (1991) *Thermo-Chemical Properties of Inorganic Substances*, 2nd Ed., Berlin: Springer-Verlag, Berlin, pp. 1836–1837.

Koo, J.H., Kneer, M., et al., (1992a) A cost-effective approach to evaluate high-temperature ablatives for military applications, *Naval Engineers Journal*, 104(3), 166–177.

Koo, J.H., Lin, S., et al., (1992b) Performance of high-temperature polymer composite ablatives under a hostile environment, *Science of Advanced Materials and Process Engineering Series*, 37, SAMPE, Covina, CA, 506–520.

Koo, J.H., Miller, M., et al., (1993) Evaluation of fiber-reinforced composites ablatives for thermal protection, *Science of Advanced Materials and Process Engineering Series*, 38, SAMPE, Covina, CA, pp. 1085–1098.

Koo, J.H., et al., (2003), Nanocomposites rocket ablative materials: Processing, characterization, and performance, *Proc. of 2003 SAMPE ISSE*, 48, SAMPE, Covina, CA, pp. 1156–1170.

Koo, J.H., Pilato, L., and Wissler, G.E. (2005) Polymer nanostructured materials for high-temperature applications, *SAMPE J.*, 41(2), 7.

Koo, J.H., (2006) *Polymer Nanocomposites: Processing, Characterization, and Applications*, New York: McGraw-Hill.

Koo, J.H., and Pilato, L.A., (2006a) Thermal properties and microstructures of polymer nanostructured materials, In M.J. Schulz, A. Kelkar, and M.J. Sundaresan, Eds., *Nanoengineering of Structural, Functional, and Smart Materials*, Boca Raton, FL: CCR Press, pp. 409–441.

Koo, J.H., Ho, W.K., and Ezekoye, O.A., (2006b) A review of numerical and experimental characterization of thermal protection materials - Part I. Numerical modeling, AIAA-2006-4936, *42nd AIAA/ASME/SAE/ASEE Joint Propulsion Conference*, Sacramento, CA, July 9–12.

Koo, J.H., Ho, W.K., Bruns, M., and Ezekoye, O.A., (2007a) A review of numerical and experimental characterization of thermal protection materials - Part II. Material properties characterization, AIAA-2007-2131, *48th AIAA/ASME/ASCE/AHS Structures, Structural Dynamics, and Materials Conference*, Honolulu, HI, Apr. 23–26.

Koo, J.H., Ho, W.K., Bruns, M., and Ezekoye, O.A., (2007b) A review of numerical and experimental characterization of thermal protection materials - Part III. Experimental testing, AIAA-2007-5773, *43rd AIAA/ASME/SAE/ASEE Joint Propulsion Conference*, Cincinnati, OH, July 8–11.

Koo, J.H., Pilato, L., and Wissler, G., (2007c) Polymer nanostructured materials for propulsion systems, *Journal of Spacecraft and Rockets*, 44(6), 1250–1262.

Koo, J.H., Ezekoye, O.A., et al., (2009) Characterization of polymer nanocomposites for solid rocket motor – Recent progress, *Proc. of SAMPE 2009 ISSE*, SAMPE, Covina, CA.

Koo, J.H., et al. (2010) Flammability studies of a novel class of thermoplastic elastomer nanocomposites, *Journal of Fire Sciences*, 28(1), 49–85.

Koo, J.H., Miller, M.J., Weispfenning, J., and Blackmon, C., (2011a) Silicone polymer composites for thermal protection system: fiber reinforcements and microstructures, *Journal of Composite Materials*, 45(13), 1363–1380.

Koo, J.H., Miller, M.J., Weispfenning, J., and Blackmon, C., (2011b) Silicone polymer composite for thermal protection of naval launching system, *Journal Spacecraft and Rockets*, 48(6), 904–919.

Koo, J.H., Ezekoye, O.A., Lee, J.C., Ho, W.K., and Bruns, M.C., (2011c) Rubber-clay nanocomposites based on thermoplastic elastomers, In M. Galimberti, Ed., *Rubber-Clay Nanocomposites*, Hoboken, NJ: Wiley and Sons, pp. 489–521.

Kumar, S., Kumar, A., Shukla, A., Devi, G.R., and Gupta, A.K., (2005) Thermal-diffusivity measurement of 3D-stitched C–SiC composites, *Journal of the European Ceramic Society*, 29(3), 489–495.

Laine, R.M., (2005) Nanobuilding blocks based on the [OSiO] (x=6, 8, 10) octasilsesquioxanes, *Journal of Materials Chemistry*, 15, 3725–3744.

Laub, B., and Venkatathy, E., (2003) Thermal protection system technology and facility needs for demanding future planetary missions, *Proc. of International Workshop on Planetary Probe Atmospheric Entry and Descent Trajectory Analysis and Science*, Lisbon, Portugal.

Lee, J.C., Koo, J.H., and Ezekoye, O.A., (2009a) Flammability studies of thermoplastic polyurethane elastomer nanocomposites, AIAA-2009-2544, *50th AIAA/ASME/ASCE/AHS/ASC Structures, Structural Dynamics, and Materials Conference*, Palm Spring, CA May 4–7.

Lee, J.C., Koo, J.H., Ezekoye, O.A, et al., (2009b) Heating rate and nanoparticle loading effects on thermoplastic polyurethane elastomer nanocomposite kinetics., AIAA-2009-4096, *AIAA Thermophysics Conference*, San Antonio, TX, June 22–25.

Lee, J.C., Koo, J.H., and Ezekoye, O.A., (2009c) Thermoplastic polyurethane elastomer nanocomposites: Density, hardness, and flammability properties correlations, AIAA-2009-5273, *AIAA Joint Propulsion Conference*, Denver, CO, Aug. 2–5.

Lee, J.C., (2010) *Characterization of ablative properties of thermoplastic polyurethane elastomer nanocomposites*, Ph.D. Dissertation, The University of Texas at Austin.

Lee, J.C., Koo, J.H., and Ezekoye, O.A. (2011) Thermoplastic polyurethane elastomer nanocomposite ablatives: Characterization and performance, AIAA-2011-6051, *47th AIAA/ASME/SAE Joint Propulsion Conference*, San Diego, CA, Aug. 1–4.

Lee, Y.J., and Joo, H., (2004) Ablation characteristics of carbon fiber reinforced carbon (CFRC) composites in the presence of silicon carbide, *Surface and Coatings Technology*, 180–181, 286–289.

Lincoln, D.M., Vaia, R.A., Brown, J.M., and Benison Tolle, T.H. (2000) Revolutionary nanocomposite materials to enable space systems, *Proc. of 21st Century Aerospace Conference IEEE*, Big Sky, MT, vol. 4, 183–192.

Liu, Y., Lu, Z., Chen, X., Wang, D., Liu, J., and Hu, L., (2009) Study on Phenolic-Resin/Carbon-fiber Ablation Composites Modified with Polyhedral Oligomeric Silsesquioxanes, *Proc. of 2009 4th IEEE International Conference on Nano/Micro Engineered and Molecular Systems*, Shenzhen, China, Jan. 5–8.

Lombardi, M., Fino, P., Malicelli, G., and Montanaro, L., (2012) Exploring composites based on PPO blend as ablative thermal protection systems. Part I: The role of layered fillers, *Composite Structures*, 94, 1067–1074.

LR1504 Prepreg datasheet. http://www.lewcott.com/ablatives.html.

Marra, F., Pulci, G., Tirillo, J., Bartuli, C., and Valente, T., (2010) Numerical simulation of oxy-acetylene testing procedure of ablative materials for re-entry space vehicles, *Proc. IMechE Vol. 225 Part L: J. Materials: Design and Applications*, pp. 32–40.

Mittal, V., Ed., (2010a) *Polymer Nanotube Nanocomposites: Synthesis, Properties, and Applications*, Wiley, Hoboken, NJ: Wiley & Sons.

Mittal, V., Ed., (2010b) *Optimization of Polymer Nanocomposites Properties*, Weinheim, Germany: Wiley-VCH.

Mittal, V., Ed., (2012) *Characterization Techniques for Polymer Nanocomposites*, Weinhein, Germany: Wiley-VCH.

Monti, M., Natali, M., Petrucci, R., Puglia, D., Terenzi, A., Valentini, L., and Kenny, J.M., (2011) Advanced fiber reinforced composites based on nanocomposite matrices, In L. Nicolais and A. Borzacchiello, Eds., *Wiley Encyclopaedia of Composites*, 2nd ed., doi:10.1002/9781118097298.weoc025.

Morgan, A.B., and Wilkie, C.A., Eds., (2007), *Flame Retardant Polymer Nanocomposites*, Hoboken, NJ: Wiley & Sons.

Natali, M., Monti, M., Kenny, J., and Torre, L., (2011a) Synthesis and thermal characterization of phenolic resin/silica nanocomposites prepared with high shear rate-mixing technique, *Journal of Applied Polymer Science*, 120, 2632–2640.

Natali, M., Monti, M., Kenny, J., and Torre, L., (2011b) A nanostructured ablative bulk moulding compound: Development and characterization, *Composite Part A: Applied Science and Manufacturing*, 42, 1197–1204.

Natali, M., Monti, M., Puglia, D., Kenny, J., and Torre, L., (2012) Ablative properties of carbon black and MWNT/phenolic composites: a comparative study, *Composites Part A: Applied Science and Manufacturing*, 43(1), 174–182.

Patton, R.D., Pittman, Jr. C.U., Wang, L., and Hill, J.R., (1999a) Ablation, mechanical and thermal conductivity properties of vapor grown carbon fiber/phenolic matrix composites, *Composites Part A: Applied Science and Manufacturing*, 30(9), 1081–91.

Patton, R.D., Pittman, Jr., C.U., Wang, L., and Hill, J.R., (1999b) Vapor grown carbon fiber composites with epoxy and poly (phenylene fulfide) matrices, *Composites Part A: Applied Science and Manufacturing*, 30, 1081–1091.

Pavli, A.J., (1968) Experimental evaluation of several advanced ablative materials as nozzle sections of a storable propellant rocket engine, NASA TM X-1559.

Peterson, D.A., Winter, J.M., and Shinn, A.M., Jr., (1969) Rocket engine evaluation of erosion and char as functions of fabric orientation for silica-reinforced nozzle materials, NASA TM X-1721.

Philip, S.H., Gonzales, R.I., Blanski, R.L., and Viers, B.D., (2002) Hybrid inorganic/organic reactive polymers for sever environment protection, *Proc. of 47th SAMPE ISSE*, SAMPE, Covina, CA, May 12–16.

Pinnavaia, T.J., and Beall, G.W., Eds., (2000) *Polymer-Clay Nanocomposites*, New York: Wiley.

Pulci, G., Tirillo, J., Marra, F., Fossati, F., Bartuli, C., and Valente, T. (2010) Carbon-phenolic ablative materials for re-entry space vehicles: Manufacturing and properties, *Composites Part A: Applied Science and Manufacturing*, 41, 1483–1490.

Rahatekara, S.S., Zammarano, M., Matko, S., Koziol, K.K., Windle, A.H., Nyden, M., et al. (2010) Effect of carbon nanotubes and montmorillonite on the flammability of epoxy nanocomposites, *Polymer Degradation and Stability*, 95, 870–879.

Safadi, B., Abdrews, R., and Grulke, E. A., (2002) Multiwalled carbon nanotube polymer composites: Synthesis and characterization of thin films, *Journal of Applied Polymer Science*, 84, 2660–2669.

Shih, Y.C., Cheung, F.B., and Koo, J.H., (2003) Numerical study of transient thermal ablation of high-temperature insulation materials, *Journal of Thermophysics and Heat Transfer*, 17(1), 53–61.

Srikanth, I., Daniel, A., Kumar, S., Padmavathi, N., Singh, V., Ghosal, P., Kumar, A., and Devi, G.R., (2010) Nano silica modified carbon–phenolic composites for enhanced ablation resistance, *Scripta Materialia*, 63, 200–203,

Sutton, P., and Biblarz, O., (2000) *Rocket Propulsion Elements*, New York: Wiley-IEEE.

Tanaka, K., Adachi, S., and Chujo, Y., (2009) Structureproperty relationship of octa-substituted POSS in thermal and mechanical reinforcements of conventional polymers, *Journal of Polymer Science: Part A: Polymer Chemistry*, 47, 5690–5697.

Tate, J.S., Jacobs, C.J., and Koo, J.H., (2011) Dispersion of MWCNT in phenolic resin using different dispersion techniques and evaluation of thermal properties, *Proc. of 2011 SAMPE ISSE*, Long Beach, CA, May 23–26.

Thostenson, E.T., Li, C., and Chou, T.W., (2005) Nanocomposites in context, *Composites Science and Technology*, (65), 491–516.

Torre, L., Kenny, J.M., and Maffezzoli, A.M., (1998a) Degradation behaviour of a composite material for thermal protection systems Part I-Experimental characterization, *Journal of Materials Science*, 33, 3137–3143.

Torre, L., Kenny, J.M., and Maffezzoli, A.M., (1998b) Degradation behaviour of a composite material for thermal protection systems Part II Process simulation, *Journal of Materials Science*, 33(12), 3145–3149.

Torre, L., Kenny, J.M., Boghetich, G., and Maffezzoli, A.M., (2000) Degradation behaviour of a composite material for thermal protection systems Part III Char characterization, *Journal of Materials Science*, 35(18), 4563–4566.

Vaia, R.A., Price, G., Ruth, P.N., Nguyen, H.T., and Lichtenhan, J.L., (1999) Polymer/layer silicate nanocomposites as high performance ablative materials, *Applied Clay Sciences*, 15, 67–92.

Wang, W.J., Perng, L.H., Hsiue, G.H., and Chang, F.C., (2000) Characterization and properties of new silicone-containing epoxy resin, *Polymer*, 41(16), 6113–6122.

Warga, J.J., (1979) Low cost fabrication techniques for solid rocket nozzles, *Proc. of National Aeronautics and Space Engineering and Manufacturing Meeting*, Los Angeles, CA, October 5–9, pp. 700796.

Wilson, D., Beckley, D., and Koo, J.H. (1994) Development of silicone matrix-based advanced composites for thermal protection, *High Performance Polymer*, 6(2), 165–181.

Wu, C.S., Liu, Y.L., and Chiu, Y.S., (2002) Epoxy resins possessing flame retardant elements from silicon incorporated epoxy compounds cured with phosphorus or nitrogen containing curing agents, *Polymer*, 43(15), 4277–4284.

Xiao, J., Chen, J-M., Zhou, H-D., and Zhang, Q., (2007) Study of several organic resin coatings as anti-ablation coatings for supersonic craft control actuator, *Materials Science and Engineering A*, 452–453, 23–30.

Yu, Q-C., and Wan, H., (2012) Ablation Capability of Flake Graphite Reinforced Barium-phenolic Resin Composite under Long Pulse Laser Irradiation, *Journal of Inorganic Materials*, 27(2), 157–161 (in Chinese).

Zanetti, M., Camino, G., and Mülhaupt, R., (2001) Combustion behavior of EVA/flourohectorite nanocomposites, *Polymer Degradation and Stability*, 74, 413–417.

Zhao, Z., and Gou, J., (2009) Improved fire retardancy of thermoset composites modified with carbon nanofibers, *Science and Technology of Advanced Materials*, 10, 015005.

Zheng, S., Wang, H., Dai, Q., Kuo, X., Ma, D., and Wang, K., (1995) Morphology and structure of organosilicon polymer-modified epoxy resins, *Macromolecular Chemistry and Physics*, 196(1), 269–78.

Chapter 10

Antifriction Nanocomposites Based on the Chemically Modified Ultra-High Molecular Weight Polyethylene

Lyudmila A. Kornienko and Sergey V. Panin
Tomsk State University, Institute of Strength Physics and Materials Science RAS, Tomsk, Russia
kornienkola@ispms.tsc.ru, paninsergey@yahoo.co.uk

In this chapter, the mechanical and tribotechnical properties of the chemically modified ultra-high molecular weight polyethylene (UHMWPE) have been investigated. Permolecular structure and wear track surfaces of UHMWPE specimens with various volume fractions of added UHMWPE and high-density polyethylene (HDPE) grafted with maleic anhydride (SMA) and vinyltrimethoxysilane (VTMS) were studied. It is shown that adding grafted UHMWPE exerts low influence onto the viscoelastic properties of the polymer–polymeric composite. On the other hand, adding grafted HDPE brings about visible improvement in the technological properties (flowability, ductility) at nearly constant strength. Chemical modification of UHMWPE makes it possible to increase wear resistance considerably (by 2–3 times) that is related to features of crystallization and formation of permolecular structure. It is shown that the mechanical and tribotechnical properties of

Characterization of Nanocomposites: Technology and Industrial Applications
Edited by Frank Abdi and Mohit Garg

ISBN 978-981-4669-02-3 (Hardcover), 978-1-315-36489-6 (eBook)
www.panstanford.com

nanocomposites based on the block copolymers are essentially different from those of nanocomposites based on the nonmodified UHMWPE. It is established that crystallization and formation of permolecular structure in the investigated heterogeneous materials are defined by the uniformity of nanofiller distribution.

According to the authors, the chemical modification realized by the grafting of polar monomers does not allow to increase the adhesion between nanofiller and high-molecular matrix (UHMWPE). The wear resistance of the UHMWPE-based nanocomposites is mostly determined by crystallization conditions and the type of permolecular structure (spherulitic or lamellar) formed during crystallization.

10.1 Introduction

Polymeric composites having low friction coefficient and high wear resistance in combination with reasonable stress–strain characteristics and appropriate processability are the objects of intensive scientific investigations [1–4]. Particular type of filler is determined in correlation with application area and operating conditions of the given antifriction materials. Application of nanocomposites provides new possibilities, as a decrease in the reinforcing particle sizes brings about an essential change in almost all of the physical and chemical properties of the initial components and composite as a whole [5–8].

Ultra-high molecular weight polyethylene (UHMWPE) is of special importance among the family of polymeric materials owing to its high corrosion and chemical resistance [9, 10], and also biocompatibility, which open opportunities for application of this polymer in medicine and, first of all, in endoprosthesis (implants) manufacturing. The basic problem in the development of UHMWPE based on antifriction composites is its low adhesion to nanofillers because of the absence of polar groups. Finding ways to increase adhesion between the UHMWPE and nanofillers is an actual problem [4, 11]. In the present work an attempt has been made to enhance adhesion between UHMWPE and nanofillers by means of chemical modification of the polymer by adding UHMWPE grafted with maleic anhydride (SMA) and high-density polythene (HDPE) grafted with (SMA) and a more complex multipolar component—vinyltrimethoxysilane (VTMS, $C_5H_{12}O_3Si$), for the design of antifriction nanocomposites. Grafted copolymers with functional groups are effective compatibilizers to increase interphase adhesion [12]. The mechanism of compatibilizer action consists in the following: the grafted copolymers based on UHMWPE and HDPE have thermodynamic affinity with filled nonpolar polymer of the same type and can be well combined with it.

At the same time, a compatibilizer due to the presence of active functional groups can form strong chemical bonds with the filler, which allows to realize enforcement effect of fillers onto the physical-mechanical and tribotechnical properties [12–14]. The approach based on modification of a material by strengthening bonds at the polymer–nanofiller interface can be an effective and perspective way to improve the technological and operational properties of composite polymeric materials.

The present research includes the analysis of physical-mechanical and antifrictional properties of UHMWPE modified by adding copolymers such as UHMWPE-g-SMA and HDPE-g-SMA, as well as HDPE-g-VTMS, and nanocomposites on their basis.

10.2 Experimental Techniques

UHMWPE (GUR-4120 by Ticona, Germany) was used in this work. The molecular weight of UHMWPE is 3.0 million. Copolymers of HDPE-g-SMA and HDPE-g-VTMS were manufactured by the reacting extrusion in a melted state. The degree of grafting was $\sim$1.5%. Powder was obtained by the mechanical treatment of the granules at room temperature. The powder of UHMWPE was grafted in a solid-phase state by using a planetary mill in ozone atmosphere. The grafting degree was noticeably lower. The mixing of binder powders (UHMWPE) with UHMWPE-g-SMA, HDPE-g-SMA, and HDPE-g-VTMS was carried out in the small-size industrial-type planetary ball mill MP/0.5 $\times$ 4. Specimens were manufactured by hot pressing at a pressure of 10 MPa and temperature 190°C with a cooling rate 3–4° C/min. The density of the specimens was measured by hydrostatic weighting. Tensile tests were performed at the electromechanical testing machine INSTRON-5582. The tribological properties were measured at the "SMT-1" testing machine at dry sliding friction. Loading was 160 N while roller revolution rate was 100 min^{-1}. The size of the specimens was $7 \times 7 \times 10$ mm^3.

The wear resistance of the materials at dry friction mode was determined at "block-on-ring" scheme according to ASTM G99 and DIN 50324. The surfaces of the specimens' wear tracks were studied with the help of optical profilometer Zygo New View 6200. The area of the wear tracks was measured by means of the Rhino Ceros 3.0 software by manual allocation of their contours and subsequent automatic calculation. Investigation of phase structure and degree of crystallinity was carried out by Shimadzu XRD-6000 diffractometer with CuK_α radiation and combined analyzer SDT Q600. Infrared spectra were studied by using the NIKOLET 5700 FTIR spectrometer. The permolecular structure was examined by means of a scanning electron microscope. Fracture surfaces of specimens with a notch fractured exactly after extracting specimens from liquid nitrogen tank was performed by LEO EVO 50 microscope at accelerating voltage of 20 kV.

10.3 Results and Discussion

In the first part of this chapter, the physical-mechanical and tribotechnical characteristics of UHMWPE composites with the various volume fractions of UHMWPE-g-SMA and HDPE-g-SMA (1–20 wt%) added to initial UHMWPE powder were investigated. The physical-mechanical characteristics of UHMWPE are summarized in Table 10.1.

It follows from Table 10.1 that the elastoplastic characteristics of UHMWPE (elastic modulus, yield point, and elongation) do not essentially change with increasing volume fraction of graft UHMWPE added. The density and shore hardness of this composite remain nearly constant with an increase in the content of the filler. In the case of the UHMWPE and HDPE-g-SMA mixture, the density and hardness also remain unchanged with an increase in the amount of the filler. On the other hand, the elastoplastic properties increase in comparison with those of pure UHMWPE (Table 10.2). The changes

Table 10.1 Mechanical properties of UHMWPE modified by grafted UNMWPE

Vol. fraction $UHMWPE_{grafted}$, wt%	Density d, g/cm^3	Elasticity modulus E, MPa	Shore A hardness	Yield point, $\sigma_{0.2}$, MPa	Ultimate strength, σ_B, MPa	Elongation at fracture, ε, %
0	0.916	549.42	97.9	19.99	30.94	474.25
1	0.912	–	98.0	–	–	–
3	0.916	594.26	98.1	17.62	28.27	441.99
5	0.913	539.48	97.9	18.86	28.41	429.86
10	0.912	563.37	98.0	19.99	32.00	477.00
20	0.916	574.42	97.9	19.31	32.52	441.66

Table 10.2 Physical-mechanical properties of UHMWPE modified with grafted PE

Vol. fraction of PE_{graft}, wt%	Density d, g/cm^3	Elasticity modulus E, MPa	Shore A hardness	Yield point, $\sigma_{0.2}$, MPa	Ultimate strength, σ_B, MPa	Elongation at fracture, ε, %
0	0.932	570.90	97.3	19.19	34.31	470.91
1	0.919	595.61	97.2	23.30	40.44	507.66
3	0.934	552.86	97.1	24.43	38.43	458.54
5	0.931	582.83	97.0	24.78	38.79	486.58
10	0.932	586.76	97.0	24.92	39.19	486.66

in ultimate elongation are most noticeable: the polymer becomes more ductile (by 15–30%).

The wear intensity of all the above-mentioned compositions was measured. The wear diagrams (1, 2) for UHMWPE modified by adding different volume fractions of UHMWPE-g-SMA are given in Fig. 10.1.

In both cases, wear is decreased on UHMWPE modification. The wear of UHMWPE composite specimens changes poorly with increasing volume frqactions of graft component (from 1 to 20 wt%). Figure 10.3 shows the wear intensity of UHMWPE modified with grafted UHMWPE and HDPE. It is seen that the wear resistance of UHMWPE modified with grafted HDPE is higher (approximately 3 times) than that of UHMWPE modified with grafted UHMPE (2.6 times).

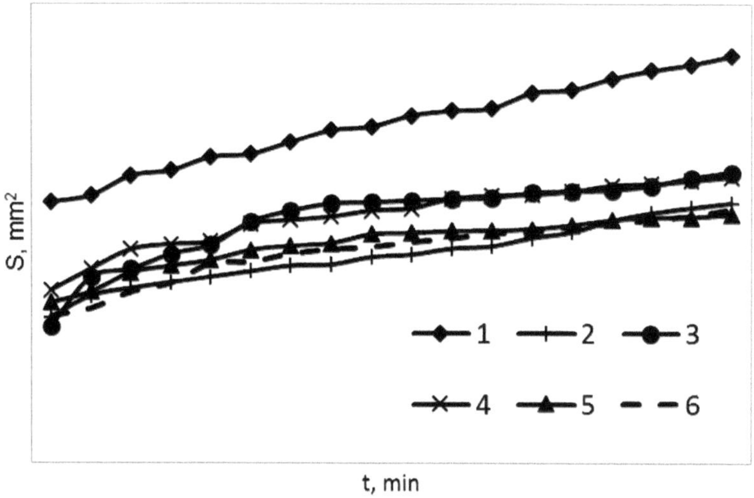

Figure 10.1 Wear diagrams of the pure UHMWPE (1) and of UHMWPE modified by adding different volume fractions of UHMWPE-g-SMA: (2) 1, (3) 3, (4) 5, (5) 10, and (6) 20 wt%.

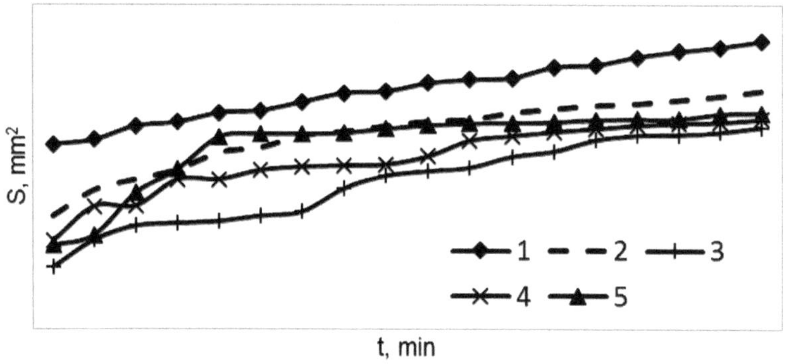

Figure 10.2 Wear diagrams of UHMWPE (1) modified by adding different volume fractions HDPE-g-SMA: (2) 1, (3) 3, (4) 5, and (5) 10 wt%.

To define the character of structural changes in UHMWPE on its chemical modification investigations of permolecular structure and wear tracks topography were carried out by varying copolymer volume fractions. It was determined that a spherulitic structure is formed in the modified UHMWPE during crystallization (Fig. 10.4).

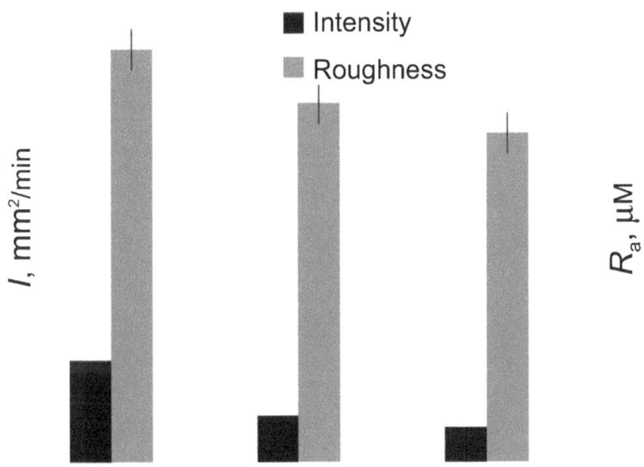

Figure 10.3 Dependance of wear intensity and roughness of wear track surfaces in pure (1) and modified UHMWPE on the steady-state wearing stage: (2) 10 wt% UHMWPE-g-SMA, (3) 10 wt% HDPE-g-SMA.

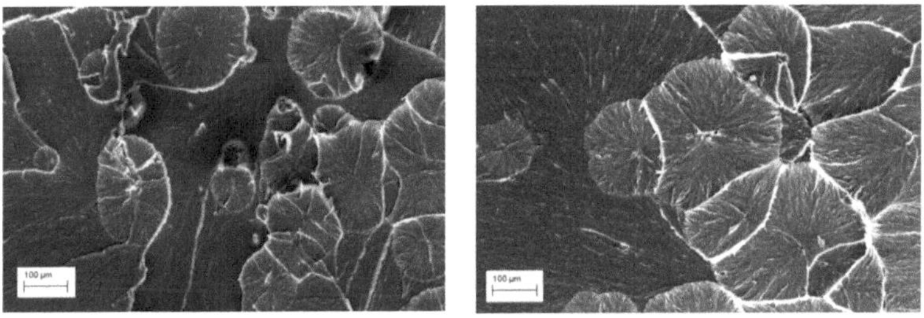

(a) (b)

Figure 10.4 Permolecular structure of UHMWPE filled with 10 wt% PE-g-SMA (a) and 10 wt% UHMWPE-g-SMA (b).

It should be noted that finer spherulitic structure is formed when the modifying is carried out with HDPE-g-SMA (Fig. 10.4a). As far as our estimates have shown, the average spherulite size in the first case (Fig. 10.4a) is 20–25% less than in the second one (Fig. 10.4b). The internal structure of the spherulites (packaging of molecules) is similar in both the cases.

The study of the crystallinity degree of polymers by differential scanning calorimetry (DSC) and X-ray diffraction analysis has shown

Table 10.3 Results of structural analysis of modified UHMWPE

Specimen	Crystallinity, χ (DSC), %	Crystallinity, χ (X-ray), %	CSR, nm
UHMWPE	48.6	48.0	56
UHMWPE + 10% HDPE-g-SMA	68.6	66.5	38
UHMWPE + 10% UHMWPE-g-SMA	50.9	52.0	52

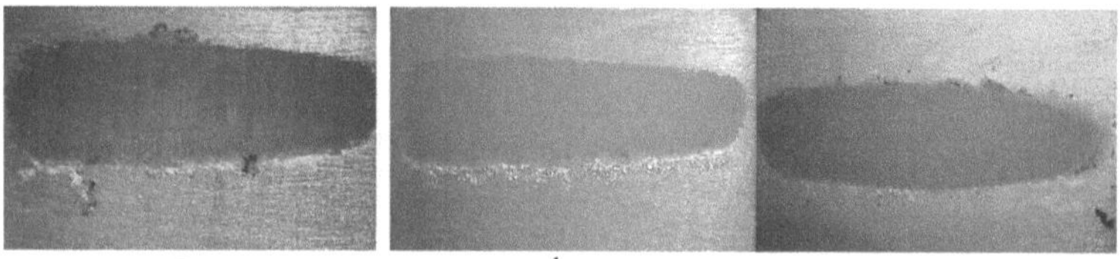

Figure 10.5 Images of wear track surfaces in pure (a) and modified UHMWPE at the steady-state wearing stage: (b) 10 wt% of UHMWPE-g-SMA, and (c) 10 wt% of HDPE-g-SMA. Magnification: ×50.

that the crystallinity (χ) is higher by 17% in UHMWPE specimens modified by adding grafted HDPE (Table 10.3). The size of coherent scattering regions (CSRs) in these specimens is less than that in the polymer filled with grafted UHMWPE.

Analysis of the roughness (R_a) of wear tracks have shown that it deacreases in the modified UHMWPE specimens (Figs. 10.3 and 10.5). It has the lowest value in UHMWPE filled with grafted PE. These data are in good agreement with tha data on wear intensity.

In our opinion, greater wear resistance in chemically modified UHMWPE is mostly related to the change in the crystallinity degree as a consequence of the size of spherulites. This can be proven by IR spectra (Fig. 10.6). The spectra of pure UHMWPE and those of UHMWPE modified by adding grafted UHMWPE and HDPE are identical. One can notice only some changes in the intensity of the C–O peaks in the region of 1100 cm^{-1} due to the nucleation of anhydride groups in the copolymer.

The presence of free functional groups in block copolymers (UHMWPE–UHMWPE$_{graft}$ and UHMWPE–HDPE$_{graft}$) defines the crystallization process and, as a result, the formation of the

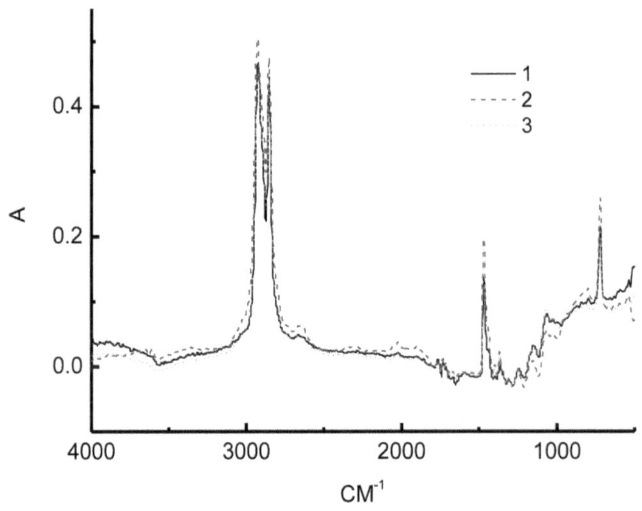

Figure 10.6 Infrared spectra of pure UHMWPE (1) and of UHMWPE modified by grafted UHMWPE (2) and PE (3).

permolecular structure of UHMWPE. Free functional groups are activators (centers) of crystallization and are responsible for the formation of the spherulitic structure. The permolecular structure of UHMWPE largely defines the frictional properties of the polymer, as shown in Refs. 14–16.

A neat part of the research is dedicated to composites on the basis of block copolymers (described above) filled with nanopowders and nanofibers. To perform a comparative analysis of the effect of chemical modification of UHMWPE on the tribotechnical properties of polymeric nanocomposites, the same fillers and compositions as studied in Refs. 14 and 17 were selected. The role of chemical modification of UHMWPE in changing the adhesion strength of nanofillers was analyzed.

The mechanical and friction characteristics of nanocomposites based on block copolymers UHMWPE–UHMWPE$_{graft}$ and UHMWPE–HDPE$_{graft}$ with various volume fractions of grafted copolymers (1–20 wt%) were studied. Carbon (external Ø 60 nm, length 2–3 μm) and Al_2O_3 (Ø 1–10 nm) nanofibers and copper (~90 nm) and SiO_2 (20–30 nm) nanoparticles with volume fraction 0.5 wt% were used as nanofillers. The mechanical characteristics of UHMPE

Table 10.4 Mechanical properties of block copolymers of UHMWPE-$UHMWPE_{graft}$ filled with 0.5 wt% C

Volume fraction of $UHMWPE_{graft}$, wt%	Density d, g/cm^3	Elasticity modulus E, MPa	Shore A hardness	Yield point, $\sigma_{0.2}$, MPa	Ultimate strength, σ_U, MPa	Strain at failure, ε, %
0	0.928	487.14	98.4	19.64	27.91	399.42
3	0.927	486.76	98.6	16.09	24.17	427.0
5	0.928	572.07	97.5	17.81	29.88	454.42
10	0.924	582.84	98.5	18.58	29.27	457.50
20	0.938	484.01	98.5	18.92	28.5	440.00

specimens filled with various fractions of carbon nanofibers and UHMWPE-g-SMA measured at stretching tests are shown in Table 10.4. For comparison, specimens of UHMWPE-g-SMA modified by adding nanofillers (100 wt% UHMWPE–$UHMPE_{graft}$) were also manufactured and tested.

As follows from Table 10.4, the elastic modulus, yield point, ultimate strength, and elongation at failure of nanocomposites based on block copolymer UHMWPE–$UHMWPE_{graft}$ with various compatibilizer volume fractions ($UHMWPE_{graft}$) increase slightly compared with similar properties of nanocomposites based on pure (unmodified) UHMWPE given in the first row of the table. The same trend is observed for nanocomposites based on block copolymer UHMPE–PE_{graft} (Table 10.5). However, the strain at failure, ε, of the nanocomposites based on UHMPE block copolymers is higher than that for those based on chemically nonmodified UHMWPE. As

Table 10.5 Mechanical properties of UHMWPE-based block copolymers: UHMWPE- PE_{graft} filled with 0.5 wt% C

Volume fracture PE_{graft}, wt%	Density d, g/cm^3	Elasticity modulus E, MPa	Shore A hardness	Yield point, $\sigma_{0.2}$, MPa	Ultimate strength, σ_U, MPa	Strain at failure, ε, %
0	0.939	487.14	98.4	19.64	34.31	399.42
3	0.941	512.50	98.3	21.95	40.44	449.49
5	0.933	496.19	98.5	22.83	38.43	417.42
10	0.935	487.75	98.5	22.51	38.79	466.33
20	0.939	526.86	98.5	23.39	39.19	458.00

far as experimental studies have shown, the tendency is similar for all composites with nanofillers Al_2O_3, SiO_2, and Cu. It is found that block copolymers filled with SiO_2 and Cu nanoparticles show a larger improvement in the mechanical characteristics, in contrast with composites filled with C and Al_2O_3 nanofibers.

The wear resistance of nanocomposites based on block copolymers in comparison with the tribotechnical properties of nanocomposites based on nonmodified UHMWPE has been analyzed. The wear diagrams for composites based on UHMWPE–$UHMWPE_{graft}$ block copolymers with the addition of Al_2O_3 nanofibers and SiO_2 nanoparticles (0.5 wt%) are given in Figs. 10.1 and 10.2.

Similar wear digrams for nanocomposites were obtained by adding carbon nanofibers and copper nanoparticles. It follows from Figs. 10.7 and 10.8 that (i) the wear of the nanocomposites based on block copolymers looks similar to the wear of nanocomposites based on pure (unmodified) UHMWPE, and (ii) the wear of the first two types of composites is slightly higher than that of nanocomposites based on 100 wt% $UHMWPE_{graft}$ matrix (curve 7, Figs. 10.7 and

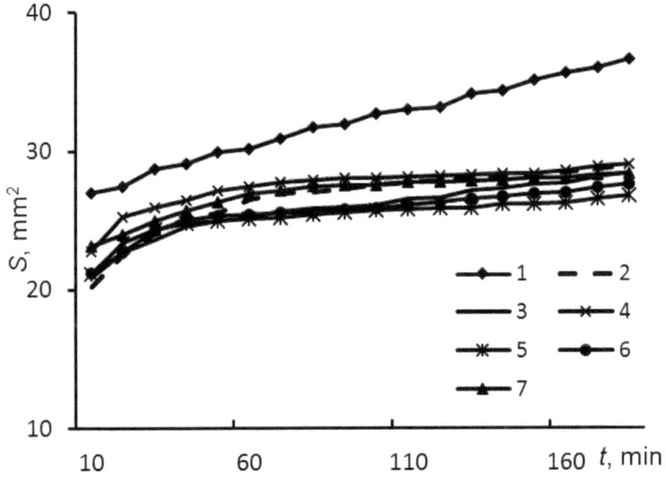

Figure 10.7 Wear diagrams of pure UHMWPE (1), UHMWPE + 0.5 wt% Al_2O_3 nanocomposite (2), block copolymer UHMWPE nanocomposites with different volume fractions of $UHMWPE_{graft}$: (3) 3, (4) 5, (5) 10, (6) 20 wt%; $UHMWPE_{graft}$ + 0.5 wt% Al_2O_3 nanocomposite (7).

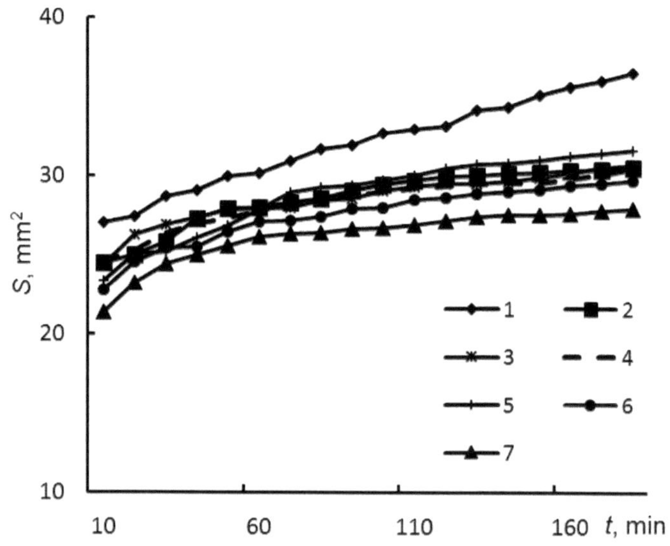

Figure 10.8 Wear diagrams of pure UHMWPE (1), UHMWPE + 0.5 wt% SiO_2 nanocomposite (2), block copolymer UHMWPE-based nanocomposites with different volume fractions of $UHMWPE_{graft}$: (3) 3, (4) 5, (5) 10, (6) 20 wt%; $UHMWPE_{graft}$+ 0.5 wt% SiO_2 nanocomposite (7).

10.8). The type of nanofiller (fibers or particles) does not affect the wearing pattern of all composites studied in this chapter.

The roughness of the wear track surface of nanocomposites based on pure UHMWPE, block copolymer UHMWPE-$UHMWPE_{graft}$, and $UHMWPE_{graft}$ is illustrated in Fig. 10.9. It is seen that the roughness of the wear tracks of nanocomposites based on pure UHMWPE and that of the block copolymers is nearly identical, while for nanocomposites based on $UHMWPE_{graft}$ the roughness of the wear tracks surface is slightly lower (columns 6, 9, 12, 15).

The diagram to characterize wear intensity at steady-state stage for all investigated nanocomposites is presented in Fig. 10.10. Analysis of the data described above shows that the wear resistance of chemically modified UHMWPE increases by 2.5–3 times in comparison with pure polymer (columns 1, 2, 3). The filling of block copolymers with nanoparticles and nanofibers is not accompanied by an increase in wear resistance (columns 5, 11, 14).

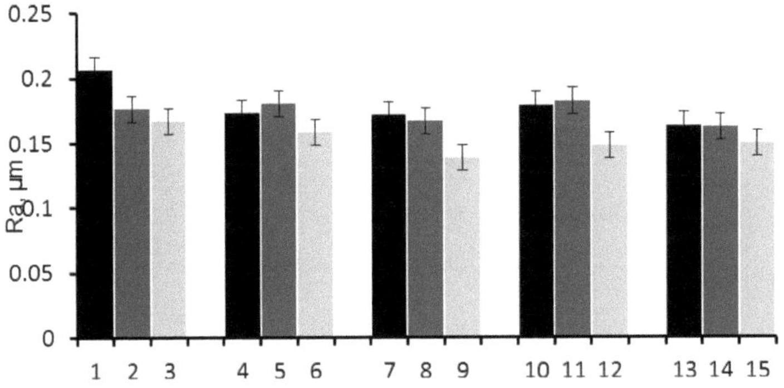

Figure 10.9 Roughness of wear tracks surface at the steady-state wearing stage of pure UHMWPE (1), block-copolymer UHMWPE-UHMWPE$_{graft}$ (2), UHMWPE$_{graft}$ (3) and nanocomposites on its basis: 0.5 wt.% SiO_2 (4, 5, 6), 0.5 wt.% Al_2O_3 (7, 8, 9), 0.5 wt.% Cu (10, 11, 12) and 0.5 wt.% C (13, 14, 15).

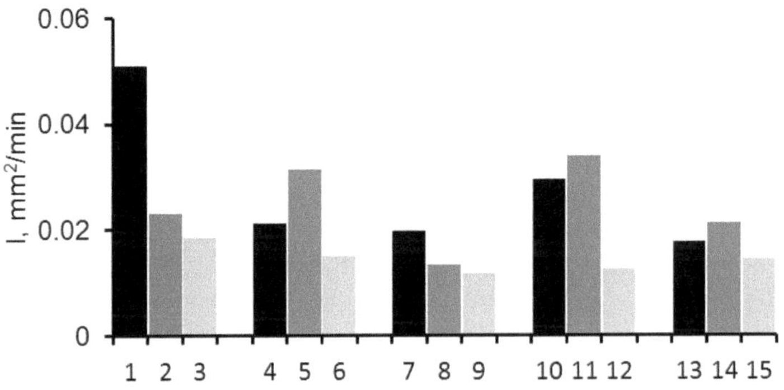

Figure 10.10 Wear intensity of the pure UHMWPE (1), UHMWPE-UHMWPE$_{graft}$ block copolymer (2), UHMWPE$_{graft}$ (3), and nanocomposites filled with 0.5 wt% SiO_2 (4, 5, 6), 0.5 wt% Al_2O_3 (7, 8, 9), 0.5 wt% Cu (10, 11, 12), and 0.5 wt% C (13, 14, 15).

At the same time, the wear resistance of nanocomposites based on 100% UHMWPE$_{graft}$ increases by 1.5–2 times in comparison with chemically modified UHMWPE-UHMWPE$_{graft}$ (columns 3, 6, 9, 12, 15) and is higher than that of nanocomposites based on pure UHMWPE (columns 1, 4, 7, 10, 13).

Table 10.6 Crystallinity (χ, %) of nanocomposite specimens based on UHMWPE

Specimen	Pure	0.5 wt% SiO_2	0.5 wt% Al_2O_3	0.5 wt% Cu	0.5 wt% C
UHMWPE	48.6	53.3	55.8	53.5	50.1
UHMWPE - $UHMWPE_{graft}$	56.5	50.7	54.5	52.8	49.3
UHMWPE-PE_{graft}	60.0	56.0	58.0	55.5	52.4

To clarify the reason behind structural changes in nanocomposites based on chemically modified UHMWPE, studies of permolecular structure by means of scanning electron microscopy, differential scanning calorimetry, and infrared spectroscopy were performed. Table 10.6 presents the results of measuring the degree of crystallinity for specimens of pure UHMWPE, chemically modified UHMWPE (UHMWPE-$UHMWPE_{graft}$, UHMWPE-PE_{graft}), and nanocomposites based on it.

Table 10.6 shows that the degree of crystallinity of nanocomposites based on block copolymers is lower than that of the block copolymers, and is comparable with the crystallinity of nanocomposites based on pure UHMWPE. The infrared spectra of these composites are identical with those of block copolymers (Fig. 10.11). One can assume that the chemical bonds between nanofillers and grafted polar groups do not arise. This can be proven by IR spectra of UHMWPE-$UHMWPE_{graft}$ and the corresponding nanocomposites (Fig. 10.12). The spectrum of pure UHMWPE (curve 1) is presented in Fig. 10.12 for the sake of comparison. The intensities of the C–O peaks in the range of 1100 cm^{-1} for grafted UHMWPE and its nanocomposites are the same.

It might be concluded that in block copolymers the presence of grafted polar groups gives rise to an increase in crystallinity compared with pure UHMWPE (Table 10.6) and, as a result, in the improvement of the polymer's wear resistance. If this takes place on adding nanofillers, crystallization in the polymer is determined by the presence and uniformity of the distribution of the latter in the matrix. As a result, the wear resistance of nanocomposites based on chemically modified UHMWPE is close to that of the nanocomposites based on pure UHMWPE.

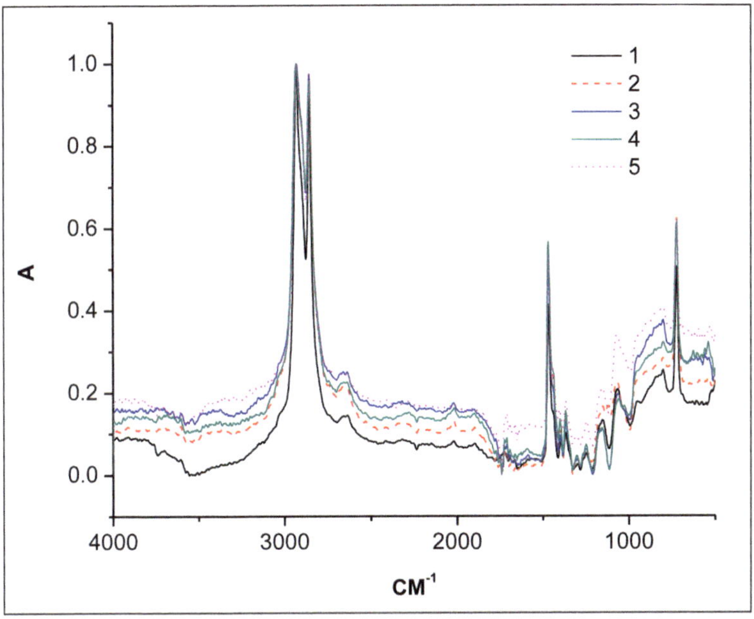

Figure 10.11 Infrared spectra of the UHMWPE-UHMWPE$_{graft}$ block copolymer (1) and nanocomposites based on filled UHMWPE + UHMWPE$_{graft}$: + 0.5 wt% SiO_2 (2), + 0.5 wt% Al_2O_3 (3), + 0.5 wt% Cu (4), and + 0.5 wt% C (5).

In Fig. 10.13 SEM micrographs of the permolecular structure of nanocomposites based on pure (Fig. 10.13a,b) and chemically modified UHMWPE (Fig. 10.13c,d) are illustrated. It is seen that in both types of composites, a structure similar to the spherulitic structure is formed.

In the center of spherulites, some inclusions in the form of agglomerates of nanoparticles are usually observed. It is confirmed by the microanalysis of inhomogeneities of the fracture surface of the composites. For both types of composites, the polymer melt crystallization is initiated, first of all, from nanofillers. Consequently, the wear resistance of nanocomposites based on chemically modified UHMWPE as well as of pure polymer is largely determined by the structure (packing of molecules) which is formed during the crystallization process. The conditions of the crystallization of melted UHMWPE in addition to thermodynamics are defined by the

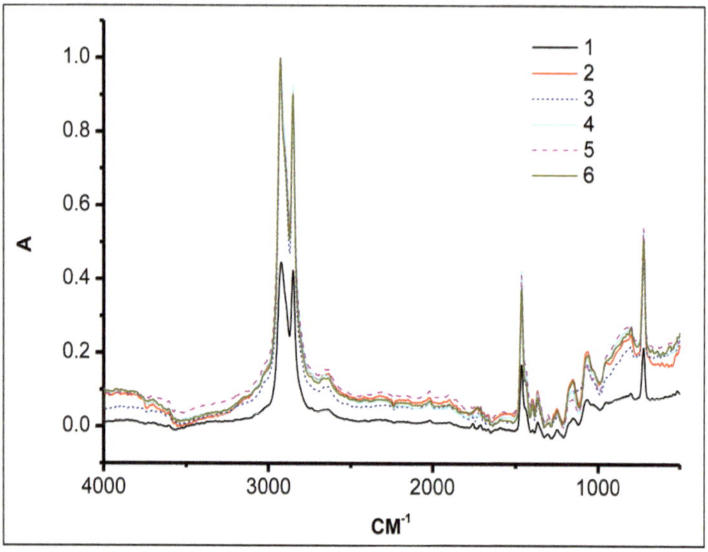

Figure 10.12 Infrared spectra of pure UHMWPE (1), $UHMWPE_{graft}$ block copolymer (2), and nanocomposites based on $UHMWPE_{graft}$: + 0.5 wt% SiO_2 (3), + 0.5 wt% Al_2O_3 (4), + 0.5 wt% Cu (5), and + 0.5 wt% C (6).

presence of nano- or microfillers, and unbonded polar groups. The formed structure determines the degree of the matrix destruction under friction.

The next part of the research was dedicated to an attempt to change the adhesion properties of UHMWPE to nanofillers by grafting more complex multipolar component: vinyltrimethxysilane (VTMS-$S_5N_{12}O_3Si$). A comparative analysis of the influence of a compatibilizer type onto the tribotechnical properties of polymer nanocomposites in order to determine the effect of different compatibilizers on the adhesion properties of UHMWPE and fillers as well as to estimate the role of filler size (nano, micro) in the formation of permolecular structure of composites based on high-molecular matrix was carried out. The mechanical and frictional characteristics of block copolymer UHMWPE-$HDPE_{graft}$ with various volume fractions of grafted component (1–20 wt%) were investigated for a comparative analysis of the effect of different types of compatibilizers on the adhesion properties of the high-

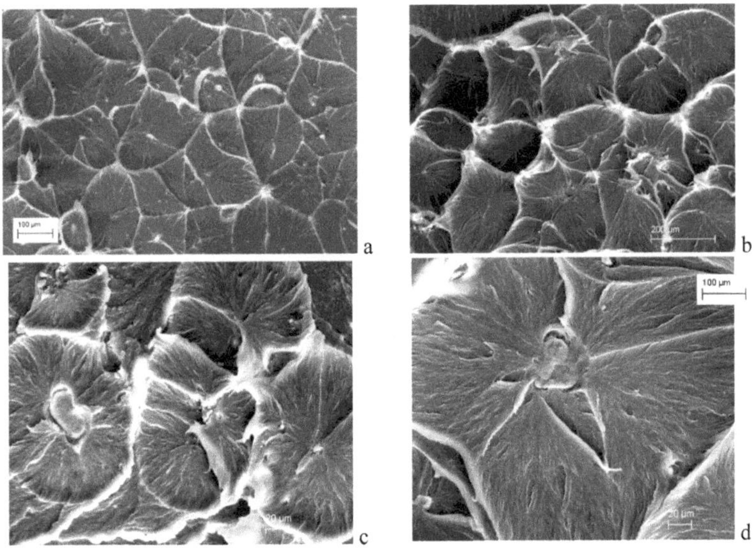

Figure 10.13 SEM micrographs of the permolecular structure of nanocomposites: UHMWPE + 0.5 wt% C (a, b) and UHMWPE-$UHMWPE_{graft}$+ 0.5 wt% C (b, d).

molecular polymer to nanofillers. Carbon nanofibers (C) and Al_2O_3, copper nanoparticles (Cu), and SiO_2 with volume fraction 0.5 wt% were used as nanofillers. Table 10.7 shows the mechanical properties measured under tensile tests of UHMWPE specimens modified by adding various volume fractions of HDPE-g-VTMS.

Table 10.7 Mechanical characteristics of UHMWPE modified by HD-PE-g-VTMS

Volume fracture of PE_{graft}, wt%	Density d, g/cm^3	Elastic modulus E, MPa	Shore hardness HSA	Yield point, $\sigma_{0.2}$, MPa	Ultimate strength, σ_U, MPa	Strain at failure, ε, %
0	0.939	549.42	97.9	19.99	30.94	474.25
3	0.942	471.52	98.6	19.54	29.73	476.25
5	0.938	453.93	98.5	19.46	29.37	483.50
10	0.945	468.44	98.0	16.39	24.01	493.58
20	0.941	454.85	98.5	16.82	27.43	512.50

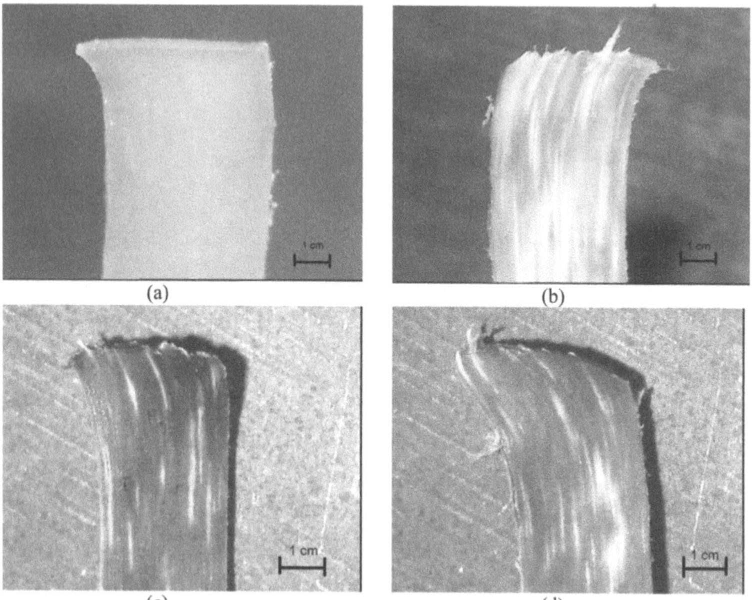

(a) (b) (c) (d)

Figure 10.14 Images of fractured specimens: pure UHMWPE (a), UHMWPE + 10% HDPE-g-VTMS block copolymer (b), and nanocomposites based on UHMWPE + 10 wt% HDPE-g-VTMS + 0.5 wt% C (c), and UHMWPE + 10 wt% HDPE-g-VTMS + 0.5 wt% Cu (d).

It is follows from the table that on adding HDPE grafted with VTMS, the modulus, yield point, and ultimate strength decrease with increasing content of the compatibilizer, and the elongation at failure increases. The optimum volume fraction of the compatibilizer in the block copolymer is about 10 wt%. Therefore, a further comparative analysis of nanocomposites based on this block copolymer was made. It was shown that the wear intensity (I) measured at the steady-state stage of wear for such a block copolymer is minimum.

The fractured specimens of pure UHMWPE (Fig. 10.15a) and block copolymer UHMWPE-HDPE grafted with VTMS (Fig. 10.15b), where heterogeneity (layering, peeling, lamination, foliation) of the block copolymer component is visible, are presented in Fig. 10.15. These conclusions are confirmed by the data of infrared spectra. There are peaks in the range of 1100–1200 sm^{-1}, corresponding to Si–O bonds [18]. However, the component that corresponds to HDPE

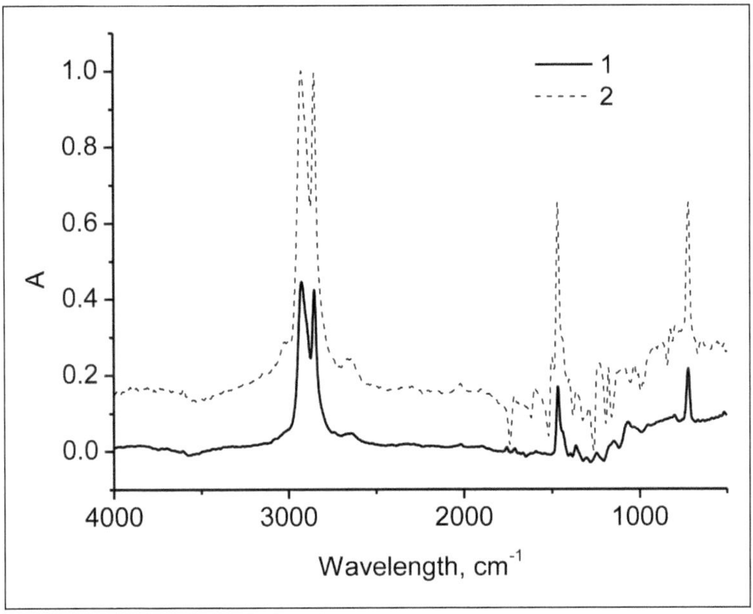

Figure 10.15 Infrared spectra of pure UHMWPE (1) and UHMWPE + 10% HDPE-g-VTMS block copolymer (2).

grafted with a complex monomer is deformed uniformly together with matrix UHMWPE: no sign of losing the adhesion between the components of the mixture is seen.

Analysis of the tribotechnical properties of nanocomposites based on UHMWPE + 10% HDPE-g-VTMS block copolymer has shown the following. Similar to nanocomposites based on block copolymers with HDPE-g-SMA with nanofibers and nanoparticles, approximately an equal increase in the wear resistance of the compositions is observed (Fig. 10.16).

The mechanical properties of nanocomposites based on UHMWPE + 10% HDPE-g-VTMS block copolymer are presented in Table 10.8. A comparative analysis of the mechanical and tribotechnical properties of nanocomposites based on UHMWPE + 10% HDPE-g-VTMS and UHMWPE + 10% HDPE-g-SMA block copolymers have shown that they are close to each other. This is clearly illustrated in Fig. 10.17, which characterizes wear intensity

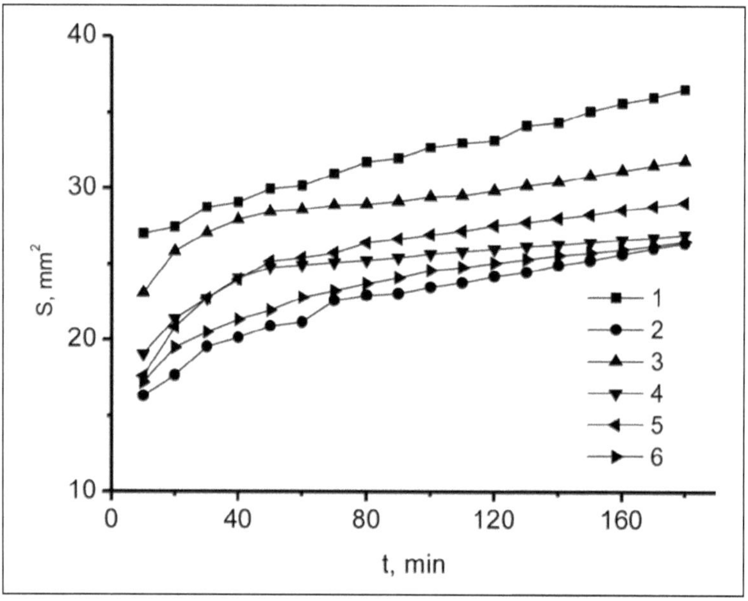

Figure 10.16 Wearing diagrams of pure UHMWPE specimens (1), and nanocomposites based on UHMWPE + 10% HDPE-g-VTMS block copolymer with 0.5 wt% SiO_2 (2), 0.5 wt% Al_2O_3 (3), 0.5 wt% Cu (4), and 0.5 wt% C (5).

at the steady-state stage for the three types of nanocomposites. It is seen that the wear intensity of nanocomposites based on chemically modified UHMWPE is close to that of nanocomposites based on pure, high molecular weight polymer. The roughness of the wear track surfaces for all three types of nanocomposites at a steady-state wearing stage is also the same.

Analysis of the permolecular structure of nanocomposites based on UHMWPE + 10% HDPE-g-VTMS block copolymer has shown that it is similar to that of nanocomposites based on UHMWPE + 10% HDPE-g-SMA (spherulitic). One should note that there is a feature in the formation process of the permolecular structure for composites modified by HDPE-g-VTMS to be studied. As was pointed above, there are heterogeneities in the polymer structure (delamination, etc.). It is similar for nanocomposites based on it (Fig. 10.14c,d).

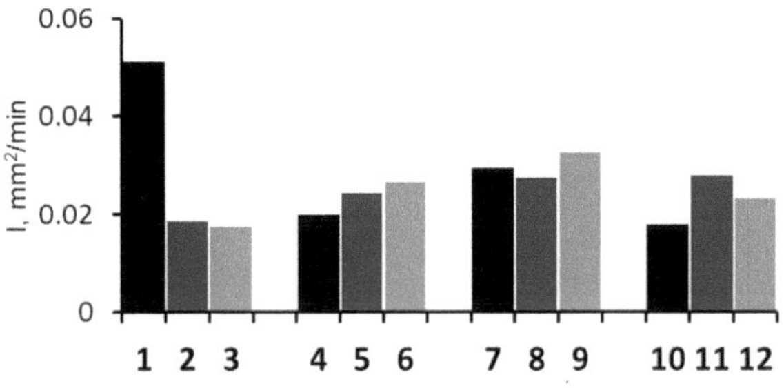

Figure 10.17 Wear intensity of the pure UHMWPE specimens (1), UHMWPE + 10% HDPE-g-VTMS block copolymer (2), UHMWPE + 10% HDPE-g-SMA (3), and nanocomposites based on it with 0.5 wt% Al_2O_3 (4, 5, 6), 0.5 wt% Cu (7, 8, 9), and 0.5 wt% C (10, 11, 12).

Table 10.8 Mechanical characteristics of nanocomposites based on UHMWPE + 10% HDPE-g-VTMS block copolymer filled with 0.5 wt% of filler

Filler	Density d, Г/cm^3	Elastic modulus E, MPa	Shore A hardness	Yield point $\sigma_{0.2}$, MPa	Ultimate Strength σ_U, MPa	Strain at failure ε, %
0	0.939	549.42	97.9	19.99	34.31	474.25
SiO_2	0.937	512.64	98.6	20.94	31.29	442.07
Al_2O_3	0.937	534.12	98.5	21.48	29.44	430.33
Cu	0.943	489.67	98.5	20.84	26.39	451.67
C	0.937	480.44	98.5	22.29	22.72	374.33

In doing so, heterogeneity in the spherulitic structure formation is visible on a fractured surface (Fig. 10.18).

The infrared spectra of nanocomposites based on this block copolymer are identical to those of the matrix UHMWPE + 10% HDPE-g-VTMS (Fig. 10.19). In other words, there are no additional bonds of free radicals with nanofillers. While the uneven distribution of copolymer HDPE-g-VTMS in UHMWPE slightly decreases the crystallinity of compositions, it is not critical, though, to decrease the tribotechnical properties of composites based on the specified matrix (Fig. 10.18, Table 10.9). As was shown above, the type of formed permolecular structure (packing molecules) plays a key role

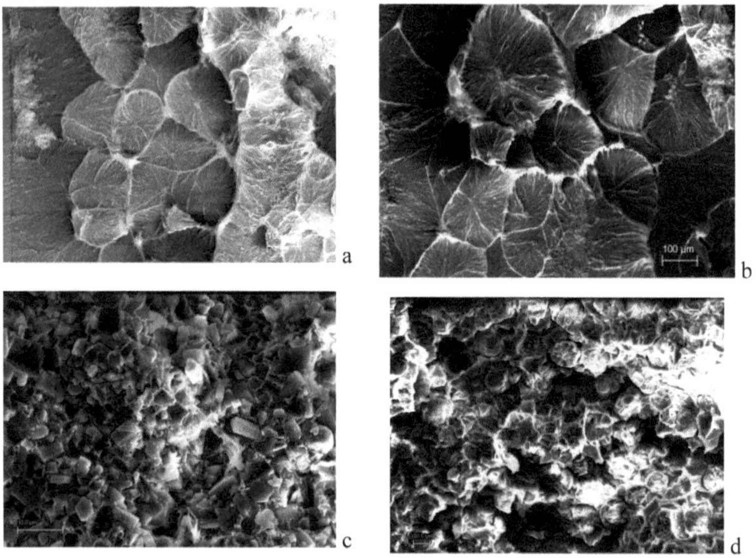

Figure 10.18 SEM micrographs of the permolecular structure of nanocomposites based on UHMWPE + 10% HDPE-g-VTMS with 0.5 wt% C (a) and 0.5 wt% Al_2O_3 (b), and microcomposites with 40 wt% AlO(OH) (c) and 40 wt% Al_2O_3 (d).

Table 10.9 Crystallinity (χ, %) of specimens of composite materials based on UHMWPE + 10% HDPE-g-VTMS

Specimen	Pure	0.5 wt% SiO_2	0.5 wt% Al_2O_3	0.5 wt% Cu	0.5 wt% C
UHMWPE	48.6	53.3	55.8	53.5	50.1
UHMWPE + 10% HDPE-g-VTMS	42.5	42.5	40.0	39.3	38.9

in the destruction of the surface layers of the high-molecular matrix at tribounits.

In this chapter, the tribotechnical properties and the permolecular structure of composites based on UHMWPE + 10% HDPE-g-VTMS and UHMWPE + 10% HDPE-g-SMA block copolymers by adding 40 wt% of micrometer-size fillers—particles (AlO(OH) of 2 µm and Al_2O_3 of 50 µm—were investigated to illustrate the key role of the permolecular structure in the formation of the

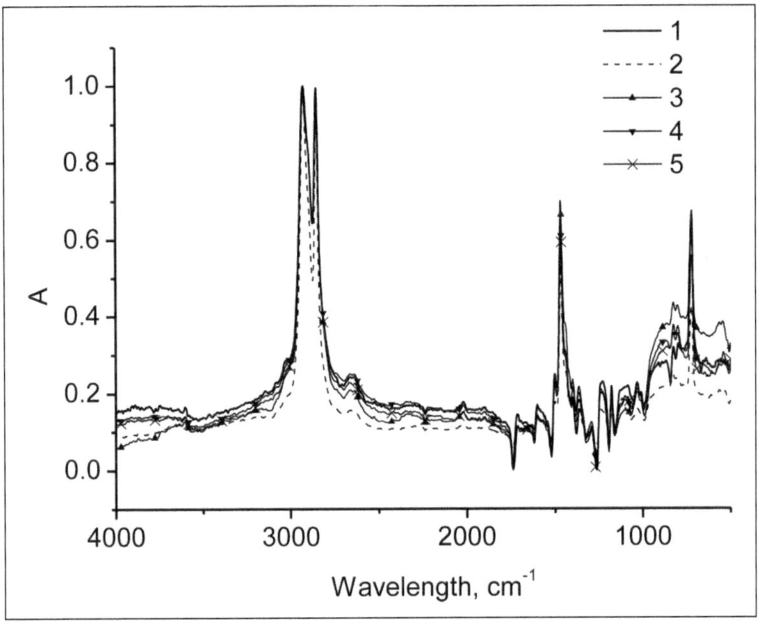

Figure 10.19 Infrared spectra of UHMWPE + 10% HDPE-g-VTMS block-copolymer (1) and nanocomposites on its base with 0.5 wt% SiO_2 (2), 0.5 wt% Al_2O_3 (3), 0.5 wt% Cu (4) and 0.5 wt% C (5).

tribotechnical properties of composites based on UHMWPE [17]. It was shown that the wear intensity of these composites increases compared to the nanocomposites and that the crystallinity of the structure falls sharply down to 16.6% and 23.1%, respectively. The amorphous component in the permolecular structure prevails. Permolecular structures of the examined microcomposites based on the chemically modified high-molecular matrix are presented in Fig. 10.18c,d. It is seen that the micrometer-size filler suppresses the formation of an ordered permolecular structure. The adhesion bond of the "filler matrix" is also weak [3].

Using the data obtained in this study, as well as in the papers [14, 15, 17, 19], it is possible to propose the following mechanism of wear of polymer nanocomposites based on UHMWPE.

There is an "oscillation" of solid particles of the nanofiller that plays the role of the dry-film lubricant in the soft matrix destroyed within the wear track surface. At the same time, smoothing of counterface surface can take place as a result of filling microrelief cavities by products of a plastic matrix fracture [11]. It can be also assumed that the nano-size particles of filler, having very low adhesion to the polymer matrix, provide a polish to the counterface surface that causes wear intensity reduction. Later, filler particles "oscillating" like fixed springs in the area of the isostructurized matrix provide high wear resistance and low friction coefficient of the polymer composite. The destruction degree of a composite surface layer is determined by type of the packing of molecules. So, permolecular structure is crucial in the formation of the tribotechnical properties of nanocomposites based on high-molecular matrix.

The complex research of tribotechnical properties of composites based on chemically modified UHMWPE shows that in order to obtain high tribotechnical characteristics of composites based on high-molecular matrix, it is essential to select optimum conditions of composition crystallization for the purpose of forming ordered (crystallized) permolecular structure. The latter can be achieved by means of high-energy effect, with the size and volume of the filler, and by chemical modification.

10.4 Conclusions

- The wear resistance of chemically modified UHMWPE at the steady-state stage of wear increases by 2.5–3 times.
- Growth of UHMWPE wear resistance due to increase in polymer crystallinity degree is caused by chemical modification with grafted copolymers.
- Grafted polar groups of various chemical natures (SMA, VTMS) determine the conditions of melt crystallization and, as a result, the type of permolecular structure of the composition formed.
- Nanocomposites based on chemically modified UHMWPE (UHMWPE-UHMWPE$_{graft}$ and UHMWPE-PE$_{graft}$ block copolymers) have mechanical and tribitechnical properties

close to those of nanocomposites based on pure (chemically unmodified) UHMWPE.

- The adhesion properties of the high-molecular matrix to nanofiller do not change on chemical modification of UHMWPE by adding into the composition HDPE grafted with monomers of difference structures.
- The permolecular structure of nanocomposites based on chemically modified UHMWPE plays a key role in the formation of tribotechnical characteristics.
- The mechanism of wear of polymer nanocomposites based on UHMWPE is proposed. The nanoparticles "oscillating" in the subsurface layer of destructured high-molecular matrix provide high wear resistance and low friction coefficient.

References

1. Mashkov Yu. K., Kropotin O. V. (2009). *Tribofizika i strukturnaya modifikatsiya materialov tribosistem* [Tribophysics and Structural Modification of Tribosystem Materials]. Omsk: OmGTU Publishing House, p. 322
2. Pesetsky S. S., Bogdanovich S. P., Myshkin N. K. (2007). Tribotechnical properties of nanocomposites obtained by dispersion of fillers in polymer melts. *J. Friction Wear.* **28**(5):500–524.
3. Myshkin N. K., Petrokovets M. I., Kovalev A. V. (2005). Tribology of polymers: adhesion, friction, wear and mass-transfer. *Tribol. Int.* **38**:910–921.
4. Buznik V. M., Fomin V. M., Alkhimov A. P. (2005). *Metallopolimernye nanokompozity (poluchenie, svoistva, primenenie)* [Metal-polymeric Nanocomposites: Synthesis, Properties, Application]. Novosibirsk: Sibir. Otd. Ross. Akad. Nauk, p. 258.
5. Galetz M. C., Blar T., Ruckdaschel H., Sandler K. W., Alstadt V. (2007). Carbon nanofibre-reinforced ultrahigh molecular weight polyethylene for tribological applications. *J. Appl. Polym. Sci.* **104**:4173–4181.
6. Zhou J., Yan F. (2005). Improvement of the tribological behavior of ultra-high-molecular-weight polyethylene by incorporation of poly(phenyl p-hydroxyzoate). *J. Appl. Polym. Sci.* **96**:2336–2343.

7. Ruan S. L., Gao P., Yang X. G., Yu T. X. (2003). Toughening high performance ultrahigh molecular weight polyethylene using multiwalled carbon nanotubes. *Polymer.* **44**(19):5643–5654.

8. Vaobaug Z., Vongeheng F., Lu W., Viaobo L. (2004). Processing and properties of MWNT/HDPE composites. *Carbon.* **42**(2):271–277.

9. Sinha S. R., Briscoe B. J. (2009). *Polymer Tribology.* London: Imperial College Press.

10. Shi W. (2004). Preliminary investigation into the load bearing capacity of ion beam surface modified UHMWPE. *J. Mater. Sci. Lett.* (39):3183–3186.

11. Cooper J., Dowson D., Fisher J. (1993). Macroscopic and microscopic wear mechanisms in ultra-high molecular weight polyethylene. *Wear.* **162**:378–384.

12. Pruth E. V., Zelenetsky A. N. (2001). Chemical modification and blending polymers in the extruder reactor. *Uspekhi Chimii.* **70**(1):72–87.

13. Krasnov A. P., Said-Galiev A. E., Afonicheva O. V., et al. (2007). Friction behavior of blends of incompatible UHMWPE polymer and polymethylmethacrylate obtained in the media of supercritical carbon dioxide. *J. Frict. Wear.* **28**(3):288–295.

14. Panin S. V., Kornienko L. A., Wannasri S., et al. (2010). Comparative analysis of nano and microfillers influence of oxidized Al on friction-mechanical properties of UHMWPE. *J. Frict. Wear.* **31**(5):353–360.

15. Sirota A. G. (1974). *Modification of Polymer's Structure and Properties.* Leningrad: Khimiya.

16. Panin V. E., Panin S. V., Kornienko L. A., et al. (2010). The effect of UHMWPE mechanical activation on its mechanical and tribotechnical properties. *J. Frict Wear.* **31**(2):168–176.

17. Panin S. V., Kornienko L. A., Wannasri S., et al. (2010). Comparison of the effectiveness of UHMWPE modification by nanofibers (C, Al_2O_3) and nanoparticles (Cu, SiO_2) at obtaining antifriction composites. *J. Frict. Wear.* **31**(6): 685–696.

18. Smith A. (1982). *Applied Infrared Spectroscopy.* Moscow: Mir Publishing House.

19. Panin S. V., Kornienko L. A., Poowadin T., Sergeev V. P., Ivanova L. R., Shilko S. V. (2010). Friction and wear of ion-implanted ultrahigh-molecular weight polyethylene for implants. *Frict. Lubr. Mach. Mech.* (10):3–11.

Chapter 11

Modeling of Mechanical Properties in Nanoparticle Reinforced Polymers Using Atomistic Simulations

Samit Roy and Avinash Reddy Akepati
Department of Aerospace Engineering and Mechanics, The University of Alabama, 209 Hardaway Hall, Box 870280, Tuscaloosa, AL 35401, USA
sroy@eng.ua.edu, arakepati@crimson.ua.edu

This chapter is aimed towards understanding damage progression and failure in advanced nanostructured composite materials. The molecular dynamics (MD) program LAMMPS (large-scale atomic/molecular massively parallel simulator) is a versatile code with numerous functionalities which can be used to simulate polymer systems. Although the results for molecular simulation models, in particular polymer models, are system size dependent, the ease with which large molecular systems can be evaluated, especially with parallel molecular dynamics codes such as LAMMPS, makes it a more attractive choice compared to other approaches such as ab initio calculations, which are limited to the calculation of a few 100 atoms.

In the context of fracture behavior in polymers, the critical value of the J-integral (J_I) at crack initiation is related to the fracture toughness of the material, where the subscript I denotes the fracture

Characterization of Nanocomposites: Technology and Industrial Applications
Edited by Frank Abdi and Mohit Garg

ISBN 978-981-4669-02-3 (Hardcover), 978-1-315-36489-6 (eBook)
www.panstanford.com

mode ($I = 1, 2, 3$). Therefore, the J-integral could be used as a suitable metric for estimating the crack driving force as well as fracture toughness of the material as the crack begins to initiate. However, for the conventional macroscale definition of the J-integral to be valid at the nanoscale, in terms of the continuum stress and displacement fields and their spatial derivatives, requires the construction of local continuum fields from discrete atomistic data, and using these data in the conventional contour integral expression for atomistic J-integral. One such methodology is proposed by Hardy that allows for the local averaging necessary to obtain the definition of free energy, deformation gradient, and Piola–Kirchoff stress as fields (and divergence of fields) and not just as total system averages. Further, the atomistic J-integral takes into account the effect of reduction in J from continuum estimates due to the fact that the free energy available for crack propagation is less than the internal energy at sufficiently high temperatures when entropic contributions become significant.

As a case study, the feasibility of computing the dynamic atomistic J-integral over the MD domain at finite temperatures using Hardy estimates of continuum fields based on localization functions is evaluated for a graphene nanoplatelet having a central crack. For model verification, the values of atomistic J-integral are compared with results from linear elastic fracture mechanics (LEFM) for isothermal crack initiation at 0 K. The atomistic computational results are in very good agreement with LEFM data up to the point of onset of a nonlinear region near the crack tip at elevated strain levels. Computational results related to the path-independence of the atomistic J-Integral are also presented. Further, a novel approach that circumvents the complexities of direct computation of entropic contributions is also discussed. Good agreement is observed between the atomistic J and the LEFM results at 0 K, with predictable discrepancies at 300 K due to entropic effects. Work is currently to study the effect of nanographene on the fracture properties of polymers. Elastic moduli data from MD simulations of crosslinked neat EPON 862 resin are also presented and compared with results available in the literature.

11.1 Introduction

Nanoparticle reinforcements are nanometer ($\sim 10^{-9}$ m) thick particles that interact intimately with the polymer matrix molecules to provide strengthening of the polymer matrix at the nanoscale. The size of the polymer molecules are roughly on the order of the nanoparticle thickness, depending on the degree of polymerization. Moving up the scale, we find that the carbon fiber diameter is usually on the order of 5 to 10 μm, although its length may extend up to several meters. The resulting composite (matrix reinforced with fiber) is in the macro- or engineering scale of analysis. The combination of fiber reinforcements in the polymer matrix gives the macroscale composite orthotropic material characteristics. The nanoscale interaction between polymer molecules and nanoparticle is a key factor in determining the macroscale strength of the composite.

In recent years, numerous efforts have been directed towards modeling nanocomposites in order to better understand the reasons behind the enhancement of mechanical properties, even by the slight addition (a few weight percent) of nanomaterials (Roy et al., 2007; Liu et al., 2006). In order to better understand the local influence of nanoparticle on the mechanical properties of the composite, it is required to perform nanoscale analysis. In this chapter, a methodology to study fracture at the atomistic level is discussed. The fracture toughness of a material is related to the critical value of J-integral and therefore could be used as a metric for estimating the crack driving force as well as the fracture toughness of the material. Evaluation of J-integral at the atomistic level using information obtained from molecular dynamics (MD) simulations is discussed in this chapter along with its application for modeling fracture properties in nanographene reinforced EPON 862 polymer systems. Phenomena such as nanoscale debond and delamination can be more accurately addressed using this approach. Also, the effect of dispersion and orientation of nanoreinforcements on fracture properties can also be better understood by such an atomistic scale analysis.

11.2 Atomistic J-Integral Evaluation Methodology

Our goal is to be able to use MD simulations to compute a performance metric, such as the atomistic J-integral of a nanomaterial, in order to quantify the influence of nanofillers (such as graphene platelets) on phenomenon such as delamination crack propagation in a laminated composite. In this section our objective is to validate this methodology for a nanographene sheet with a defect before applying it to a more complex nanographene reinforced polymeric system. Earlier efforts to calculate J-integral for graphene sheets using atomistic data have been reported in (Khare et al., 2007), but the J-integral computation is performed by summing over a discrete number of atoms, and the entropic contribution at higher temperatures is not considered in that work. In this chapter, the feasibility of computing the J-integral within a purely MD domain at finite temperatures using Hardy estimates of *continuum fields* for a polymer system is evaluated. The J-integral evaluation scheme discussed below can be subsequently applied to polymeric systems to evaluate the fracture metrics, such as work of separation, which can then be employed in a hierarchical multiscale model. In conventional macroscale fracture mechanics, the J-integral vector, defined as the divergence of the Eshelby stress tensor (refer to Eq. 11.1), has been used to quantify the crack driving force available from thermomechanical loading as well as material inhomogeneities.

$$J = \int_{\partial\Omega} \underline{S} N dA = \int_{\partial\Omega} (\Psi N - \underline{F}^T \underline{P} N) dA \qquad (11.1)$$

In Eq. 11.1, S is the Eshelby stress tensor, Ψ is the free energy density, $\underline{F}$ is the deformation gradient tensor, $\underline{P}$ is the first Piola–Kirchhoff stress tensor and **N** is the outward normal to the surface $\partial\Omega$ along which the J-integral is evaluated. At a temperature of 0 K and under an equilibrium state, Eq. 11.1 reduces to Eq. 11.2, where W is the stored energy density and $\underline{H}$ is the displacement gradient.

$$J = \int_{\partial\Omega} (WN - \underline{H}^T \underline{P} N) dA \qquad (11.2)$$

The critical value of J_k at crack initiation is related to the fracture toughness of the material, where the subscript k denotes components of the J-vector ($k = 1, 2, 3$) and these components are related to the fracture modes (Rigby and Aliabadi, 1993). Therefore, the J-integral could be used as a suitable metric for estimating the crack driving force as well as the fracture toughness of the material as the crack begins to initiate. For the conventional macroscale definition of J-integral to be valid at the nanoscale in terms of the continuum stress and displacement fields (and their spatial derivatives) requires the construction of local continuum fields from discrete atomistic data, and using these data, the conventional contour integral expression for J, is given by Eq. 11.1 (Jones et al., 2011, Jones and Zimmerman, 2010). One such methodology is proposed by Hardy (Hardy, 1982) that allows for the local averaging necessary to obtain the definition of free energy, deformation gradient, and Piola–Kirchoff stress as fields (and divergence of fields) and not just as total system averages. Here it is assumed that ensemble average $\langle * \rangle$ is approximated by the time average of the quantity over a sufficiently long period of time. A detailed description for the evaluation of each term required to calculate the atomistic J-integral is given below.

The stored energy density field (W) can be expressed as a sum of atomistic potentials of each atom using Eq. 11.3.

$$\begin{aligned} W(\boldsymbol{X}, t) &= \sum_{\alpha=1}^{M} (\phi^{\alpha}(t) - \phi_{\mathrm{X}}^{\alpha}) \psi(\boldsymbol{X} - \boldsymbol{X}^{\alpha}) \\ &= \sum_{\alpha=1}^{M} \phi^{\alpha}(t)\, \psi(\boldsymbol{X} - \boldsymbol{X}^{\alpha}) - W(\boldsymbol{X}) \end{aligned} \quad (11.3)$$

In Eq. 11.3, ϕ^{α} is atomistic potential of atom α, $\phi_{\boldsymbol{X}}^{\alpha}$ is the reference potential energy density of at 0 K, ψ is the Hardy localization function, M is the number of atoms in the localization box, and $W(\boldsymbol{X})$ is a constant. Dividing the MD domain into localization boxes, the atomistic potential at the centroid of the box can be defined using Eq. 11.4 and the field distribution of the potential over the domain is given by Eq. 11.5, where $N_{\mathrm{I}}(\boldsymbol{X})$ are the interpolation functions.

$$\phi_I(\boldsymbol{X}_I, t) = \sum_{\alpha=1}^{M} \psi_{I\alpha}(\boldsymbol{X}^{\alpha} - \boldsymbol{X}_{\mathrm{I}})\, \phi^{\alpha}(t) \quad (11.4)$$

$$\phi(\boldsymbol{X}, t) = \sum_{I=1}^{N} N_I(\boldsymbol{X}) \phi_{\mathrm{I}}(\boldsymbol{X}_I, t) \quad (11.5)$$

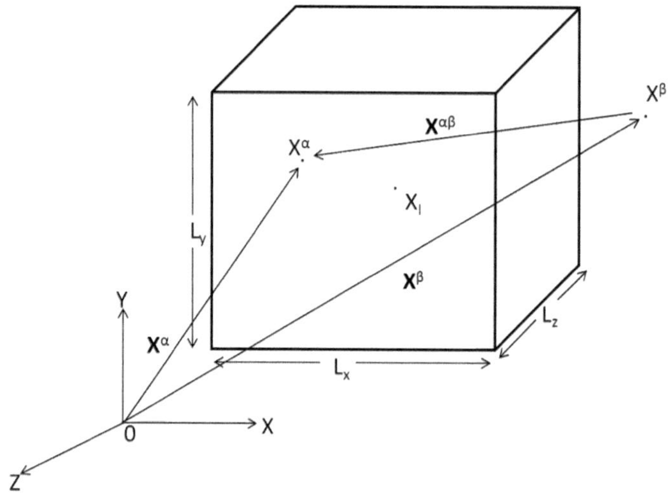

Figure 11.1 Illustration of a localization box with one atom inside and another outside the box.

Substituting Eq. 11.5 in Eq. 11.3 gives the final expression for calculating the stored energy density field:

$$W(\boldsymbol{X}, t) = \sum_{I=1}^{N} \sum_{\alpha=1}^{M} N_I(\boldsymbol{X}) \psi_{I\alpha} (\boldsymbol{X}^\alpha - \boldsymbol{X}_I) \phi^\alpha (t) - W(\boldsymbol{X}) \quad (11.6)$$

As originally proposed by Hardy (1982), the localization function depends on the location of the atoms in the localization box. The function has to satisfy the properties of $\psi > 0$ and $\int_\Omega \psi dV = 1$ within the localization box. Referring to Fig. 11.1, localization function used in this chapter is given by Eq. 11.7.

$$\psi (\boldsymbol{X} - \boldsymbol{X}^\alpha) = \begin{cases} \dfrac{1}{L_x L_y L_z} & \text{if} \quad \begin{matrix} X_I - \frac{L_x}{2} \leqslant X^\alpha \leqslant X_I + \frac{L_x}{2} \\ Y_I - \frac{L_y}{2} \leqslant Y^\alpha \leqslant Y_I + \frac{L_y}{2} \\ Z_I - \frac{L_z}{2} \leqslant Z^\alpha \leqslant Z_I + \frac{L_z}{2} \end{matrix} \\ 0 & \end{cases} \quad (11.7)$$

where, L_x, L_y, and L_z are the dimensions of the localization box and X_I, Y_I, and Z_I are the coordinates of the centroid, as illustrated in Fig. 11.1. In a similar way, the displacement gradient can be evaluated using Eq. 11.8, where $\boldsymbol{u}(\boldsymbol{X}, t)$ is the displacement vector.

$$\underline{H}(\boldsymbol{X}, t) = \nabla_X \boldsymbol{u}(\boldsymbol{X}, t)$$
$$= \sum_{I=1}^{N} \sum_{\alpha=1}^{M} \nabla_X N_I(\boldsymbol{X}) \psi_{I\alpha} (\boldsymbol{X}^{\alpha} - \boldsymbol{X}_I) \, \boldsymbol{u}^{\alpha}(t) \tag{11.8}$$

The so-called bond function $B_{\alpha\beta} = \int_0^1 \psi(\lambda(\boldsymbol{X}_{\alpha} - \boldsymbol{X}) + (1-\lambda)(\boldsymbol{X}_{\beta} - \boldsymbol{X}))d\lambda$ as defined by Hardy (1982) can then be used to compute the first Piola–Kirchoff stress along the contour, as given by Eq. 11.9:

$$\underline{P}(\boldsymbol{X}, t) = -\frac{1}{2} \sum_{\alpha=1}^{M} \sum_{\beta \neq \alpha}^{M} f^{\alpha\beta} \otimes X^{\alpha\beta} B^{\alpha\beta}(\boldsymbol{X}) \tag{11.9}$$

where $\boldsymbol{f}^{\alpha\beta}$ is the vector representing the force between atoms α and β, and $\boldsymbol{X}^{\alpha\beta}$ is the vector representing the difference in their positions. In Eq. 11.1, the free energy density (Ψ) is given by

$$\Psi = \mathrm{U} - \mathrm{T}\eta = -\frac{k_{\mathrm{B}} T}{V} Log Z \tag{11.10}$$

where U is the internal energy density, T is the temperature, η is the entropy density, V is the volume of the ensemble, k_{B} is Boltzmann's constant and Z is the partition function of the atoms occupying the region Ω. Note that our general definition of free energy density Ψ includes the entropy term and therefore is valid for finite temperature applications of the atomistic J-integral. The conventional definition of J-integral does not take into account the entropic contribution to the free energy and, consequently, may lead to significant over-estimation of the J-integral at the atomistic level especially at finite temperature. Jones et al. (2011) used a local harmonic (LH) approximation based on the Cauchy-Born model to compute the partition function (Z) for an idealized material with defect-free crystalline structure at finite (non-zero) temperature. However, the LH approximation is not valid for an amorphous (disordered) crosslinked polymer network (Jones et al., 2011), and an alternate approach (such as virtual internal bond) may be necessary.

11.3 Numerical Evaluation of Atomistic J-Integral

Numerical integration through Gaussian quadrature was employed to evaluate atomistic J-integral using the equations discussed in the

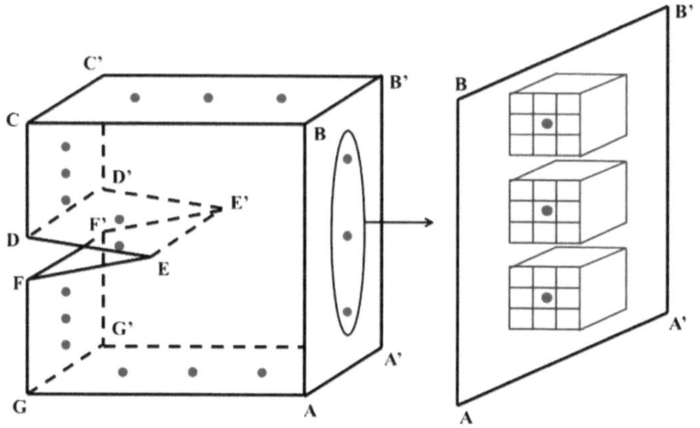

Figure 11.2 Gaussian quadrature points along the integration contour and localization boxes on one side showing subcells.

previous section. Figure 11.2 shows that the Gaussian quadrature points along the integration contour. Gaussian integration points were considered on each side of the contour, and at each integration point, a localization box was created which was further divided into subcells, as illustrated in Figs. 11.2 and 11.3.

As discussed earlier, the discrete atomistic values of potentials and displacement gradients are converted into field quantities using the finite element interpolation functions for a nine-node element (see Fig. 11.3).

From Eq. 11.2, the J-integral for the right-hand side of the contour (path AB) is given by

$$J_{AB} = h\int_a^b (W\hat{N} - \underline{H}^T \underline{P}\hat{N})dY$$

Referring to Fig. 11.3, assuming, $Y = C_1\xi + C_2$ and applying boundary conditions,

$$a = C_1(-1) + C_2; \quad b = C_1(1) + C_2$$

$$C_1 = \frac{b-a}{2}; \qquad C_2 = \frac{b+a}{2}$$

$$dY = C_1 d\xi = \left(\frac{b-a}{2}\right) d\xi$$

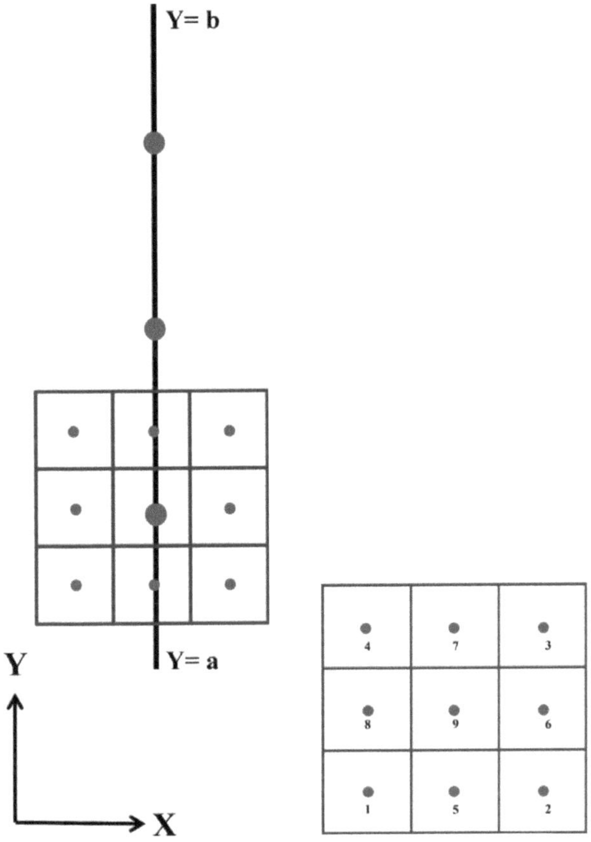

Figure 11.3 Gaussian quadrature points with 3 × 3 computational subcells.

$$J_{AB} = h\int_{-1}^{1}(W\hat{N} - \underline{H}^T\underline{P}\hat{N})C_1 d\xi$$

$$J_{AB} = C_1 h\sum_{k=1}^{N_G}\left[W(\xi_k)\,\hat{N} - \underline{H}^T\,(\xi_k)\,\underline{P}\,(\xi_k)\,\hat{N}\right]w_k$$

$$J_{AB} = C_1 h\sum_{k=1}^{N_G}\left[W(Y_k)\,\hat{N} - \underline{H}^T\,(Y_k)\,\underline{P}\,(Y_k)\,\hat{N}\right]w_k \quad (11.11)$$

where, $Y_k = C_1\xi_k + C_2$ and N_G is the number of Gaussian quadrature points along path AB. Similarly, the J-integral can be calculated using this procedure at all Gaussian quadrature points along each segment of the contour (AB-BC-CD-DE-EF-FG-GA) and then summed.

11.3.1 *Preliminary Results of Atomistic J-Integral Calculation Using 4420 Atoms Graphene Sheet*

A single graphene platelet was modeled using MD (LAMMPS) and was subjected to a far-field uniaxial stress, and the atomistic J-integral was computed on the contour as shown in Fig. 11.4. In this example, the graphene platelet is modeled using harmonic style for bonds and angles, and OPLS style for dihedrals. The graphene platelet, measuring 10.5 nm $\times$ 10 nm, consisted of 4420 carbon atoms. The graphene platelet was subjected to a far-field uniaxial stress normal to the crack, and the atomistic J-integral was computed on the contour as shown in Fig. 11.4. In order to establish proof-of-concept for atomistic J-integral computation, the MD simulations were carried out at temperature $T = 0$ K, that is, without any entropic contribution to the free energy. Pressure barostatting was not used in these simulations. The localization box size used for these computations varied between 0.4 nm and 0.8 nm. The initial length of the pre-existing (starter) crack was assumed to be 2.2 nm in the graphene platelet. The coarse time-step for the problem (Δt) was set equal to 4 femtoseconds. Figure 11.4 shows the MD simulation of the initiation of a nanoscale dynamic mode I crack in the graphene platelet. The atomistic J-integral results are shown in Fig. 11.5 for a graphene platelet and compared with linear elastic fracture mechanics (LEFM) predictions. The figure shows that the quadratic dependence of the atomistic J-integral on mode I stress intensity factor (K_{I}) is in good agreement with LEFM predictions. The figure also verifies that J is identically zero in the absence of any singularity within the integration contour. The calculated value starts deviating from LEFM at around $K_{\mathrm{I}} = 0.4$ MP·m$^{1/2}$ because of nonlinear effects due to bond stretching near the crack tip. The critical value of J_{I} (J_{IC}) for graphene at 0 K was calculated by (Jin and Yuan, 2005) using molecular mechanics to be between 3 and 5 J/m^2 depending on the length of the crack. The value of J_{IC} reported in this chapter is approximately 1 J/m^2, as shown in Fig. 11.5. This discrepancy is probably due to the difference

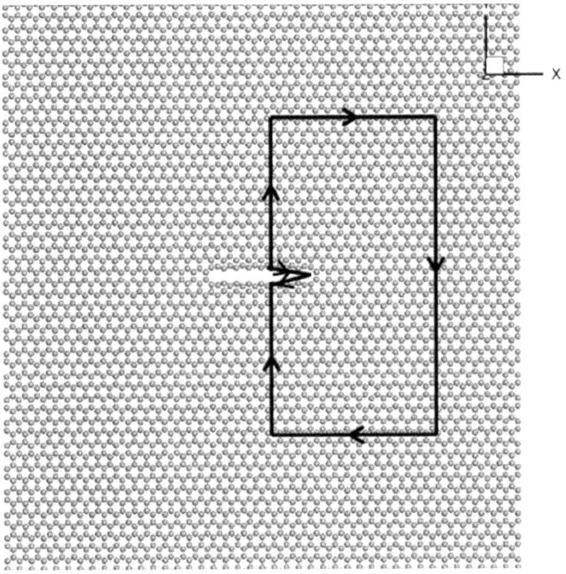

Figure 11.4 Graphene sheet with nanoscale crack and contour for J-Integral computation.

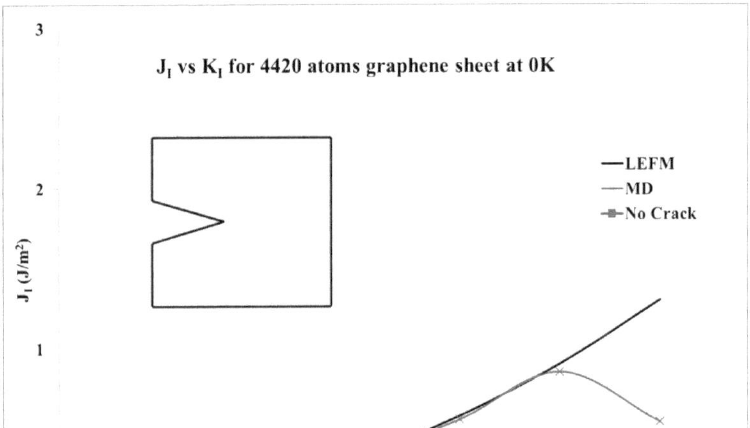

Figure 11.5 Comparison of calculated atomistic J_I with LEFM values at 0 K.

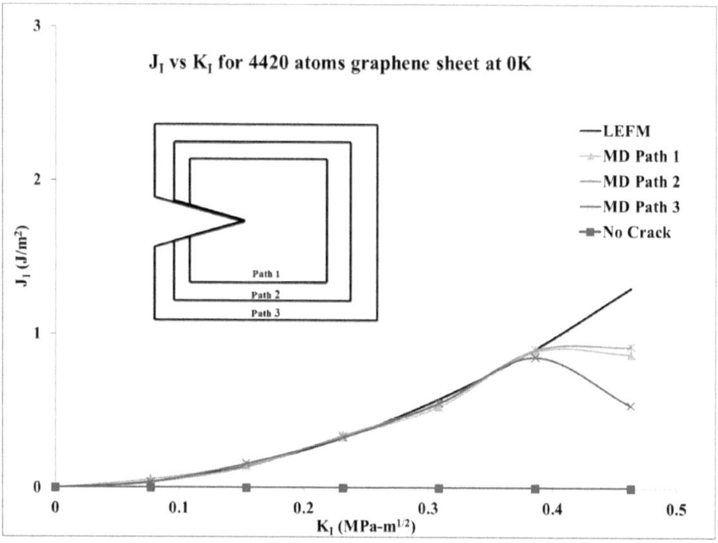

Figure 11.6 Path independence of atomistic J_{I} at 0 K.

in the size of the nanographene system used, and the force fields used by (Jin and Yuan, 2005) are different from the current work. Path independence of J_{I} was also verified using three different contours and the results are shown in Fig. 11.6. It is clear that the calculated value of J_{I} is path independent. This result facilitates the use of calculating J_{I} at finite temperatures using a cohesive zone-based technique, which will be discussed later.

The J-integral evaluation methodology that was verified for a graphene platelet at 0 K was used for the finite temperature case at 300 K, as depicted in Fig. 11.4. Though the calculated value of atomistic J-integral is in good agreement with LEFM predictions at 0 K, there is a disagreement between those values at significantly higher temperatures due to entropic contributions, as delineated by the term $T\eta$ term in Eq. 11.10. Hence, simple LEFM predictions cannot be used at higher temperatures and MD results would be more accurate. Work is in progress to evaluate J-integral for a graphene platelet at 300 K. Figure 11.7 shows the preliminary results for atomistic J_{I} for graphene sheet at 300 K. It can be observed that the calculated values of atomistic J_{I} are significantly

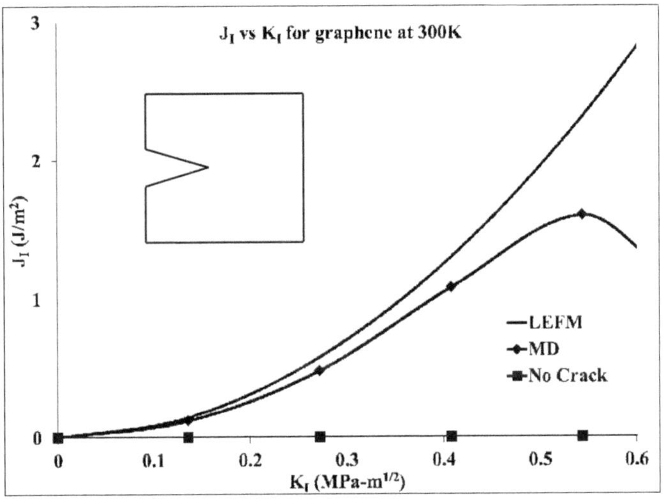

Figure 11.7 Comparison of calculated atomistic J_{I} with LEFM values at 300 K.

lower than the predicted LEFM values, as one would expect when entropic contributions are taken into account as given by Eq. 11.10.

11.4 Molecular Dynamics Simulations of EPON 862-DETDA Epoxy System

Several simulations were run using the LAMMPS software package on the EPON 862 epoxy model that was originally created by Dr. Greg Odegard's group at Michigan Technological University (MTU). Figure 11.8 shows the coarse-grained MD model. In the figure, the legend on the right is for the different types of atoms. Each atom or united atom model is given a different color that corresponds to the numbers one to thirteen. These numbers correspond to the specific atoms or united atoms described in Table 11.1 In the table, the EPON atoms are those that originated in the uncrosslinked epoxy system, and the DETDA atoms are those that originated in the DETDA hardener crosslinking agent before the system was crosslinked.

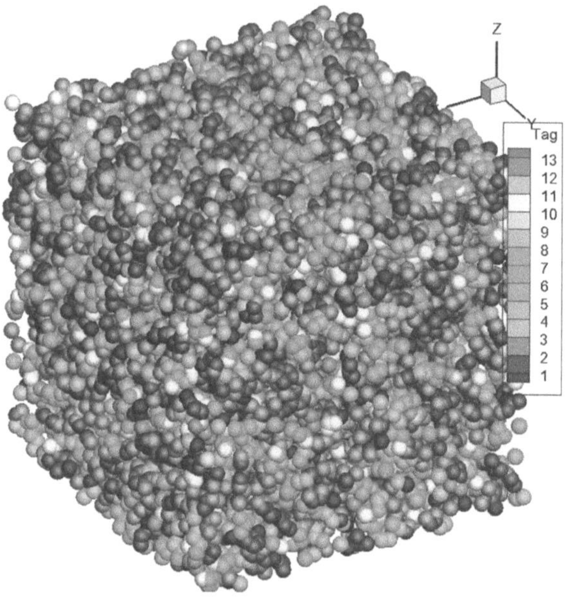

Figure 11.8 Three-dimensional view of the EPON 862-DETDA epoxy system.

11.4.1 *Materials Characterization*

11.4.1.1 Comparison of mechanical properties with MTU's simulation results

Simulations were run to calculate material characteristics of the system, i.e., bulk modulus (K), shear modulus (G), and Young's modulus (E). These simulations were run in conformance with MTU simulations in order to obtain the same material data for EPON 862, as described by Bandyopadhyay et al. (2011). A crosslink density of 76% was assumed for all cases. The simulations that were completed were tensile and compressive tests that strain the system $\pm 0.3\%$ in the x, y, and z directions, respectively, and positive and negative shear tests that strain the system $\pm 0.3\%$ along the diagonal of the epoxy cube. After the simulations were completed, linear regression analysis was performed on the LAMMPS output to obtain material data, such as tensile and shear modulus, bulk modulus, and Poisson's ratio (ν). These material properties are displayed in

Table 11.1 Legend for atoms in the EPON 862-DETDA epoxy system

EPON
1. CH_2 (central group between 2 benzenes in EPON 862)
2. C of benzene
3. H of benzene
4. C of benzene connected to O
5. CH_2 connected to O (not the CH_2 connected to O in epoxy ring)
6. CH_2 of epoxy ring
7. CH of epoxy ring
8. O of epoxy ring
DETDA
9. CH_3 in DETDA
10. N in NH_2
11. H in NH_2
12. Carbon in CH_3 directly connected to benzene in DETDA
13. H in CH_3 directly connected to benzene in DETDA

Table 11.2 Material characteristic results compared to the MTU material characteristic results

	K (GPa)	*E* (GPa)	*G* (GPa)	*ν*
Computed at UA	2.73	2.214	0.811	0.365
Computed MTU	2.66	2.258	0.831	0.359
Percentage error	+2.61	−1.93	−2.38	+1.62

Table 11.2 along with the percent error compared with the MTU results. As it can be observed from Table 11.2 the results obtained using MD simulations are in good agreement with those obtained at MTU.

11.4.1.2 Comparison with MTU's experimental results

Experimental results for the Young's modulus of the EPON 862 DETDA system were obtained from MTU. The tested specimens had post cure times of 0, 0.2, 2, 20, and 200 hours. A variety of simulations were run to compare with the experimental results. Three simulations were run at 300 K (NVT) (10%, 5%, and 1.5% strain in the *x* direction), and four simulations at 0 K (2% and 5%

Table 11.3 Experimental Young's moduli compared to various MD simulation Young's moduli

Experimental results	Young's modulus (GPa)
Room temperature experimental values for MTU post cure time 0 h	1.7 ± 1.3141
Room temperature experimental values for MTU post cure time 0.2 h	1.63 ± 0.6734
Room temperature experimental values for MTU post cure time 2 h	1.62 ± 0.8765
Room temperature experimental values for MTU post cure time 20 h	1.34 ± 1.2351
Room temperature experimental values for MTU post cure time 200 h	1.94 ± 0.6731
Simulation results	
10% Strain11 300 K cubic fitted curve	2.509
1.5% Strain11 300 K linear fitted curve	4.924
5% Strain11 300 K linear fitted curve	2.622
10% Strain11 300 K linear fitted curve	1.488
2% Strain11 0 K (0 to 0.5% strain)	49.094
5% Strain11 0 K (0 to 0.5% strain)	46.652
5% Strain22 0 K (0 to 0.5% strain)	57.745
5% Strain33 0 K (0 to 0.5% strain)	48.827

strain in the x direction, 5% strain in the y direction, and 5% strain in the z direction). For the 10% strain simulations at 300 K a cubic curve and a linear curve were fitted to the results to get Young's modulus, and for the 5% and 1.5% strain simulations at 300 K only a linear curve was plotted to attain Young's modulus. For the simulations at 0 K only a linear curve was fitted from 0 to 0.5% strain to attain Young's modulus because this was the linear part of the curve before the system began to yield. Table 11.3 shows the different experimental results and the different simulation results. As seen in Table 11.3, the simulations at 300 K are much closer to the experimental results. As might be expected, the simulations at 0 K indicate that the system is significantly stiffer at 0 K compared to 300 K.

11.5 Future Work

Work is currently underway to quantify and verify the difference due to the entropic contribution at even higher temperatures. All the simulations till now have been done using optimized potentials for liquid simulations (OPLS) force field. Work is currently underway to use ReaxFF force field (van Duin, 2001) for MD simulations. ReaxFF is a bond order-based force field which can model chemical reactions including formation and breaking of chemical bonds. This force field is useful for fracture simulation because it can model crack propagation by breaking chemical bonds, which is a more accurate representation of the physical phenomenon occurring at atomic scale. This would give a better understanding of the effect of nanographene on fracture properties of EPON 862. As mentioned earlier, our ultimate goal is to be able to use MD simulations to compute parameters such as the partition function, free energy density, bond density function (for bonded as well as nonbonded interactions), Piola–Kirchoff stress, etc., in order to quantify the influence of nanofillers such as graphene nanoplatelets on macroscale phenomenon such as delamination crack propagation in a nanoreinforced laminated polymer composite. As mentioned in an earlier section, the local harmonic (LH) approximation is not valid for an amorphous (disordered) crosslinked polymer network. A methodology for the derivation of a proper partition function (Z) for a flexible polymer crosslinked network through the use of statistical mechanics of phantom networks is described in (Weiner, 1983) and will be adapted for evaluating J-integral for polymers at finite temperatures. In the event, the partition function computation proves to be computationally intractable for a large amorphous polymeric system, an alternate approach using a cohesive zone-based technique will be employed for computation of J-integral at the nanoscale, as outlined by Klein and Gao (1998). Atomistic J-integral evaluation scheme discussed earlier can then be applied to polymeric systems to evaluate performance metrics such as work of separation at the polymer–nanoparticle interface. These data can subsequently be incorporated in a multiscale model for simulating failure at the macroscale. The multiscale analysis will be performed using a concurrent coupling scheme as discussed in by Roy and Nair (2011). A major advantage of the coupled simulation over pure MD simulation of crack propagation is that the spurious effect of stress waves reflected from the boundaries of the MD domain impinging on the moving crack tip is mitigated. The nanoscale fracture toughness data obtained will subsequently be hierarchically incorporated at the microscale to predict fuzzy fiber/matrix debonding or at the macroscale to predict inter-ply delamination in a composite laminate reinforced with interlaced layers of CNT forest.

11.6 Discussion

It is envisioned that a better understanding of the strength and toughness enhancement mechanisms at the nanoscale will lead to optimization of materials processing variables at the macroscale, which in turn will lead to the manufacture of nanocomposites more efficiently and at lower cost. The proposed multiscale analysis will help provide insight into the nanoscale interactions that are responsible for toughening/strengthening mechanisms in nanocomposite systems at the macroscale. The work presented in this chapter will help in evaluating properties such as work of separation at the nanoscale and incorporate them at higher length scales using multiscale modeling methods.

Acknowledgments

The authors would like to acknowledge sponsorship of this research by NASA Fundamentals Aeronautics Research Program, Aerospace Sciences Project, Contract No. NNX11AI32A, with Dr. Brett Bednarcyk as technical monitor. The authors would also like to thank Dr. Gregory Odegard and his research group at Michigan Technological University in Houghton for their help with the EPON 862-DETDA polymer modeling.

References

Roy, S., Hussain, F., Narasimhan, K., Vengadassalam, K., and Lu, H. (2007). E-glass/polypropylene nanocomposites: manufacture, characterization, thermal and mechanical properties, *Polymer Polymer Compos.*, **15**(2): 91–102.

Saether, E., Yamakov, V., and Glaessgen, E. (2007). A statistical approach for the concurrent coupling of molecular dynamics and finite element methods. AIAA paper, AIAA-2007-2169, Honolulu, HI.

Valavala, P. K., Clancy, T. C., Odegard, G. M., and Gates, T. S. (2007). Nonlinear multiscale modeling of polymer materials, *Int. J. Solids Struct.*, **44**: 1161–1179.

Buryachenko, V. A., Roy, A., Lafdi, K., Anderson K. L., and Chellapilla, S. (2005). Multi-scale mechanics of nanocomposites including interface: experimental and numerical investigation, *Compos. Sci. Technol.*, **65**: 2435–2465.

Riddick, J. C., Frankland S. J. V., and Gates, T. S. (2006). Multiscale analysis of delamination of carbon fiber-epoxy laminates with carbon nanotubes. AIAA paper, AIAA-2006-1676, Newport, RI.

Awasthi, A. P., Lagoudas, D. C., and Hammerand, D. C. (2006). Modeling of interface behavior in carbon nanotube composites. AIAA paper, AIAA-2006-1677, Newport, RI.

Abraham, F. F., Broughton, J. Q., Bernstein, N., and Kaxiras, E. (1999). Concurrent coupling of length scales: methodology and application, *Phys. Rev. B*, **60**(4): 2391–2403.

Ogata, S., Lidorikis, E., Shimojo, F., Nakano, A., Vashishta, P., and Kalia, R. K. (2001). Hybrid finite-element/molecular-dynamics/electronic-density-functional approach to materials simulations on parallel computers, *Comput. Phys. Commun.*, **138**: 143–154.

Wagner, G. J., and Liu, W. K. (2003). Coupling of atomistic and continuum simulations using a bridging scale decomposition, *J. Comput. Phys.*, **190**: 249–274.

Ma, J., Lu, H., Wang, B., Hornung, R., Wissink A., and Komanduri, R. (2006). Multiscale simulation using generalized interpolation material point (GIMP) method and molecular dynamics (MD), *J. Comput. Phys.*, **14**(2): 101–117.

Khare, R., Mielke, S. L., Paci, J. T., Zhang, S., Schatz, G. C., and Belytschko, T. (2007). Two quantum mechanical/molecular mechanical coupling schemes appropriate for fracture mechanics studies. AIAA paper, Honolulu, HI.

Liu, W. K., Karpov, E. G., and Park, H. S. (2006). *Nano Mechanics and Materials: Theory, Multiscale Methods and Applications*. John Wiley & Sons Ltd., U.K.

Khare, R., Mielke, S. L., Paci, J. T., Zhang, S., Ballarini, R., Schatz, G. C., and Belytschko, T. (2007). Coupled quantum mechanical/molecular mechanical modeling of the fracture of defective carbon nanotubes and graphene sheets, *Phys. Rev. B*, **75**(075412): 1–12.

Rigby, R. H., and Aliabadi, M. H. (1993). Mixed-mode J-integral method for analysis of 3D fracture problems using BEM, *Eng. Anal. Boundary Elem.*, **11**: 239–256.

Jones, R. E., Zimmerman, J. A., Oswald, J., and Belytschko, T. (2011). An atomistic J-integral at finite temperature based on Hardy estimates of continuum fields, *J. Phys.: Condens. Matter*, **23**: 1–12.

Jones, R. E., and Zimmerman, J. A. (2010). The construction and application of an atomistic J-integral via Hardy estimates of continuum fields, *J. Mech. Phys. Solids*, **58**: 11318–1337.

Hardy, R. (1982). Formulas for determining local properties in molecular dynamics simulations: shock waves, *J. Chem. Phys.*, **76**(1): 622–628.

Jin, Y., and Yuan, F. G. (2005). Atomistic simulations of J-integral in 2D graphene nanosystems, *J. Nanosci. Nanotechnol.*, **5**(12): 2099–2107.

Bandyopadhyay, A., Valavala, P. K., Clancy, T. C., Wise K. E., and Odegard, G. M. (2011). Molecular modeling of crosslinked epoxy polymers: the effect of crosslink density on thermomechanical properties, *Polymer*, **52**(3): 2445–2452.

Van Duin, A. C. T., Dasgupta, S., Lorant, F., and Goddard, W. A. (2001). ReaxFF: a reactive force field for hydrocarbons, *J. Phys. Chem. A*, **105**(41): 9396–9409.

Weiner, J. H. (1983). *Statistical Mechanics in Elasticity*. John Wiley & Sons.

Klein, P., and Gao, H. (1998). Crack nucleation and growth as strain localization in a virtual-bond continuum, *Eng. Fract. Mech.*, **61**: 21–48.

Roy, S., and Nair, A. (2011). Concurrent multi-scale modeling of nanoparticle reinforced polymers using statistical coupling of MD and GIMPM. Proceedings of the 52nd AIAA-SDM Conference, Denver, CO.

Chapter 12

Prediction of Effect of Waviness, Interfacial Bonding, and Agglomeration of Carbon Nanotubes on Their Polymer Composites

Mohit Garg, Frank Abdi, and Jerrold Housner
AlphaStar Corporation, 5150 E. Pacific Coast Highway, Suite 650, Long Beach, CA 90804, USA
jhousner@alphastarcorp.com

Carbon nanotubes (CNTs) show superior mechanical and physical properties and seem to be promising ideal reinforcing material for polymer matrix composites of high strength and low density. In most of the experimental results till date, however, only modest or negligible improvements in stiffness and strength have been observed. The reason for less than expected improvements has been blamed on weak interfacial bonding of CNTs to the matrix, waviness in CNTs, and agglomeration of CNTs in resin at higher particle volume fractions. In this chapter, the effect of interfacial bonding, waviness, and agglomerations on the nanocomposite stiffness, Poisson's ratio and strength are considered via numerical simulations. The proposed methodology combines the Mori-Tanaka effective field method, micromechanics, classical laminate theory (CLT)

Characterization of Nanocomposites: Technology and Industrial Applications
Edited by Frank Abdi and Mohit Garg

ISBN 978-981-4669-02-3 (Hardcover), 978-1-315-36489-6 (eBook)
www.panstanford.com

and multiscale progressive failure analysis (MS-PFA) to predict the desired material properties of nanocomposites without the need for time consuming and expensive testing. The results demonstrate the robustness of the methodology with the comparison against test data from literature without relying on extensive finite element modeling of CNTs within polymer matrix material.

12.1 Introduction

Both experimental and theoretical studies conclude that carbon nanotubes (CNTs) show superior mechanical and physical properties and seem to be promising ideal reinforcing material for polymer matrix composites of high strength and low density. In most of the experimental results till date, however, only modest or negligible improvements in stiffness and strength have been observed. Theoretical models, however, just started to look into the reason for the cause of over predictions compared to the test data. The reason for less than expected improvements has been identified to be weak interfacial bonding of CNTs to the matrix, waviness in CNTs, and agglomeration of CNTs in resin at higher particle volume fractions. Efforts have been made to address and model the interfacial bonding, waviness and agglomeration issues [1–4]. In this chapter, a methodology has been developed that is able to account for all three issues simultaneously. The proposed methodology combines the Mori–Tanaka effective field method, micromechanics, classical laminate theory (CLT), and multiscale progressive failure analysis (MS-PFA) to predict the desired material properties of nanocomposites without the need for time-consuming and expensive testing and detailed finite element models. The effect of interface has been presented earlier by the authors on multiscale carbon fiber reinforced composite material system on interlaminar shear strength using a short-beam shear specimen from ASTM D 2344 [3, 5, 6]. In this chapter, all three effects will be combined and predictions will be made for carbon nanotube composites with higher fiber volume fraction. The methodology is capable of predicting the strength of the advanced nanocomposites.

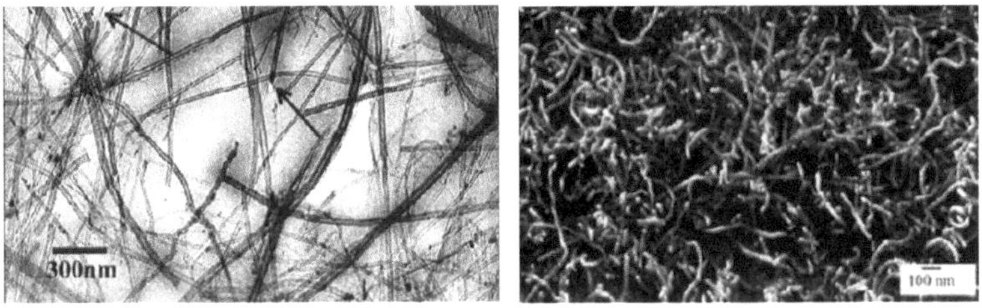

(a) 1 wt% MWCNT/polystyrene resin (b) 50 wt% MWCNTs/polyvinyl alcohol

Figure 12.1 (a) Transmission electron microscope (TEM) image and (b) scanning electron microscope (SEM) image of randomly oriented wavy carbon nonotubes in resin. Reprinted from Ref. [1], Copyright (2003) with permission from Elsevier.

12.2 Experiment

Fisher et al., (2003) presented analytical data [1] for multiwalled carbon nanotubes (MWCNTs) infused in polyester resin. According to the reference [1], the nanotubes were uniformly distributed in epoxy and constituted 5 wt% in the end product [7]. The MWCNTs were approximately 3000 nm in length and 30 nm in diameter. The waviness (a/λ) was assumed to be 0.05 and wavelength ratio (λ/d) to be 100, where a, d, and λ are amplitude, MWCNT diameter, and wavelength, respectively [1]. The wave amplitude is assumed to be 150 nm. Figures 12.1 and 12.2 show the transmission electron

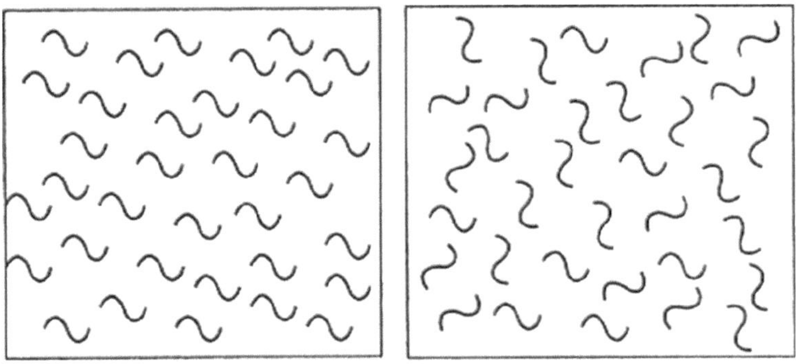

(a) Aligned wavy nanotubes (b) Randomly oriented wavy nanotubes

Figure 12.2 Uniformly dispersed wavy nanotubes orientation modeled. Reprinted from Ref. [8], Copyright (2002) with permission from Elsevier.

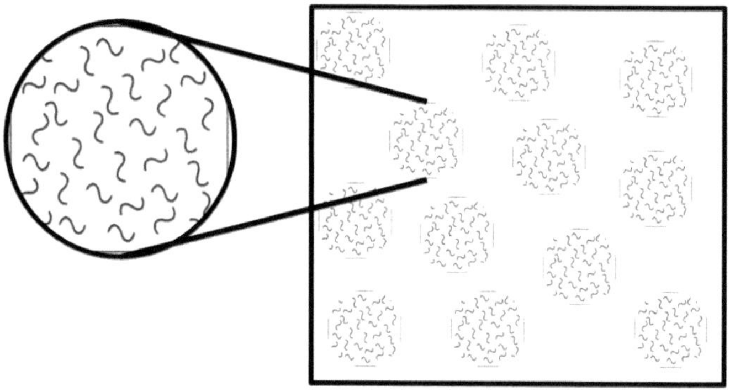

Figure 12.3 Agglomerate (uniformly dispersed) containing wavy nanotubes in matrix modeled.

microscope image and a schematic of resin infused with wavy MWCNTs [1]. Figure 12.3 shows schematic of agglomerated wavy nanotubes reinforcing the resin. Note that the agglomerations are assumed to be uniformly dispersed and are spherical in shape. The test data (modulus) was obtained from literature [7]. Young's modulus from test for resin infused with randomly oriented wavy MWCNTs was approximately 4 GPa. No strength data was provided in the literature [1].

Tables 12.1 and 12.2 list the typical tested and effective material properties for MWCNTs and epoxy from the literature [1–3, 9].

The authors will be using above constituent properties and methodology (discussed next) to predict the effect of MWCNT waviness, interface effect, and agglomeration on resin material properties.

Table 12.1 Typical modulus and strength properties for MWCNTs [1–3]

MWCNT material properties	Symbol	Value	Units
Density	Rhop	2.16	g/cm^3
Modulus	Ep	450	GPa
Poisson's ratio	vp	0.19	-
Tensile strength	SpT	63	GPa
Compressive strength	SpC	21.1	GPa

Table 12.2 Modulus and strength properties for epoxy [1, 9]

Matrix material properties	Symbol	Value	Units
Density	Rhom	1.2	g/cm^3
Modulus	Em	2.25	GPa
Poisson's ratio	vm	0.38	
Tensile strength	SmT	139.7	MPa
Compressive strength	SmC	260	MPa
Shear strength	SmS	80	MPa

Note: Shear strength is obtained by dividing tension strength by sqrt(3).

12.3 Methodology

The proposed methodology for fatigue durability simulation of nanocomposite material system consists of utilizing the closed-form solution from Mori–Tanaka, classical laminate theory (CLT), and multiscale multiphysics progressive failure analysis of a unit cell or a representative volume element (RVE).

12.3.1 *Waviness Modeling*

First, the wavy MWCNTs are broken down into small straight small tubes. The orientation of each tube is determined assuming a sign wave, as shown in Fig. 12.4. The material properties of the resin reinforced with straight MWCNTs are determined using Mori–Tanaka formulation. Then a simple FE model is generated where each element is assigned unidirectional ply properties obtained from Mori–Tanaka formulation, but different orientation to the adjacent elements determinate from the sine function. The simple FE model is then used to compute the effective modulus, Poisson's ratio.

12.3.2 *Nanocomposite: Closed-Form Analytical Solution*

Effective nanocomposite (or enhanced matrix) material properties, where randomly dispersed MWCNTs of a known weight fraction in neat matrix, are determined analytically. The analysis approach

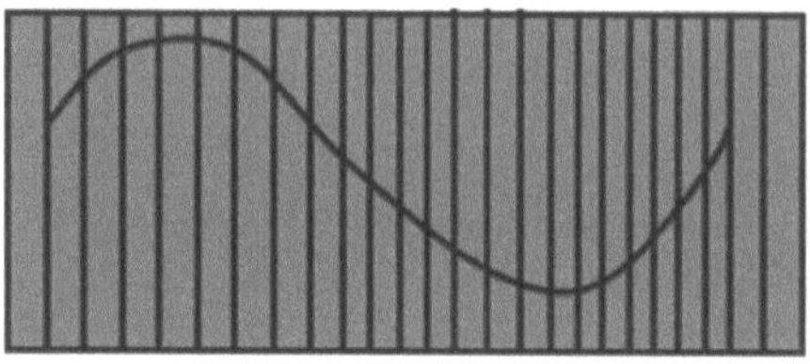

Figure 12.4 Wavy MWCNTs model.

uses well-known Mori–Tanaka formulation [10, 11] for calculating the anisotropic (aligned nanoparticles) nanocomposite properties from isotropic matrix and in situ isotropic/anisotropic nanoparticle properties (stiffness, aspect ratio and volume fraction). For a composite material reinforced with aligned nanoparticles, the Tandon and Weng (1984) prediction of the moduli E_{11} (aligned particle direction), E_{22} (transverse to the aligned particle direction), the in-plane shear modulus G_{12}, and the out-of-plane shear modulus G_{23} of the composite are

$$E_{11} = {}^{E_m}\!/\!\left(1 + f_p\left(A_1 + 2\upsilon_m A_2\right)/A\right) \tag{12.1}$$

$$E_{22} = {}^{E_m}\!/\!\left(1 + f_p\left(-2\upsilon_m A_3 + \left(1 - \upsilon_m\right)A_4 + \left(1 + \upsilon_m\right)A_5{}^{*}A\right)/\left(2A\right)\right) \tag{12.2}$$

$$G_{12} = {}^{G_m}\!/\!\left(1 + f_p\left(\mu_m/\left(\mu_p - \mu_m\right)\right) + 2\left(1 - f_p\right)H_{1212}\right) \tag{12.3}$$

$$G_{23} = {}^{G_m}\!/\!\left(1 + f_p\left(\mu_m/\left(\mu_p - \mu_m\right)\right) + 2\left(1 - f_p\right)H_{2323}\right) \tag{12.4}$$

where A and A_i are constants depending on the components of the Eshelby tensor and the matrix/nanoparticles properties, and Hijkl are the Cartesian components of the Eshelby tensor [12].

In this chapter, a closed-form analytical solution for the complete set of anisotropic elastic properties of the nanocomposite derived by Tandon and Weng (1984) by combining the Eshelby theory [12] and the Mori–Tanaka model [10], as shown in Eqs. 12.1–12.4, has been developed.

The in situ strength properties of the nanocomposites are calculated based on stress and strain-based micromechanics theories developed earlier in the literature [13, 14]. At this stage we have calculated the effective material properties of the enhanced matrix where the nanoparticles are randomly dispersed. Note that for fibrous nanoparticles the above methodology calculates aligned anisotropic properties. In order to predict random isotropic properties for fibrous nanoparticles composite we follow the process described in Section 12.3.2 below.

12.3.3 *Conventional Composite: Multiscale Micromechanics*

A unit cell–based coupon model is then subjected to multiscale multiphysics progressive failure analysis (PFA). The unit cell with quasi-isotropic $[45/0/\text{-}45/90]_{2s}$ lay-up is subjected to five in-plane loading conditions: longitudinal tension and compression, transverse tension and compression, and in-plane shear. Then modulus and strength values are recorded from each simulation. Each ply is assigned unidirectional enhanced matrix properties that were previously derived analytically using the closed-form Mori–Tanaka approach. The combined unit cell and PFA simulation of a quasi-isotropic ply lay-up gives equivalent effective 2D and 3D random (transversely isotropic and isotropic) stiffness, strength and Poisson's ratio of the enhanced matrix (nanocomposite), as shown in Fig. 12.5.

The constituent material properties can be estimated using limited test data by reverse engineering the constituent material properties. MCQ-Nano, a commercial nanocomposite material characterization and qualification software [15], was used to predict the constituent properties. MCQ-Nano uses the above mentioned methodology to predict the aligned, transversely isotropic and isotropic nanocomposite thermoelectromechanical material properties. The formulation in MCQ-Nano also accounts for the manufacturing variables such as particle and void volume ratio.

12.3.4 *Progressive Failure Analysis (PFA)*

Progressive failure analysis is the preferred approach for failure analysis in composite materials. In this approach, representative volume element (RVE) independent of finite element model (FEM) failure in each ply constituting the laminate, under investigation,

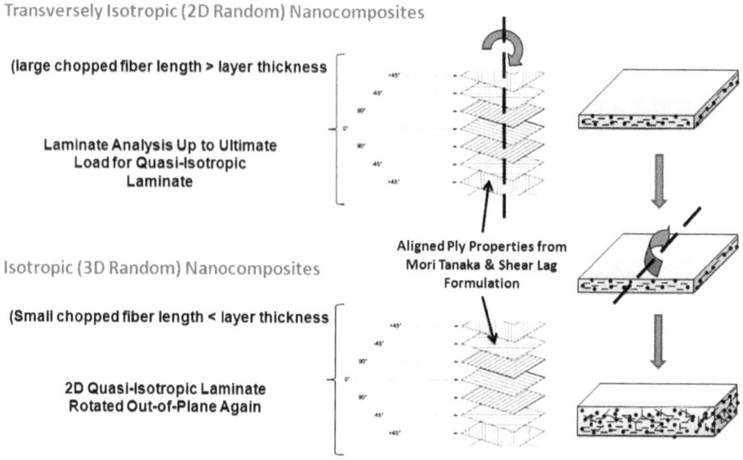

Figure 12.5 Classical laminate theory and multiscale multiphysics progressive failure analysis approach to determine 2D and 3D random nanocomposite properties.

is assessed on the basis of the ply limit strength/strain properties, as shown in Fig. 12.5. Some failure criteria are maximum stress- or strain-based and some are interactive. The damaged plies are then either removed from the unit cell model or given a negligible stiffness to simulate a broken ply and study its effect on overall laminate. Later, once all the plies in the unit cell are damaged, it is assumed to be completely broken. The stresses in individual plies are obtained using classical laminate theory, which are then compared against the limit stresses or strains parameters using default failure criteria [9]. The predictions from progressive failure analysis then give the laminate modulus, Poisson's ratio, and strength in addition to other thermal and electrical properties that consists of randomly oriented particles in 2D or 3D space, are shown in Fig. 12.5.

12.3.5 *Material Modeling Assumptions*

The following assumptions have been made in order to predict the material properties for resin reinforced with wavy MWCNTs:

(1) Matrix and MWCNTs material properties are assumed to be isotropic and linear elastic.
(2) The void content was assumed to be 0%.
(3) The following properties were assumed for the interphase (Table 12.3 and Fig. 12.6) between the matrix and the MWCNTs. Note that Young's modulus for the interphase is assumed to be 1/10th of the control matrix (Table 12.2).

Table 12.3 Assumed effective interphase properties (thickness = 1 nm)

Interphase material properties	Symbol	Value	Units
Density	Rhom	1.2	g/cm^3
Modulus	Em	0.225	GPa
Poisson's ratio	vm	0.49	–
Tensile strength	SmT	139.7	MPa
Compressive strength	SmC	260	MPa
Shear strength	SmC	80	MPa

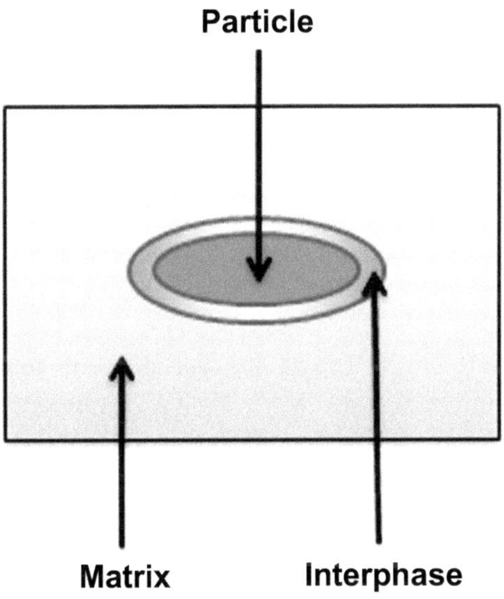

Figure 12.6 Modeling interphase effects for progressive failure analysis.

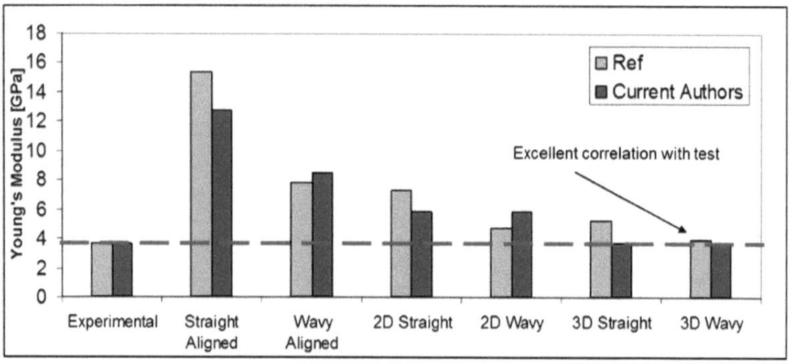

Figure 12.7 Comparison of predicted versus tested Young's modulus for the nanocomposite: 5 wt% MWCNT in epoxy; $L = 3000$ nm; $D = 30$ nm; and $L/d = 100$; wavelength ratio $= \lambda/d = 100$; waviness $= a/\lambda = 0.05$; amplitude (a) $= 150.0$ nm; $\lambda =$ wavelength. Reprinted from Ref. [1], Copyright (2003) with permission from Elsevier.

12.4 Results and Discussion

12.4.1 *Modulus Prediction/Test Validation of Control and Nanocomposite*

Using the above methodology in material characterization and qualification Nano (MCQ-Nano) tool and constituent test data listed under Tables 12.1 and 12.2, the modulus of straight and wavy MWCNTs reinforced nanocomposite are predicted when they are aligned or randomly oriented in two- and three-dimensional space. Figure 12.7 shows the comparison between various predictions for straight and wavy MWCNTs reinforced composite and test data from literature [1].

Figure 12.8 below shows the variation of Young's modulus of epoxy reinforced with wavy MWCNTs randomly oriented in three-dimensional space for different wave amplitude (a) values ranging between 0 nm and 700 nm. Figure 12.8 indicates that the nanocomposite Young's modulus decreases with increase in MWCNTs waviness.

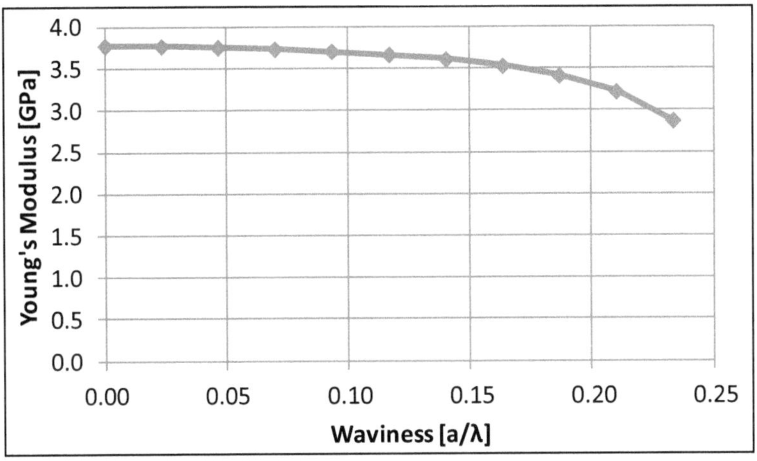

Figure 12.8 Predicted Young's modulus for the randomly oriented 3D MWCNTs in epoxy: 5 wt% MWCNT in epoxy; $L = 3000$ nm; $D = 30$ nm; $L/d = 100$; wavelength ratio $= \lambda/d = 100$; waviness $= a/\lambda = 0.05$; amplitude $(a) = 0.0$ to 700.0 nm; $\lambda =$ wavelength.

12.4.1.1 Effect of weak interphase and agglomeration

Figure 12.9 shows the effect of an interphase on Young's modulus of the epoxies reinforced with MWCNTs. The interphase was assumed to be in between the MWCNT outer wall and matrix. The interphase is assumed to be plastically yielded due to thermal mismatch between the nanoparticles and matrix. The assumed properties are listed in Table 12.3 for the interphase. The interphase is also assumed to be perfectly bonded to the particle and the matrix. The modulus drops due to 1 nm thick interphase layer from 3.758 GPa to 2.732 GPa (27.3% decrease). The predictions clearly indicate the importance of interphase layer. The interphase layers in resin reinforced with MWCNTs are usually weaker because the carbon atoms in the CNTs outer wall cannot bond with the epoxy surrounding the tube.

Figure 12.9 also shows the effect of uniform and non-uniform nanoparticle dispersion on the Young's modulus of the epoxies reinforced with MWCNTs. For non-uniform dispersion of nanoparticles, the authors assumed large spherical shape particles consisting of agglomerates of nanoparticles in neat matrix. The authors assumed

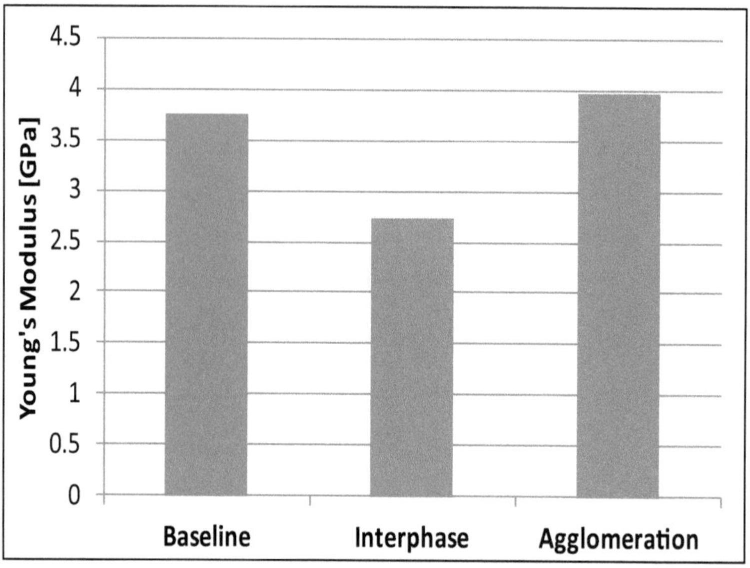

Figure 12.9 Predicted Young's modulus for the 3D randomly oriented MWCNTs in epoxy, baseline, baseline with interphase of 1 nm thickness, and baseline with agglomeration (no interphase); 5 wt% MWCNT in epoxy; $L =$ 3000 nm; $D = 30$ nm; $L/d = 100$; wavelength ratio $= \lambda/d = 100$; waviness $= a/\lambda = 0.05$; amplitude $(a) = 0.0$ to 700.0 nm; $\lambda =$ wavelength.

the ratio of agglomerates to the nanocomposite volume to be 0.5 and that of nanoparticles to agglomerates to be 1.0. All other variables were same as the baseline case. Figure 12.9 indicates that assumed spherical shaped agglomeration actually helps increase Young's modulus slightly (5.4% increase). Note that agglomerations in nanocomposites usually occur at a critical particle weight fraction depending on the infusing techniques of the manufacturer.

12.4.2 *Strength Property Prediction*

Using the above methodology in material characterization and qualification Nano (MCQ-Nano) tool and constituent test data listed under Tables 12.1 and 12.2, the tension, compression, and shear strength of straight and wavy MWCNTs reinforced nanocomposite are predicted when they are aligned or randomly oriented in two- and three-dimensional space. Figure 12.10 shows the comparison

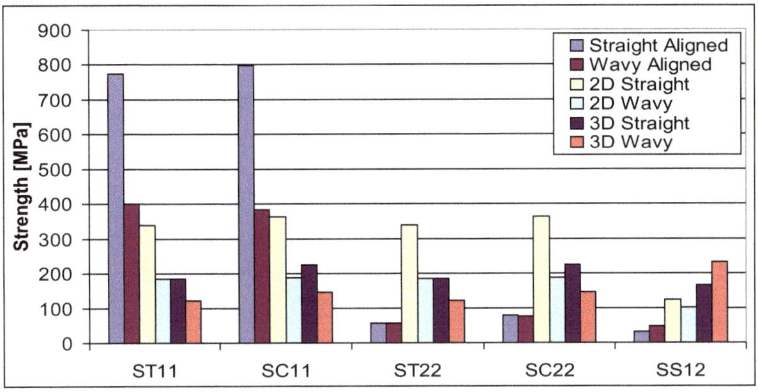

Figure 12.10 Predicted strength for the epoxy: 5 wt% MWCNT in epoxy; $L = 3000$ nm; $D = 30$ nm; $L/d = 100$; wavelength ratio $= \lambda/d = 100$; waviness $= a/\lambda = 0.05$; amplitude $(a) = 1.5\text{E-}4$ nm; $\lambda =$ wavelength.

between various predictions for straight and wavy MWCNTs reinforced composite.

Figure 12.11 below shows the variation of tension, compression and shear strength of epoxy reinforced with wavy MWCNTs randomly oriented in three-dimensional space for different wave amplitude (a) values ranging between 0 nm and 700 nm. Figure 12.11 indicates that the nanocomposite tension strength decreases with increase in MWCNTs waviness, the compression decreases slightly and then increases again with increases again with change in wave amplitude and shear strength is maximum when wave amplitude is 0.12. The increase in shear strength is due to the amount of fiber oriented in that direction. The tension strength increases slightly again as the amount of fiber is oriented again along the principal directions.

12.4.2.1 Effect of weak interphase and agglomeration

Figure 12.12 shows the effect of an interphase on the tension, compression, and shear strength of the epoxies reinforced with MWCNTs. The interphase was assumed to be in between the MWCNT outer wall and matrix. The interphase is assumed to be plastically yielded due to thermal mismatch between the nanoparticles and

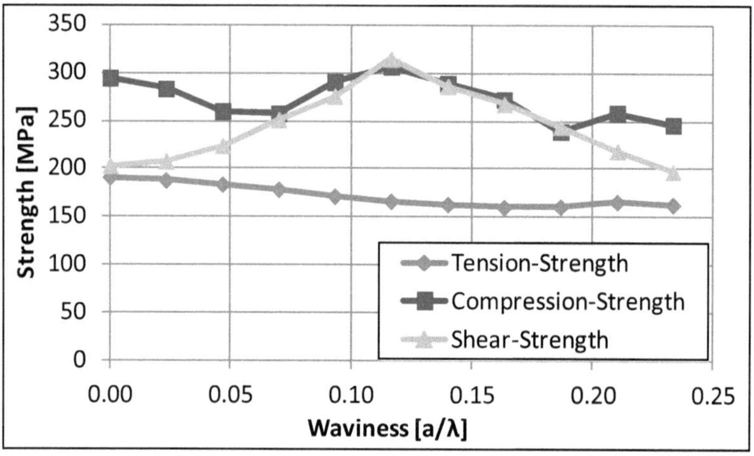

Figure 12.11 Predicted tension, compression, and shear strength of randomly oriented 3D MWCNTs in nanocomposite: 5 wt% MWCNT in epoxy; $L = 3000$ nm; $D = 30$ nm; $L/d = 100$; wavelength ratio $= \lambda/d = 100$; waviness $= a/\lambda = 0.05$; amplitude $(a) = 0.0$ to 700.0 nm; $\lambda =$ wavelength.

matrix. The assumed properties are listed in Table 12.3 for the interphase. The interphase is also assumed to be perfectly bonded to the particle and the matrix. The tension and shear strength drop slightly (−6.3% and −10.0%, respectively) because of the interphase; however, the compression strength increases by about 21.6%.

Figure 12.12 also shows the effect of uniform and non-uniform nanoparticle dispersion on Young's modulus of the epoxies reinforced with MWCNTs. For non-uniform dispersion of nanoparticles, the authors assumed large spherical shape particles consisting of agglomerates of nanoparticles in neat matrix. As for Young's modulus earlier, the authors assumed the ratio of agglomerates to the nanocomposite volume to be 0.5 and that of nanoparticles to agglomerates to be 1.0. All other variables were same as the baseline case. Figure 12.12 indicates that assumed spherical shaped agglomeration, unlike Young's modulus, decreases the tension, compression, and shear strengths significantly compared to the baseline (−37.5%, −10.3%, and −71.2%, respectively).

In summary, the above results (Figs. 12.7, 12.8, 12.10, and 12.11) indicate that waviness significantly affects the modulus and strength

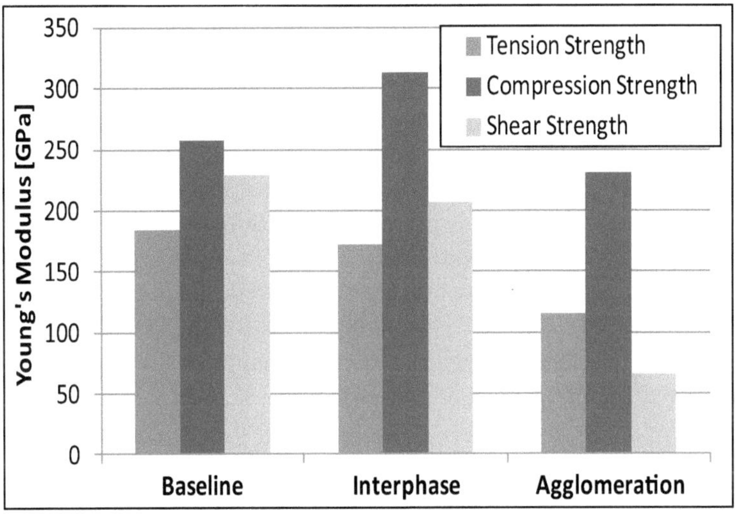

Figure 12.12 Predicted tension, compression, and shear strength of the randomly oriented 3D MWCNTs in epoxy, baseline, baseline with interphase of 1 nm thickness, and baseline with agglomeration (no interphase); 5 wt% MWCNT in epoxy; $L = 3000$ nm; $D = 30$ nm; and $L/d = 100$; wavelength ratio $= \lambda/d = 100$; waviness $= a/\lambda = 0.05$; amplitude $(a) = 0$ to 700 nm; $\lambda =$ wavelength.

of the nanocomposites. However, the shear strength increases with increase in waviness to 0.12 (Fig. 12.11). Figures 12.9 and 12.12 indicate that the interphase in between the CNTs and matrix can significantly affect both modulus and strength of the nanocomposite (epoxy reinforced with MWCNTs). On the other hand, uniformly dispersed spherical agglomeration helps increase the modulus, but adversely affects the tension, compression, and shear strength of epoxies reinforced with wavy MWCNTs. The trends predicted are in general agreement with those of experimentalists in literature.

12.5 Conclusions

The authors described and applied a novel computational simulation approach to predict the static properties (modulus and strength) of nanocomposite polymeric materials used in aerospace and automobile applications. The constituent (MWCNTs and matrix) properties were used to predict the material properties of epoxy reinforced with 5 wt% MWCNTs. A combined approach of Mori–Tanaka, micromechanics, classical laminate theory, and multiscale progressive failure analysis were used to account for CNT waviness, interphase, and agglomeration in nanocomposites. The predicted Young's modulus value compared well with the reported test data. The predictions were extended to the tension, compression, and shear strength, which are in agreement with the general observation of other researchers. The validation with experimental results suggests the methodology is robust and may possibly be used for other nanocomposite material combinations.

References

1. Fisher, F. T., Bradshaw, R. D., and Brinson, L. C. (2003). Fiber waviness in nanotube-reinforced polymer composites-I: modulus predictions using effective nanotube properties. *Compos. Sci. Technol.*, **63**(11), 1689–1703.
2. Garg, M., Pantano, A., and Boyce, M. C. (2007). An equivalent orthotropic representation of the nonlinear elastic behavior of multiwalled carbon nanotbues. *J. Eng. Mater. Technol.* **129**, 431–439.
3. Garg, M., Abdi, F., and Stuart, M. (2009). Analyzing interlaminar shear strength of multi-scale composites via combined finite element and progressive failure analysis approach. SAMPE Tennessee.
4. Shi, D-L., Feng, X-Q., Huang, Y. Y., Hwang, K-C., and Gao, H. (2004). The effect of nanotube waviness and agglomeration on the elastic property of carbon nanotube-reinforced composites. *Trans. ASME*, **126**, 250.
5. Garg, M., Abdi, F., Kinloch, A. J., and Nikbin, K. (2011). Part 1: Multi-Scale Nanocomposite Fatigue Life Determination, SAMPE Fort Worth, TX, Oct 17–20, 2011.
6. Garg, M., Abdi, F., Kinloch, A. J., and Nikbin, K. (2012). Part 2: Multi-Scale Nanocomposite Fatigue Life Determination, SAMPE Baltimore, MD, May 18, 2012.
7. Schadler, L. S., Giannaris, S. C., and Ajayan, P. M. (1998). Load transfer in carbon nanotube epoxy composites. *Appl. Phys. Lett.* **73**(26), 3842.

8. Bradshaw, R. D., Fisher, F. T., and Brinson, L. C. (2002). Fiber waviness in nanotube-reinforced polymer composites-II: modeling via numerical approximation of the dilute strain concentration tensor. *Compos. Sci. Technol.*, **63**(11), 1705–1722
9. *GENOA User Manual.* www.ascgenoa.com
10. Mori, T., and Tanaka, K. (1973). Average stress in matrix and average elastic energy of mateials with misfitting inclusions. *Acta Metall.* **21**(5), 571–574.
11. Tandon, G. P., and Weng, G. J. (1984). The effect of aspect ratio of inclusions on the elastic properties of unidirectionally aligned composites. *Polymer Compos.* **5**(4), 327–333.
12. Eshelby, J. D. (1957). The determination of the elastic field of an ellipsoidal inclusion and related problems. *Proc. R. Soc. A.* **241**(1226), 376–396.
13. Courtney, T. H. (2000). *Mechanical Behavior of Materials* (2nd ed.). New York, McGraw-Hill.
14. Murthy, P. L. N., and Chamis, C. C. (1986). Integrated Composites Analysis (ICAN). Technical Paper 2515, Cleveland, Ohio, NASA, 1986.
15. Garg, M., Abumeri, G., and Abdi, F. (2010). Material properties characterization for multi-scale composites: glass/epoxy/silica nanopartcles. *J. Mater. Eval.* **68**(5), 2010.

Chapter 13

Dispersion of Nanoparticles in Polymers

Ambrose C. Taylor[a] **and David J. Bray**[b]
[a] *Department of Mechanical Engineering, South Kensington Campus, Imperial College London, London SW7 2AZ, UK*
[b] *Scientific Computing Department, SciTech Daresbury, Warrington, Cheshire WA4 4AD, UK*
a.c.taylor@imperial.ac.uk, david.bray@stfc.ac.uk

The degree of dispersion of nanoparticles can greatly affect the properties of nanocomposites. Agglomeration of the particles can cause reductions in strength or poor barrier performance. This chapter outlines the principal types of nanoparticles, and discusses some of the methods that may be used to disperse these particles in polymers. Homogeneously dispersing nanoparticles in polymer matrices can be difficult due to their high surface area and incompatability with polymers. Even if a good dispersion is initially achieved the particles may agglomerate during further processing, leaving a poor dispersion in the final material. If agglomerates are formed, these can act as defects resulting in a reduction in performance rather than any enhancement. Hence it is crucial that the degree of dispersion of the nanoparticles can be described quantitatively. The use of the area disorder parameter gives such a quantitative classification, distinguishing between good, random and poor dispersion. The area disorder is also robust enough to

Characterization of Nanocomposites: Technology and Industrial Applications
Edited by Frank Abdi and Mohit Garg

ISBN 978-981-4669-02-3 (Hardcover), 978-1-315-36489-6 (eBook)
www.panstanford.com

cope with the incompleteness of data associated with the analysis of micrographs of nanocomposite materials.

13.1 Nanocomposites

13.1.1 *Introduction*

It can be argued that the production of nanocomposites is not novel, as the use of nanoparticles to alter the properties of materials is not an exclusively modern technique. For example, the Lycurgus cup on display in London at the British Museum was manufactured in the late Roman period (4th century AD). This cup is made from glass, which looks green in reflected light but appears red in transmitted light. This curious effect is due to the light being scattered by the gold and silver nanoparticles in the glass. These particles are approximately 70 nm in diameter, and are present as a colloidal dispersion. Another example is carbon black, which was first produced 3500 years ago in China as lamp black and used as a pigment for ink. This is composed of particles of size 20 nm in diameter and above. This material is now used in millions of car tyres and printer cartridges per year.

Although nanocomposites may not be new, there has been an explosion in the level of interest in the development of these materials in recent years. This interest is linked to the increasing ability to synthesise nanometre-scale materials, and the availability of the tools which enable the imaging of structures at this scale. Thus it is now possible to characterise how the nanoparticles are dispersed and start to understand why the observed effects occur, something that was not possible previously.

The addition of nanoparticles can alter the properties of a material, but the effect of this addition will depend on several factors. These include the material, shape and size of the particles. However, one of the most critical factors is the quality of the dispersion of the particles in the matrix. The dispersion quality must be suitable for the effect that is required. For example, if high transparency is required then the nanoparticles must be well dispersed, so that there are no features larger than the wavelength of light which would scatter the light and reduce transparency. If conductivity is required, then a good dispersion is not the ideal state, as the formation of a percolating network requires that charge can jump from one conductive nanoparticle to the next. When each nanoparticle is surrounded by nonconducting polymer, then the material will not conduct. In this case some clustering may be helpful to obtain the required property.

To develop better nanocomposites it is critical to be able to quantitatively describe the quality of the dispersion of the nanoparticles within the matrix. This allows different dispersion techniques or surfactants to be objectively compared.

This chapter will briefly introduce nanoparticle reinforcement, and summarise the types of nanoparticles which are typically used in polymers. It will then summarise some of the dispersion techniques used, and discuss how the quality of dispersion can affect the properties of the nanocomposite produced. It will then outline how the quality of the dispersion can be quantified by introducing the area disorder parameter.

13.1.2 *Nanoparticle Reinforcement*

When selecting nanoparticles to alter the properties of a material, the five basic characteristics of particulate fillers defined by Rothon and Hancock [1] should be considered:

(1) What properties are being sought?
(2) What deleterious changes may also occur and can they be tolerated?
(3) How easy is the filler to handle and how might it affect processing?
(4) Are any special additives needed?
(5) What is the true cost of using the filler, is it justifiable and are there more cost-effective alternatives?

These characteristics remain true whether the particles are micro- or nanoparticles, although Rothon and Hancock originally devised this list for microparticles. As nanoparticles are very small by definition, then it must be noted that the interparticle distance for nanocomposites is also very small, especially at high volume fractions. These factors may affect the interaction between a particle and its neighbours, or between a particle and the polymer surrounding it, or may affect the structure of the polymer itself. Hence specific properties of a particle-modified polymer may vary depending on whether nanoparticles or microparticles are used. This may also lead to differences in the quality of the dispersion depending on whether nanoparticles or microparticles are used.

It should be noted that the performance benefit which the addition of nanoparticles provides could often be obtained using conventional micron-sized fillers at a much lower cost. However, some very successful applications that rely on the small size of nanoparticles have been shown, for example the addition of ceramic nanoparticles to transparent coatings gives improved scratch resistance without loss of transparency.

13.2 Nanoparticle Types

13.2.1 *Nanoparticle Definition*

The common definition of a nanoparticle is that it has at least one dimension in the nanometre range. Indeed, the term 'nano' has virtually replaced 'sub-micron', even when the latter is more appropriate. The additional condition that typically all of the phases are in the 1–20 nm range [2] is generally ignored. Nanoparticles are characterised by a large surface area to volume ratio. They can be made from a wide range of materials - metallic (e.g., gold, silver), ceramic (e.g., silica, alumina, layered silicates, silicon carbide), or organic (e.g., carbon black, rubber particles, graphene, carbon nanotubes and nanofibres). Due to the small size of nanoparticles, the numbers present in nanocomposites are enormous. For example, there are approximately 10^{17} individual particles in 1 kg of 20 nm diameter silica nanoparticles.

As many nanoparticles are smaller than the wavelength of light (about 390 to 780 nm [3]), they appear translucent when added to a translucent polymer, if they do not cluster to form larger features. Consequently they are too small to be observed using optical microscopy. High-magnification techniques, e.g., transmission electron microscopy (TEM), atomic force microscopy (AFM) and field emission gun scanning electron microscopy (FEGSEM) are required to image the individual particles. It is not normally possible to measure the mechanical properties of such small particles, and hence their properties are assumed to be the same as for the bulk material, or equal to those of a similar material. For many applications, such as the prediction of the mechanical properties of nanocomposites, this appears to be a reasonable assumption.

The shape and aspect ratio of the particles are important in determining the properties of nanocomposites. Nanoparticles can be classified by their shape, as equi-axed (e.g., spherical), rod-like and plate-like. These classes will be discussed in turn.

13.2.2 *Equi-Axed Nanoparticles*

Most equi-axed nanoparticles are spherical, but irregular particles are also available. The common particles are metallic (e.g., gold, silver), ceramic (e.g., silica, alumina, titania), or organic (e.g., rubber particles, fullerenes). The ceramic particles are typically prepared by sol-gel or flame-spraying methods. Rubber particles are typically core-shell particles, with a soft rubbery core and a hard shell of a polymer such as poly(methyl methacrylate).

13.2.3 *Nanotubes and Nanofibres*

The rod-like nanoparticles available include carbon nanotubes (CNTs) and carbon nanofibres (CNFs), which have a larger diameter. Ceramic nanotubes (e.g., zirconia, tungsten disulfide) or whiskers (e.g., silicon nitride, silicon carbide, alumina) are also used.

Carbon nanotubes were first reported by Iijima [4, 5], and two principal types have been identified. Firstly, single-walled nanotubes (SWNTs) have a single graphene layer rolled into a tube typically 1–2 nm in diameter, with hemispherical end caps [5, 6]. Secondly, multi-walled nanotubes (MWNTs) are comprised of a number of co-axial capped graphene tubes. The outer diameter of a MWNT is typically 3–10 nm. Under tension, only the outer layer of a MWNT carries the load as the van der Waals forces between the layers are too weak to transfer stress to the inner layers [7]. Variants such as double-walled nanotubes (DWNTs) are also available. The length of nanotubes varies significantly, but they can be as long as several millimetres [8].

Vapour-grown carbon nanofibres are typically 50 to 200 nm in diameter and 30 to 100 μm long [9, 10]. They have a lower aspect ratio and more defects than carbon nanotubes, but are much cheaper than carbon nanotubes, and are available in much larger quantities.

Ceramic whiskers have been produced for many years, as reported by Gordon [11]. However, aluminium borate whiskers have been combined with thermoset polymer matrices by Liang and co-workers [12, 13]. Ceramic nanotubes have been produced by coating carbon nanotubes with silicon carbide [14]. Silicon carbide nanofibres, which can be about 40 nm in diameter with lengths of up to several hundred microns, have also been produced [15, 16].

13.2.4 *Plate-Like Nanoparticles*

A range of plate-like nanoparticles are available, including layered silicate nanoclays (e.g., montmorillonite), graphite and α-zirconium phosphate. The particles are comprised of stacks of platelets which can be intercalated or exfoliated by the polymer. The resulting morphology is typically described as 'particulate', 'intercalated' or 'exfoliated', as identified by wide-angle X-ray scattering (WAXS). In an intercalated microstructure, polymer chains enter the galleries between the platelets and increase the measured platelet spacing [17]. For an exfoliated nanocomposite, the platelets are further apart, and the spacing is too large to measure using WAXS. For a particulate structure, the spacing between the platelets remains the same as for the original particles.

Nanoclays (e.g., montmorillonite, hectorite) are layered silicates with a structure of stacked platelets similar to that of mica. The platelet thickness is approximately 1 nm and the lateral dimensions of the plates can vary from 30 nm to tens of microns.

Graphite is also a layered material, and a combination of chemical and thermal treatments can generate exfoliated graphite platelets with a thickness of 20 to 100 nm. Here the graphite is intercalated by an acid treatment, followed by exfoliation by a thermal shock at a temperature of around 600°C. The platelet diameter is in the order of 10 μm, and hence these platelets have a high aspect ratio [18, 19]. Graphene, where a single layer of carbon atoms is isolated, has attracted considerable attention over recent years due to its excellent properties. There is also a growing interest in chemically-modified graphene such as hydrogenated graphene (i.e., graphane), fluorinated graphene and graphene oxide) [20].

13.2.5 *Other Nanoparticles*

Some nanoparticles do not fall into the above categories. Carbon black is mostly an aciniform (grape-like cluster) particulate produced by the incomplete combustion of oil. The current worldwide production of carbon black is 8.1 million tonnes per year, being used as a pigment in inks and toners, and as a functional filler in tyres and other rubber products [21].

13.3 Dispersing Nanoparticles

13.3.1 *Introduction*

To obtain the best properties from a polymer nanocomposite, it is considered that the nanoparticles should be well dispersed, and each particle should be properly wetted by the polymer. If agglomerates are formed instead, these can act as defects resulting in a reduction in performance rather than any enhancement, as discussed below. This is a major challenge when preparing nanocomposites, as even small volume fractions of nanoparticles involve huge numbers of nanoparticles and very small interparticle distances. Shaffer and Kinloch pointed out that:

> 'One major difficulty with small diameter nanotubes is that they become increasingly difficult to wet. By trivial estimation, even a 1 vol% loading of single-walled nanotubes ensures that all of the polymer molecules are within one radius of gyration (say 5 nm) of a nanotube. This result implies that complete wetting of high loading fractions of single-walled nanotubes will be difficult, at least by conventional means, and that even more modest

> concentrations may be brittle and hard to process due to the constraint of the matrix.' [22]

Hence, the combination of poor dispersion and poor interfacial bonding limit the optimum utilisation of carbon nanotubes for reinforcing polymers [23].

Dispersing nanoparticles in polymers is difficult because nanoparticles have a high surface area and are generally incompatible with the matrix polymer. Chemical functionalisation of the nanoparticle surfaces can help improve dispersion, as it can reduce agglomeration and can also improve the bonding between the particle and the matrix [24, 25].

Once agglomeration occurs it is very difficult to break up the agglomerates that are formed. Nanoparticles can be purchased pre-dispersed in a resin (e.g., nanopox from nanoresins [26–29], or the MX range of core-shell particles from Kaneka [30]), and these are often easier to work with than dry nanoparticles. Even when a good dispersion is obtained, there is no guarantee that this good dispersion will remain after subsequent processing. For example, the addition of a liquid rubber, which phase-separates during curing, to some systems can cause nanoparticles to agglomerate [27, 31]. Although it is normally assumed that well-dispersed individual particles are required, small agglomerates do not necessarily adversely affect the performance of such systems.

13.3.2 *Mixing*

If nanoparticles are obtained which are well and stably dispersed in a polymer or resin, then simple mixing is often enough to blend the particles into a polymer, e.g., [32]. Low-shear mixing is also sufficient for well-compatabilised systems, as the thermodynamics of these materials can ensure a good dispersion. However, if nanoparticles are supplied as an untreated powder they will generally be agglomerated, and these agglomerates must be broken down during processing. Low-shear mixing alone is not sufficient to break up such agglomerates. Higher power dispersion methods, such as high-shear mixing and sonication, are the easiest methods to improve the dispersion of nanoparticles in polymers [33].

For carbon nanotubes, Fiedler et al. [34] report that the size and shape of the impeller and the mixing speed control the dispersion. Intensive stirring of MWNTs in epoxy resin can achieve a relatively good dispersion. However, reagglomeration occurs due to frictional contacts and elastic interlocking mechanisms [35]. SWNTs have a greater tendency to reagglomerate than MWNTs, and hence higher shear forces are required to achieve a reasonable dispersion.

High-shear mixing can be achieved using a range of techniques. Hackman and Hollaway [36] used a grinding media mill, filled with 2 mm diameter glass beads. Green et al. [10] generated high-shear by extrusion of a nanoparticle-modified epoxy resin through a small orifice. In both cases the high-shear method gave better dispersion compared low-shear mixing. The glass beads were shown to break down agglomerated nanoclay particles to form a uniform material. Although neither work evaluated the dispersion at the nanometre-scale using microscopy, differences could be seen at the micro-scale [36]. Wetzel et al. [37] have also shown that agglomerates can be broken down. They showed that agglomerates of 13 nm alumina nanoparticles which were initially several micrometres in diameter could be reduced to agglomerates in the size range of 30–50 nm by processing using a dissolver and bead mill.

A three-roll mill (or calender) has been used by Yasmin et al. to disperse nanoclay in epoxy [38, 39], and by Gojny et al. to disperse nanotubes in epoxy [40]. This method is commonly used to disperse microparticles in polymers [34], although this is not a continuous process as the material must be collected and fed back into the mill after each pass. The authors assumed that the dispersion was achieved (a) by the shear forces generated dispersing the particles as smaller tactoids, and (b) by the combined shear and diffusion processes facilitating the separation and penetration of polymer between the clay platelets [38, 39]. This process requires a relatively viscous material, and hence is suitable for higher loadings of nanoparticles. According to Fiedler et al. [34], a major advantage of this method is that it allows the efficient manufacturing of larger quantities of nanocomposites, as the nanoparticle-modified polymer requires only minutes of processing. By comparison, low-shear methods may require processing times of hours.

As thermoplastics are generally processed at much higher viscosities than thermoset resins, achieving the high shear required to disperse nanoparticles is much easier for thermoplastics. Processing using twin screw extrusion can be sufficient to achieve a reasonable dispersion [41, 42], although agglomerates can remain [43].

The introduction of a solvent into the processing is also common, e.g., [44, 45]. The nanoparticles are dispersed in a solvent. This mixture is then added to the polymer, and the solvent is evaporated off using heat. Nanoclays can be incorporated into polymers using tetrahydrofuran for example [46]. However, Bokobza showed that although carbon nanotubes were well dispersed in toluene solution during processing, they can reagglomerate after incorporation into the polymer [23]. Alternatively small amounts of a reactive solvent may aid the processing and dispersion of the nanoparticles, e.g., [47].

For more complex polymer systems containing multiple polymeric components, such as structural adhesives, the polymer component into which the nanoparticles are introduced does not generally cause a noticeable difference in the final morphology [36, 48]. However, the morphology may be dependent on the cure time and temperature; plus the temperature, time and shear rate used for mixing.

13.3.3 *Sonication*

Sonication uses an ultrasonic bath or probe to apply sound energy to the liquid which contains the nanoparticles. Processing commonly combines sonication and mixing, especially when using probe sonicators as here the power is directed into only a small volume, and the low shear forces generated by the probe do not generally ensure that all of the polymer passes through this volume. The combination of probe sonication with high-speed mixing has been reported to be a relatively successful way to disperse nanoparticles. This technique has been used with nanoclays in epoxy resin [49] and in polyimide [50]. It has also been used with carbon nanotubes in epoxy [51].

The very localised introduction of energy from probe sonication can damage nanoparticles, for example causing buckling, bending, dislocations and rupture of carbon nanotubes [34, 52]. However, ultrasonic cavitation is an efficient method of dispersing carbon nanotubes into epoxy resin [9], although agglomeration can still occur at higher volume fractions. Probe sonication will also cause local heating of the sample, and hence if undertaken with a thermoset polymer when the curing agent or initiator is present, then it may be necessary to cool the sample to prevent premature curing [52]. The use of an ultrasonic bath reduces the energy density and alleviates some of these problems, but the time taken to achieve a reasonable dispersion will be much greater [48].

The processing time can affect the size of nanoclay agglomerates [53]. However, the d-spacing (the distance between the nanoclay platelets) was unaffected by the sonication time, and hence exfoliation of the nanoclay could not be achieved using sonication. Gintert et al. [50] also found that sonication was an important step in breaking up nanoclay agglomerates.

13.3.4 *Alignment*

The properties of fibre composites are improved by the alignment of the fibres, although this does produce an anisotropic material. It is also possible to align nanoparticles. This is generally achieved by physical means, for example by application of a force, cutting, extrusion or drawing, e.g., [54, 55]. Alignment can also be achieved using the application of magnetic or electric fields for carbon nanotubes, e.g., [56, 57], and for layered silicate nanoclays, e.g., [58]. For more information on the alignment of carbon nanotubes, and their dispersion, see the review by Xie et al. [33]. The best properties are obtained if the nanoparticles are well dispersed, i.e., fully wetted by the polymer.

13.3.5 *Re-agglomeration*

Dispersion is achieved in the liquid state, be it in the melt, in a solvent or in the resin before crosslinking occurs. However, it is necessary to preserve this dispersion in the solid state. Once the dispersion process, e.g., sonication or mixing, is stopped then the nanoparticles may reagglomerate due to Brownian motion. For

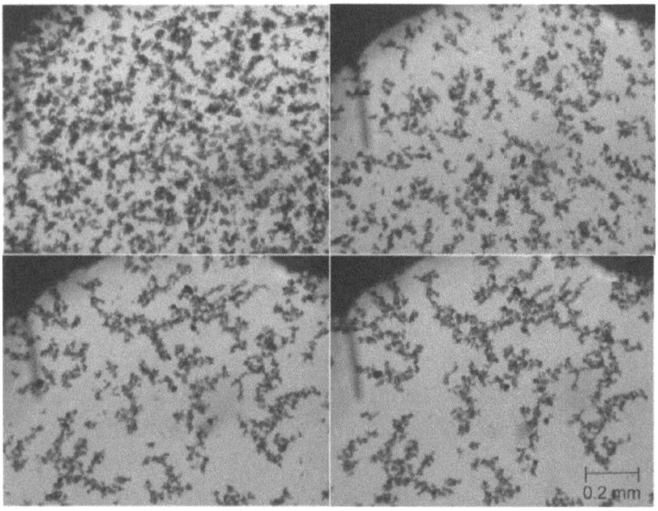

Figure 13.1 The hot stage images for samples with 0.1 wt% NT and 0 wt% R1, the nanotubes were sonicated into the epoxy. The images were taken at 22°C, 50°C, 90°C and 180°C. Reprinted with permission from Ref. [60].

thermoplastics maintaining a good dispersion is relatively simple, due to the high viscosity and relatively fast cooling. For example, at low concentrations, non-interacting nanotubes homogeneously dispersed in the polymer matrix take a very long time to re-aggregate, provided that the matrix viscosity is high enough to suppress fast Brownian motion (or crosslinked into an elastomer after dispersion) [59]. Transmission optical microscopy using a heated sample stage can be used to follow the re-agglomeration of nanoparticles during simulated processing; see Fig. 13.1, showing how higher viscosities reduce the rate of re-agglomeration [60, 61].

13.4 The Effect of Dispersion on the Properties of Nanoparticle-Modified Polymers

13.4.1 *Introduction*

The intention for adding nanoparticles to a polymer is to improve its properties, hence the resulting nanocomposite should perform better than the control polymer. This is normally the case, but not all the properties may be improved concurrently. For example, increasing the modulus can result in a reduction in strength. Hence the required balance of properties must be considered. There are also often conflicting reports of degrees of improvement between different materials, because the change in properties will depend on many factors including the degree of dispersion.

13.4.2 *Mechanical Properties*

The addition of high modulus particles increases the modulus of a polymeric material, independent of the particle size; although the stiffening effect should be higher for particles with larger aspect ratios [62]. The modulus of nanocomposites is generally independent of the dispersion. For example, Wetzel et al. [37] noted that the change in elastic properties at constant testing conditions depended on the nanoparticle volume fraction and not on the particle dispersion. However, Reynaud et al. suggested that the formation of aggregates can increase the modulus compared to the presence of well-dispersed particles due to more effective coupling with the polymer matrix [63].

There are many theoretical models available to predict the modulus of particle-modified polymers, such as the Halpin–Tsai and Mori–Tanaka models [62, 64, 65], which have been shown to broadly apply for nanoparticle-modified polymers, e.g., [64, 66], although more accurate predictions may be obtained by finite-element modelling [67]. Knowledge of the true aspect ratio of the particles, their alignment, and the adhesion between the particle and the polymer matrix is required for accurate predictions. However, the elastic properties of nanoparticles cannot generally be measured, because the small particle size makes it impossible to perform reliable measurements [68]. Hence, most authors assume that the properties (e.g., modulus, density) of the nanoparticles are equal to those of the bulk material, or of a similar material [66].

Although the modulus of nanocomposites is generally not affected by the dispersion, the formation of large agglomerates tends to reduce both the strain to failure and the tensile strength compared to the unmodified polymer. In contrast, for well-dispersed nanoparticles the strain to failure and the tensile strength may be increased or remain the same. Small agglomerates will not necessarily reduce the strength, and increased flexural strength has been observed even when agglomerates are present [37].

For CNTs, despite poor dispersion, small loadings can improve substantially the mechanical performance of the matrix [23]. However, the mechanical improvement brought about by incorporation of nanotubes remains small compared to the potential based on the properties of the nanotubes themselves. This is because the van der Waals interactions between individual nanotubes can lead to significant agglomeration, thus reducing the expected property improvements [23].

13.4.3 *Functional Properties*

The high aspect ratio and low permeability of plate-like nanoparticles provide good barrier properties when the particles are aligned and incorporated into thin films [69, 70], by creating a tortuous path for diffusion [36, 71]. Numerical predictions show that the permeability reduction is governed by the product $a.V_f$, where a is the platelet aspect ratio and V_f is the volume fraction [72]. However, this assumes good dispersion. If dispersion is poor, there will be areas of the film where there are no nanoparticles present, and hence the permeability would be no better than that of the unmodified polymer.

Studies on the fire resistance of nanocomposites have shown that nanoclays and nanotubes can improve the fire resistance, e.g., [73–75]. A significant amount of work has been performed using various thermoset polymers modified with nanoclays, e.g., [76–78]. Tests using cone calorimetry showed that the peak heat release rate was reduced by up to 50%, and the mass of char residue was increased. The effect of dispersion is generally not discussed, but the mechanism of the nanoparticles forming a barrier layer and thus reducing the peak heat release rate will be less effective if the nanoparticles are poorly dispersed.

The use of nanoparticles with a diameter less than that of the wavelength of light allows transparent nanocomposite materials to be produced. In practice, the particles do reduce the optical transparency compared to the unmodified matrix due to differences in the respective refractive indices, although using 25 nm diameter silica particles does give transmittance of 80 to 90% of the unmodified polymer value [79]. However, once again agglomeration will degrade the performance, as the agglomerates will scatter light and reduce the transparency.

13.4.4 *Electrical Properties*

As polymers are insulating materials, particles are often added to reduce the electrical resistance, e.g., [51, 80, 81]. The critical volume fraction at which the conducting inclusions form an interconnecting filler network leading to a sharp drop in the electrical resistance is known as the percolation threshold. The electrical properties are strongly affected by the filler concentration, the filler morphology (such as particle size and structure) as well as filler–filler and filler–matrix interactions which determine the state of dispersion [23].

For epoxy modified with multiwalled carbon nanotubes (MWNTs), an increase in conductivity above 0.001 wt% suggests that an infinite network of percolated nanotubes starts forming at this concentration [80]. Hence reductions in the electrical resistance can be achieved at very low volume fractions. A similar transition in conductivity occurs for carbon black, but more than 1 wt% of carbon black is required [51, 80]. As the particles must be close enough to each other to allow conduction, the degree of dispersion is important. If agglomeration does occur then much higher volume fractions of nanoparticles would be required to form a percolating network. Nanotube orientation will also influence the electrical conductivity [23], but better-dispersed samples of nanotubes will provide lower percolation thresholds [59] and higher conductivity [81].

When polymers are used as electrical insulators, then the addition of layered silicates improves the surface discharge endurance of the polymer, considerably increasing the electrical breakdown strength and the time before which electrical breakdown occurs [82, 83]. The addition of nonconducting silica or alumina nanoparticles may also reduce the conductivity of the polymer [84, 85]. Such electrical breakdown testing is sensitive to defects within the nanocomposite, as breakdown will occur at the weakest point, for example at an air bubble. Similarly, any agglomerates present may act as electrical stress concentrations and hence reduce performance or promote premature failure [86, 87].

13.4.5 *Fracture Energy*

It is well established that nanoparticles can increase the fracture energy (i.e., the toughness) of polymers. For example, Kinloch and Taylor [66, 88] compared the toughening effect of nanoclays with that due to similar microparticles, in this case mica. Although the maximum fracture energy was higher for the mica microparticles than for the nanoclays, the maximum toughness achieved at much lower volume fractions for the nanoclays. The fracture energy was dependent on type of nanoclay used, and hence on the surface treatment and the resulting morphology. The addition of nanoclays does not always increase the toughness, and may reduce it, for example with thermoset acrylic systems [89].

The effect of dispersion on fracture energy is complex. For example, the fracture energy of vinylester/epoxy systems containing nanoclay was increased more than two times compared to the unmodified polymer [90]. This was 'irrespective to the coarse dispersion of the organophilic layered silicates'. However, there was no significant effect of the type of the surface treatment used on the nanoclay, which was partly attributed to this 'coarse dispersion'.

Spherical nanoparticles have also been used to toughen polymers, for example epoxies. Even very small particles can increase the fracture energy, and 10–20 nm diameter ceramic particles have been commonly used [32, 91–93]. The toughening mechanisms of such nanocomposites have been identified as plastic shear-yielding, and debonding of the polymer from the nanoparticles, followed by plastic void-growth of the epoxy [27]. The toughening mechanisms of crack pinning, crack deflection, and immobilised polymer around the particles have all been discounted [26]. Here the particles are well dispersed.

When large agglomerates are formed and the individual particles are not wetted by the polymer, then the fracture energy of the polymer will be reduced. Kinloch and Taylor [94] showed that the addition of agglomerated nanometre-sized particles reduced the measured fracture energy of a cyanate ester polymer, as shown in Fig. 13.2a. Here the agglomerates were micron-sized (see Fig. 13.2b).

Core-shell rubber particles have also been shown to increase the toughness of polymers, e.g., [95, 96]. These particles are generally produced using a latex emulsion technique and are designed to remain well dispersed, and so the effect of poor dispersion is not clear.

The largest increases in toughness have been reported for 'hybrid' materials, which combine rubber and nanoparticle-modification of an epoxy polymer [32]. However, the addition of rubber can cause the silica nanoparticles used to agglomerate [97]. The material exhibits a considerable synergistic toughening effect, even with silica nanoparticles clustering in the epoxy matrix. It is still uncertain whether this clustering is beneficial or detrimental to the toughness.

13.4.6 *Fatigue Performance*

Nanoparticles have been shown to improve the fatigue performance of thermoset polymers, as measured using linear elastic fracture mechanics. Wetzel et al. [37] showed that the relationship between the fatigue crack propagation (FCP) rate and the stress intensity factor range for a nanocomposite followed the classic Paris–Erdogan law. These fatigue curves were shifted towards higher stress-intensity values as the nanoparticle content was increased. Silica nanoparticles have been shown to significantly improve the cyclic-fatigue behaviour of an epoxy polymer, increasing the range of applied stress-intensity factor at threshold, ΔK_{th} [98]. Manjunatha et al. [99] have shown that this improvement in fatigue performance for well-dispersed nanoparticles can be achieved in both bulk polymer samples and glass fibre reinforced polymer (GFRP) composites. Similarly, the addition of 200 nm diameter rubber particles increases fracture toughness and improves fatigue resistance [95]. Here the particles appear to be arranged in a random dispersion.

Again the effect of dispersion is not clear, and nanoparticles can still improve the performance even if not perfectly dispersed

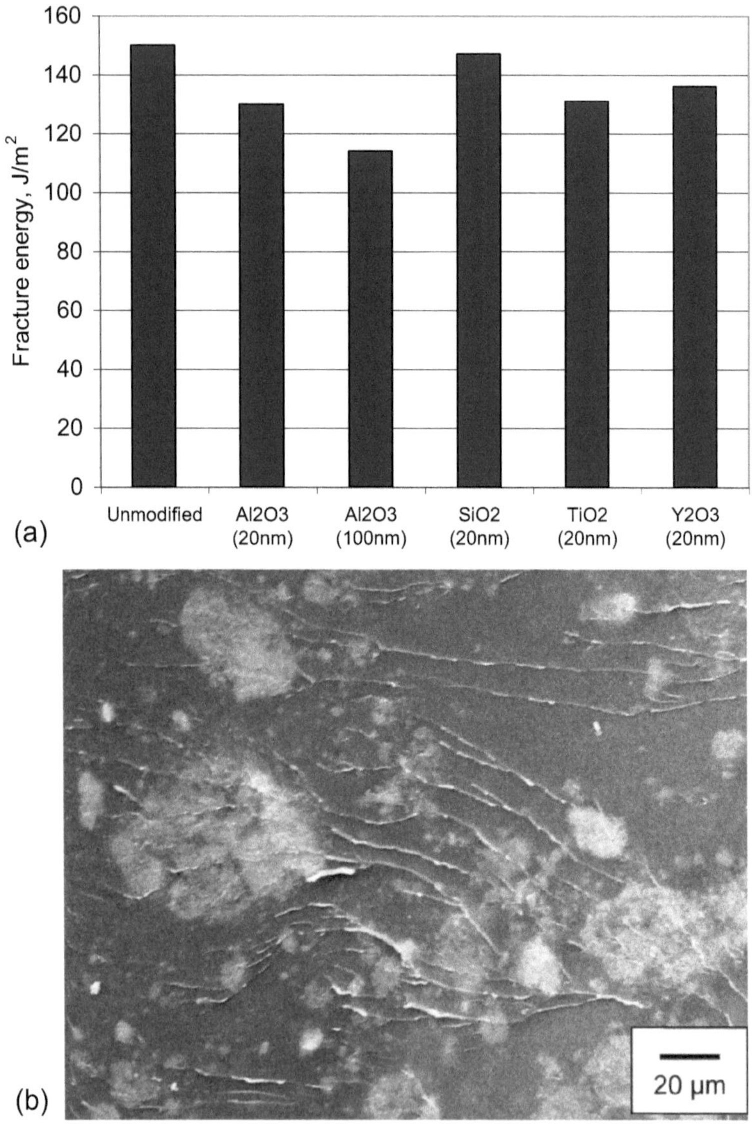

Figure 13.2 Cyanate ester polymer containing various nanoparticles, (a) fracture energy, and (b) scanning electron micrograph of fracture surface of nanocomposite.

[37, 100]. Sohn Lee [101] showed that the maximum increase in the fatigue threshold value was for a hybrid material containing rubber and agglomerated silica nanoparticles. This performed better than materials containing rubber alone or dispersed silica nanoparticles alone.

A reduction in crack growth rate per cycle has also been shown for an epoxy system with carbon nanotubes [102]. This improvement was attributed to pullout of the nanotubes that bridge across the crack as it opens. Since the dispersion quality of the surface-treated MWNT samples was superior to the unmodified MWNTs, then the number density of effective nanotubes that bridge the crack plane is expected to be greater and hence the fatigue performance is improved more, for the same weight fraction of nanotubes.

The effect of agglomeration on the fatigue performance is clearer for unnotched samples. For example, the addition of carbon nanofibres (CNF) to an epoxy polymer significantly increased the number of cycles to failure, as measured using fatigue stress versus number of cycles to failure (S-N) curves [9]. However, when the CNF content was increased further, the material failure initiated at a large agglomeration of nanofibres. The fatigue life was reduced due to the stress concentration caused by the agglomerated CNFs.

13.5 Quantifying Dispersion

13.5.1 *Introduction*

The above discussion shows how the degree of dispersion can affect the performance of a nanocomposite, and shows the importance of assessing the quality of the dispersion of nanoparticles. However, assessing this quality is difficult, as the techniques which can identify particles at the nanoscale such as AFM and TEM involve lengthy specimen preparation. Only a very small volume of the sample can be studied using these techniques, and care must be taken that the results are representative of the material as a whole. Although X-ray diffraction can be used to analyse a larger volume, it does not characterise how well the particles are dispersed. It is normally suggested that dispersion should be assessed at three size scales—(a) optically, (b) by scanning electron microscopy, and (c) by transmission electron microscopy. If the results from all three techniques indicate a good dispersion of nanoparticles, then it is fairly safe to assume that this is the case. Many authors agree that using optical microscopy is useful to investigate the dispersion of nanoparticles as it highlights the presence of agglomerates, e.g., [36, 48], although individual nanoparticles will not be distinguishable. However, this procedure is still qualitative as typically no numerical value which describes the degree of dispersion is produced. Thus there is the problem of subjectivity with the description of dispersion.

Quantitative methods to assess dispersion are crucial to allow different methods of dispersion and different materials to be compared fairly. Describing the dispersion of points within a field is a relatively common statistical problem. Dealing with particles is slightly more complex because they have a volume, and hence a hard-core approach must be used to prevent particles overlapping in space. The expected behaviour of the material can be modelled as ensembles of the geometrical arrangements of these hard-core particles. Any method that is used must also be able to cope with imperfections in the data, and these imperfections must not bias the result. For example, as dispersion is often characterised by identifying particles from micrographs, the chosen method must not be significantly biased if some of the particles are not correctly identified.

In the following sections, some of the methods which may be used to quantify dispersion in a materials science context are discussed. Finally, the authors' preferred method is described.

13.5.2 *Greyscale Method*

Greyscale analysis is carried out using transmission optical microscopy images. Thin cross-sections of the nanocomposite are prepared using petrographic techniques, and the images analysed to attach a greyscale value to each pixel in the image [103]. A histogram of the number of pixels versus the greyscale value is

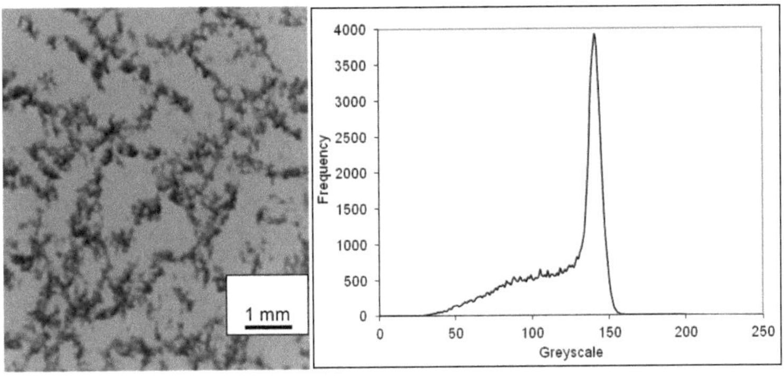

Figure 13.3 Greyscale transmission optical micrograph and corresponding histogram for carbon nanotubes in a blend of epoxy and thermoplastic. Reprinted with permission from Ref. [60].

drawn (Fig. 13.3). The mean greyscale, standard deviation from the mean and variance are then calculated.

As optical microscopy is used, the sample is generally illuminated using unfiltered visible (white) light, which covers wavelengths from about 390 to 780 nm [3], and the spatial resolution for a stereo optical microscope is typically 2 μm [104]. Thus agglomerates of nanoparticles will be visible, but individual nanoparticles will not be seen. However, although individual carbon nanotubes may not be distinguishable due to their small diameter, they will still scatter light.

The most reliable comparison according to Brooker et al. [103] is the ratio of the variance/mean. The value of this ratio takes into account any variation in the overall brightness of the image. Values greater than one indicate a tendency to cluster at this scale of measurement. Higher values indicate a greater degree of clustering; lower values may be described a showing a 'better' distribution of agglomerates or particles. The problem with this method is that it fails to discriminate between very good and very poor dispersion. If the particles are highly agglomerated, then a micrograph can contain no particles and appear the same as one containing very well dispersed nanoparticles. By comparison, other methods where individual particles are identified from higher-resolution micrographs, such as the quadat or area disorder methods, are more robust because the analysis would show clearly that no particles were present, whereas the greyscale method does not.

This method is closely related to the quadrat method described below, as it is essentially a quadrat analysis at one single scale, i.e., the pixel size, assuming that a 2-dimensional image is analysed [103].

13.5.3 *Quadrat Method*

Perhaps the simplest statistical approach that is typically used is the quadrat method [105]. The method originates from geographical methods developed during the Second World War to quantify crop production [103]. The method splits a micrograph into an array of equal-sized rectangular regions, i.e., quadrats. The quadrat size is chosen to be equal to the scale of the smallest object of interest. The number of, or the area occupied by, the objects of interest within each quadrat is determined. This is repeated for conglomerates of quadrats, which may be square areas or lines of neighbouring quadrats along orthogonal axes. The coefficient of variation of particle number is compared against the quadrat size. Deviations away from 1 (which is expected if particles are randomly dispersed) at given quadrat sizes indicate the presence of structure at particular length scales. This method is well known to be influenced by the placement and size increment of the quadrats [106], and hence is not ideal for characterising dispersion.

The quadrat method is straightforward to apply for spherical particles. For high aspect ratio particles such as carbon nanotubes, which will typically cross several quadrats unless a very coarse grid is used, the centre of the nanotube can be identified and used to describe its position. This process ensures that each nanotube is only counted once [103].

A variant of this method has been used by Khare and Burris [107], who used a square quadrat to characterise the unoccupied voids of the system. The quadrat is randomly placed on the micrograph many times and the number of particles found within it is counted. An empirical frequency distribution of particle numbers is generated, and the mode is calculated. The quadrat size is decreased until the mode equals zero, and this size is compared with that expected for a randomly dispersed material.

These methods are simple to use, but there is a common oversight that the results are not truly quantitative, in that it is not well known at what values a system can be reliably described as well or poorly dispersed [106, 108]. What is required is an analysis involving both the measurement and quantitative comparison of the sample material to standard reference behaviours.

13.5.4 *Area Disorder*

It is intuitive that a well-dispersed material will show a relatively homogeneous spread of particles across the system, i.e., the particles are completely random or better still are equally spaced in a lattice-like arrangement. On the other hand, a poorly-dispersed material contains particles which are clustered such that the system is more disordered. This degree of geometric disorder can form the basis of a statistical measure of the degree of dispersion [106].

The particles can be identified from a micrograph, by converting the micrograph into greyscale; suppressing the background noise and then using a greyscale threshold to resolve the individual particles. The method used is described in more detail by Bray et al. [108]. For each particle the centre of mass point is found, and these positions are used to generate a network of triangular cells (a Delaunay network) (Fig. 13.4), assuming periodic boundary conditions at the edges of the micrograph. Each triangle has vertices at the centre of mass points of a triplet of particles that are

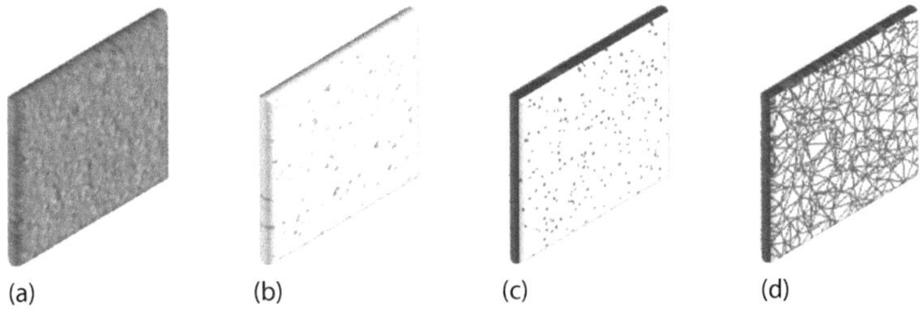

Figure 13.4 Stages of image processing: (a) Original micrograph, (b) identified particles, (c) centre of mass points, and (d) Delaunay network.

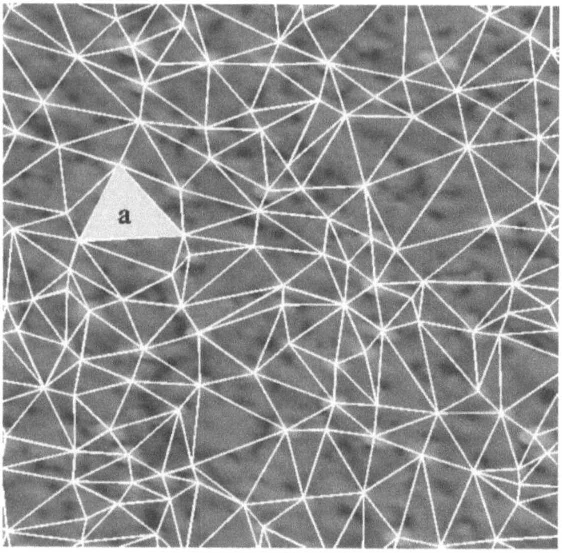

Figure 13.5 Example of the Delaunay network superimposed on an atomic force micrograph of a nanocomposite with an example triangle highlighted, where each triangle has a vertex at the centre of a nanoparticle.

considered to be mutual nearest neighbours (Fig. 13.5). Such a technique can be applied for both spherical [106, 108] and high aspect ratio particles [109]. It can also be used to quantitatively describe the dispersion of agglomerates [108]. Further, one of the more useful features of the Delaunay tessellation methods over the use of quadrats is that length-scale artefacts are not introduced. All particle separation length-scales, up to the size of the micrograph, can potentially contribute to the measurements.

The area disorder of the Delaunay network, AD_{Del}, can be calculated using the mean (σ_a) and standard deviation (μ_a) of the triangle areas using [108]:

$$AD_{Del} = (\mu_a/\sigma_a + 1)^{-1} \tag{13.1}$$

Here using the ratio of standard deviation to mean removes number density effects.

The area disorder is a dimensionless quantity with values between 0 and 1. The degree of dispersion can be characterised by calculating the mean value of the area disorder of the material.

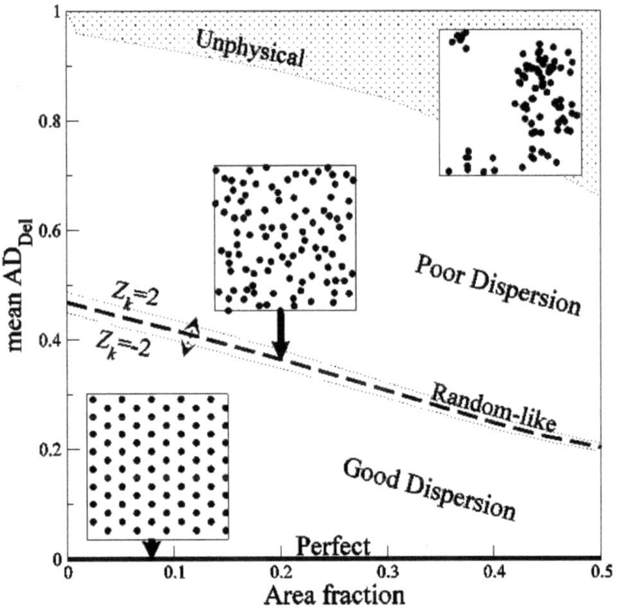

Figure 13.6 Dispersion classification diagram of mean Area Disorder versus area fraction. The insets show example dispersions for lattice-like (bottom), random (middle) or clustered (top) nanocomposites.

A lattice arrangement of particles gives the mean $AD_{Del} = 0$, and implies a perfectly dispersed system. A random arrangement gives a mean area disorder of less than 0.478, with the exact value being dependent upon the area fraction of particles in the micrograph. Note that the area fraction should equal the volume fraction of particles for a planar surface. (If the micrograph is of a thick section viewed in transmission, then there may be a small difference between the area and volume fractions as some nanoparticles will be hidden by particles above them). Larger values of the mean area disorder indicate clustering of the nanoparticles.

The degree of dispersion can be classified using the diagram shown in Fig. 13.6, which plots the mean area disorder versus the area fraction, A_f. The dashed line indicates random dispersion, which is the boundary between good dispersion below this line and poor dispersion above this line. The exact position of this division line can depend on the morphology of the particles, with the values shown here for equi-axed particles. Nevertheless it remains valid for poly-dispersed sizing and moderate aspect ratio particles. For high aspect particles, such as nanofibres, this division may need to be recalculated using measurements from a more realistic model.

As two independent micrographs are unlikely to give identical values for the mean AD_{Del} due to natural variations in the local area fraction, a large number of micrographs could be taken and analysed to calculate a precise mean value. Alternatively, and much more simply, a hypothesis test can be used as described in [108]. This effectively broadens the dashed line indicating random dispersion, as shown in Fig. 13.6 as $Z_k < 2$ and $Z_k > 2$. Within these dotted boundaries the dispersion measurement is indistinguishable from random, although the variation in value for random dispersion is small compared to the overall range of values for the mean AD_{Del}. As the number of nanoparticles analysed increases, then these boundaries contract towards the dashed line.

To allow ranking of the materials to be performed with confidence even though the exact value of the mean behaviour cannot be ascertained from a small set of micrographs, a conservative estimate for the area disorder can be determined using

$$AD_{Del}(\text{sample}) + 2U_{AD} < \text{mean } AD_{Del} < AD_{Del}(\text{sample}) - 2U_{AD} \tag{13.2}$$

where AD(sample) is an estimate based on the sample micrographs and U_{AD} is an upper estimate of the standard deviation based on model prediction [106]. Doing this provides a method by which to obtain reliable estimates for comparisons between materials.

For this measure of dispersion to be robust, it is important that the apparent dispersion of the system is not affected by the data quality. The data quality is dictated by two effects: (1) the size and number of micrographs; (2) the completeness of the data. The size of the micrograph limits the amount of structure that can be seen; i.e., large clusters of nanoparticles cannot be fully seen or may appear more uniform if too small a micrograph is used. Hence the area disorder measurement will reflect this, at small micrograph sizes, with averages differing from the true value obtained from the sampling the whole material. These deviations can be reduced by choosing better averaging techniques such as micrograph stitching, or lower magnifications as long as the individual particles can be identified.

Complete data cannot be assumed as not all of the particles can be detected due to the imperfections of the microscope or the limitations in the image analysis used to identify the particles. In addition, when samples are cut to provide a smooth surface for atomic force microscopy then on average, half of the particles present may be lost as ceramic nanoparticles are indivisible by the knife. Consideration of such incomplete data [108] shows that incompleteness increases the value of the area disorder for well-dispersed systems, such that they appear more random (Fig. 13.7). For clustered systems, the area disorder decreases such that they also appear more random. Random systems continue to appear random. However, well-dispersed materials are still classified as well dispersed, and clustered materials are still classified as clustered even if the data are incomplete [108]. Thus the area disorder is a robust measure of the degree of dispersion for nanocomposites.

13.5.5 *Discussion*

The ability to quantify dispersion is a very powerful tool as materials are statistical by nature. Assuming that there is no interaction between particles (i.e., there is no attraction or repulsion) then a material will show a random dispersion. This implies that there will be some features which may be identified qualitatively as clusters. This has been likened to the classic 'you wait ages for a bus, and then three come along at once' scenario, i.e., some clustering may simply be the result of a random process. As the volume fraction of particles is increased, the area fraction of particles on the micrographs will increase, and the particles will appear qualitatively more clustered. However, quantitative analysis is able to distinguish between the result of a random process and true clustering. Further, the use of the area disorder with the confidence interval allows the degree of dispersion in materials to be ranked successfully.

For equi-axed particles, as discussed above, there is no effect of orientation to consider. However, as the aspect ratio of the particles increases then the orientation of the particles will have an effect

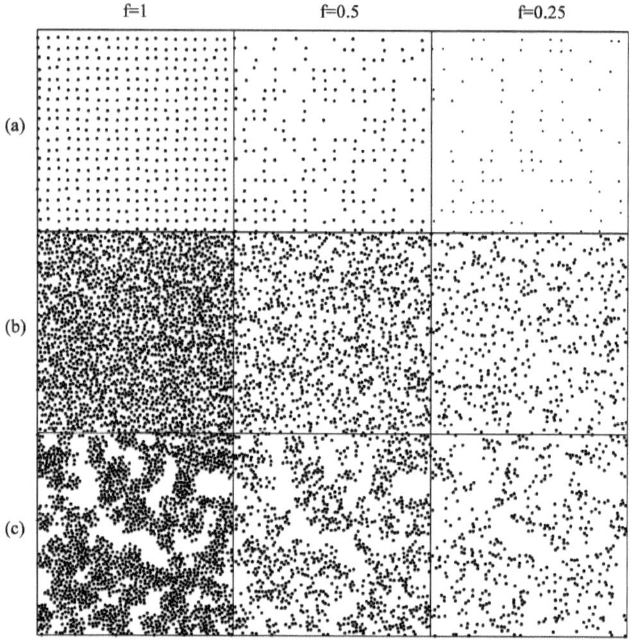

Figure 13.7 The effect of incompleteness of data on apparent dispersion when nanoparticles are arranged (a) as an approximate lattice, (b) at random, and (c) in clusters. Columns show when all ($f = 1$), half ($f = 0.5$) or a quarter ($f = 0.25$) of the nanoparticles are visible.

on the qualitative appearance of the dispersion. As the volume fraction of particles is increased, the particles will tend to become more orientated. Consider a process where particles are placed in random positions and with a random orientation into a field where they are not allowed to overlap the particles already there. If an overlap occurs, then the particle being placed is removed and tried in another random position. Initially the particles can be placed readily, but as the area fraction of particles already placed increases then it becomes more likely that a particle cannot be placed successfully first time. Indeed it becomes increasingly hard to find a free position at all, and at some point there will be no free positions available for more particles to be placed. The orientation of the particles already placed within the field will affect the orientation of those being placed, as it effectively requires less space to place a particle approximately parallel to an existing particle than to place it perpendicular. Hence there is a tendency for the particles to align even if there is no interaction between particles. The material can appear to have a crystalline structure because of this, when it is only the result of a random process. The effect of aspect ratio is significant, as higher aspect ratio particles will show a more pronounced effect. This underlines the importance of using a statistical measure of dispersion such as the area disorder to quantitatively describe the degree of dispersion in materials.

13.6 Conclusions

Nanocomposites are being extensively researched, and are being increasingly applied to commercial products as the price of nanoparticles reduces. Nanoparticles have been shown to improve the structural and functional properties of polymers. Examples include electrical percolation at very low particle volume fractions and increases in barrier properties. However, the properties of the nanocomposite can be affected by the degree of dispersion of the nanoparticles. Indeed poor dispersion can result in agglomerations of nanoparticles which lead to reductions in performance compared to the unmodified polymer rather than any improvement.

The dispersion of nanoparticles in polymer matrices can be difficult. Sonication and high-shear mixing can help, but agglomerates can remain even after aggressive processing. Even if pre-dispersed nanoparticles, available as a masterbatch in a polymer, are used the particles can agglomerate during processing or curing, and there is no guarantee that a good dispersion will be obtained in the final nanocomposite.

It is therefore crucial that the dispersion of the nanoparticles in a nanocomposite can be described quantitatively. The use of a quantitative measure removes the subjectiveness from analysing micrographs. Hence dispersion methods can be compared and the true changes in the material properties can be identified, allowing meaningful comparisons to be made. The use of the area disorder parameter gives such a quantitative classification, distinguishing between good, random and poor dispersion. The area disorder parameter is also robust enough to cope with the incompleteness of data associated with the analysis of micrographs of nanocomposite materials.

Acknowledgements

The authors would like to thank Prof F. J. Guild of Imperial College London for useful discussions.

References

1. Rothon, R. N., and M. Hancock, General principles guiding selection and use of particulate materials, in *Particulate-Filled Polymer Composites*, R. N. Rothon, ed. 1995, Longman Scientific & Technical: Harlow. p. 1–45.
2. Komarneni, S., Nanocomposites. *J. Mater. Chem.*, 1992. **2**(12): 1219–1230.
3. Held, G., Introduction to Light Emitting Diode Technology and Applications. 2009, Boca Raton: Auerbach Publications.
4. Iijima, S., Helical microtubules of graphitic carbon. *Nature*, 1991. **354**(6348): 56–58.
5. Iijima, S., and T. Ichihashi, Single-shell carbon nanotubes of 1-nm diameter. *Nature*, 1993. **363**(6430): 603–605.
6. Tjong, S. C., Structural and mechanical properties of polymer nanocomposites. *Mater. Sci. Eng., R*, 2006. **53**(3–4): 73–197.
7. Lau, K.-T., C. Gu, G.-H. Gao, H.-Y. Ling, and S. R. Reid, Stretching process of single- and multi-walled carbon nanotubes for nanocomposite applications. *Carbon*, 2004. **42**(2): 426–428.
8. Chakrabarti, S., T. Nagasaka, Y. Yoshikawa, L. J. Pan, and Y. Nakayama, Growth of super long aligned brush-like carbon nanotubes. *Jpn. J. Appl. Phys., Part 2*, 2006. **45**(24–28): L720–L722.
9. Zhou, Y., F. Pervin, and S. Jeelani, Effect vapor grown carbon nanofiber on thermal and mechanical properties of epoxy. *J. Mater. Sci.*, 2007. **42**(17): 7544–7553.
10. Green, K., D. Dean, M. Abdalla, M. V. Jose, and P. Robinson, Aligned carbon nanofiber/epoxy nanocomposites. *Polym. Mater. Sci. Eng.*, 2006. **94**: 53–54.

11. Gordon, J. E., *The New Science of Strong Materials or Why You Don't Fall Through the Floor*. 2nd ed. 1978, Harmondsworth: Penguin Books.
12. Liang, G., and X. Hu, Preparation and performance of aluminum borate whisker-reinforced epoxy composites. I. Effect of whiskers on processing, reactivity, and mechanical properties. *J. Appl. Polym. Sci.*, 2004. **92**(3): 1950–1954.
13. Tang, Y., G. Liang, Z. Zhang, and J. Han, Performance of aluminum borate whisker reinforced cyanate ester resin. *J. Appl. Polym. Sci.*, 2007. **106**(6): 4131–4137.
14. Morisada, Y., Y. Miyamoto, Y. Takaura, K. Hirota, and N. Tamari, Mechanical properties of SiC composites incorporating SiC-coated multi-walled carbon nanotubes. *Int. J. Refract. Met. Hard Mater.*, 2007. **25**(4): 322–327.
15. Zhu, Y. Q., H. W. Kroto, D. R. M. Walton, H. Lange, and A. Huczko, A systematic study of ceramic nanostructures generated by arc discharge. *Chem. Phys. Lett.*, 2002. **365**(5–6): 457–463.
16. Bechelany, M., A. Brioude, S. Bernard, G. Ferro, D. Cornu, and P. Miele, Large-scale preparation of faceted Si3N4 nanorods from beta-SiC nanowires. *Nanotechnology*, 2007. **18**(33): 335305.
17. Schadler, L. S., Polymer-based and polymer-filled nanocomposites, in *Nanocomposite Science and Technology*, P. M. Ajayan, L. S. Schadler, and P. V. Braun, eds. 2003, Wiley-VCH: Weinheim. p. 380.
18. Yasmin, A., and I. M. Daniel, Mechanical and thermal properties of graphite platelet/epoxy composites. *Polymer*, 2004. **45**(24): 8211–8219.
19. Yasmin, A., J. J. Luo, and I. M. Daniel, Processing of expanded graphite reinforced polymer nanocomposites. *Compos. Sci. Technol.*, 2006. **66**(9): 1182–1189.
20. Kim, H., A. A. Abdala, and C. W. Macosko, Graphene/polymer nanocomposites. *Macromolecules*, 2010. **43**(16): 6515–6530.
21. ICBA. International Carbon Black Association. http://www.carbon-black.org 2008 [9 April 2008].
22. Shaffer, M. S. P., and I. A. Kinloch, Prospects for nanotube and nanofibre composites. *Compos. Sci. Technol.*, 2004. **64**(15): 2281–2282.
23. Bokobza, L., Multiwall carbon nanotube elastomeric composites: a review. *Polymer*, 2007. **48**(17): 4907–4920.
24. Gojny, F. H., J. Nastalczyk, Z. Roslaniec, and K. Schulte, Surface modified multi-walled carbon nanotubes in CNT/epoxy-composites. *Chem. Phys. Lett.*, 2003. **370**(5–6): 820–824.

25. Kathi, J., and K. Rhee, Surface modification of multi-walled carbon nanotubes using 3-aminopropyltriethoxysilane. *J. Mater. Sci.*, 2008. **43**(1): 33–37.
26. Johnsen, B. B., A. J. Kinloch, R. D. Mohammed, A. C. Taylor, and S. Sprenger, Toughening mechanisms of nanoparticle-modified epoxy polymers. *Polymer*, 2007. **48**(2): 530–541.
27. Hsieh, T. H., A. J. Kinloch, K. Masania, J. Sohn Lee, A. C. Taylor, and S. Sprenger, The toughness of epoxy polymers and fibre composites modified with rubber microparticles and silica nanoparticles. *J. Mater. Sci.*, 2010. **45**(5): 1193–1210.
28. Hsieh, T. H., A. J. Kinloch, K. Masania, A. C. Taylor, and S. Sprenger, The mechanisms and mechanics of the toughening of epoxy polymers modified with silica nanoparticles. *Polymer*, 2010. **51**(26): 6284–6294.
29. Hsieh, T. H., A. J. Kinloch, A. C. Taylor, and S. Sprenger, The effect of silica nanoparticles and carbon nanotubes on the toughness of a thermosetting epoxy polymer. *J. Appl. Polym. Sci.*, 2011. **119**(4): 2135–2142.
30. Giannakopoulos, G., K. Masania, and A. C. Taylor, Toughening of epoxy using core-shell particles. *J. Mater. Sci.*, 2011. **46**(2): 327–338.
31. Mohammed, R. D., Material properties and fracture mechanisms of epoxy nano-composites, in *Mechanical Engineering*. 2007, Imperial College of Science, Technology & Medicine: London.
32. Kinloch, A. J., R. D. Mohammed, A. C. Taylor, C. Eger, S. Sprenger, and D. Egan, The effect of silica nano particles and rubber particles on the toughness of multiphase thermosetting epoxy polymers. *J. Mater. Sci.*, 2005. **40**(18): 5083–5086.
33. Xie, X.-L., Y.-W. Mai, and X.-P. Zhou, Dispersion and alignment of carbon nanotubes in polymer matrix: a review. *Mater. Sci. Eng., R*, 2005. **49**(4): 89–112.
34. Fiedler, B., F. H. Gojny, M. H. G. Wichmann, M. C. M. Nolte, and K. Schulte, Fundamental aspects of nano-reinforced composites. *Compos. Sci. Technol.*, 2006. **66**(16): 3115–3125.
35. Schmid, C. F., and D. J. Klingenberg, Mechanical flocculation in flowing fiber suspensions. *Phys. Rev. Lett.*, 2000. **84**(2): 290–293.
36. Hackman, I., and L. Hollaway, Epoxy-layered silicate nanocomposites in civil engineering. *Composites Part A*, 2006. **37**(8): 1161–1170.
37. Wetzel, B., P. Rosso, F. Haupert, and K. Friedrich, Epoxy nanocomposites - fracture and toughening mechanisms. *Eng. Fract. Mech.*, 2006. **73**(16): 2375–2398.

38. Yasmin, A., J. L. Abot, and I. M. Daniel, Processing of clay/epoxy nanocomposites by shear mixing. *Scr. Mater.*, 2003. **49**(1): 81–86.

39. Yasmin, A., J. J. Luo, J. L. Abot, and I. M. Daniel, Mechanical and thermal behavior of clay/epoxy nanocomposites. *Compos. Sci. Technol.,* 2006. **66**(14): 2415–2422.

40. Gojny, F. H., M. H. G. Wichmann, U. Kopke, B. Fiedler, and K. Schulte, Carbon nanotube-reinforced epoxy-composites: enhanced stiffness and fracture toughness at low nanotube content. *Compos. Sci. Technol.,* 2004. **64**(15): 2363–2371.

41. Chen, L., S. C. Wong, and S. Pisharath, Fracture properties of nanoclay-filled polypropylene. *J. Appl. Polym. Sci.*, 2003. **88**(14): 3298–3305.

42. Dennis, H. R., D. L. Hunter, D. Chang, S. Kim, J. L. White, J. W. Cho, and D. R. Paul, Effect of melt processing conditions on the extent of exfoliation in organoclay-based nanocomposites. *Polymer*, 2001. **42**: 9513–9522.

43. Gershon, A., D. Cole, A. Kota, and H. Bruck, Nanomechanical characterization of dispersion and its effects in nano-enhanced polymers and polymer composites. *J. Mater. Sci.*, 2010. **45**(23): 6353–6364.

44. Patton, R. D., C. U. Pittman Jr., L. Wang, and J. R. Hill, Vapor grown carbon fiber composites with epoxy and poly(phenylene sulfide) matrices. *Composites Part A*, 1999. **30**(9): 1081–1091.

45. Prolongo, S. G., M. Burón, M. R. Gude, R. Chaos-Morán, M. Campo, and A. Ureña, Effects of dispersion techniques of carbon nanofibers on the thermo-physical properties of epoxy nanocomposites. *Compos. Sci. Technol.,* 2008. **68**(13): 2722–2730.

46. Patel, S., A. Bandyopadhyay, A. Ganguly, and A. K. Bhowmick, Synthesis and properties of nanocomposite adhesives. *J. Adhes. Sci. Technol.*, 2006. **20**(4): 371–385.

47. Jana, S. C., and S. Jain, Dispersion of nanofillers in high performance polymers using reactive solvents as processing aids. *Polymer*, 2001. **42**(16): 6897–6905.

48. Brooker, R. D., B. R. K. Blackman, A. C. Taylor, and A. J. Kinloch. Nano-reinforcement of epoxy/thermoplastic blends, in *31st Annual Meeting of the Adhesion Society*. 2008. Austin, USA: Adhesion Society: Blacksburg, USA.

49. Zunjarrao, S., R. Sriraman, and R. Singh, Effect of processing parameters and clay volume fraction on the mechanical properties of epoxy-clay nanocomposites. *J. Mater. Sci.*, 2006. **41**(8): 2219–2228.

50. Gintert, M. J., S. C. Jana, and S. G. Miller, A novel strategy for nanoclay exfoliation in thermoset polyimide nanocomposite systems. *Polymer*, 2007. **48**(14): 4166–4173.

51. Sandler, J., M. S. P. Shaffer, T. Prasse, W. Bauhofer, K. Schulte, and A. H. Windle, Development of a dispersion process for carbon nanotubes in an epoxy matrix and the resulting electrical properties. *Polymer*, 1999. **40**(21): 5967–5971.
52. Lu, K. L., R. M. Lago, Y. K. Chen, M. L. H. Green, P. J. F. Harris, and S. C. Tsang, Mechanical damage of carbon nanotubes by ultrasound. *Carbon*, 1996. **34**(6): 814–816.
53. Lam, C.-K., K.-T. Lau, H.-Y. Cheung, and H.-Y. Ling, Effect of ultrasound sonication in nanoclay clusters of nanoclay/epoxy composites. *Mater. Lett.*, 2005. **59**(11): 1369–1372.
54. Ajayan, P. M., O. Stephan, C. Colliex, and D. Trauth, Aligned carbon nanotube arrays formed by cutting a polymer resin-nanotube composite. *Science*, 1994. **265**(5176): 1212–1214.
55. Baik, Y., S. Lee, Y. Jang, and S. Kim, Unidirectional alignment of Carbon nano-sized fiber using drawing process. *J. Mater. Sci.*, 2005. **40**(22): 6037–6039.
56. Chen, X. Q., T. Saito, H. Yamada, and K. Matsushige, Aligning single-wall carbon nanotubes with an alternating-current electric field. *Appl. Phys. Lett.*, 2001. **78**(23): 3714–3716.
57. Martin, C. A., J. K. W. Sandler, A. H. Windle, M. K. Schwarz, W. Bauhofer, K. Schulte, and M. S. P. Shaffer, Electric field-induced aligned multi-wall carbon nanotube networks in epoxy composites. *Polymer*, 2005. **46**(3): 877–886.
58. Koerner, H., E. Hampton, D. Dean, Z. Turgut, L. Drummy, P. Mirau, and R. Vaia, Generating triaxial reinforced epoxy/montmorillonite nanocomposites with uniaxial magnetic fields. *Chem. Mater.*, 2005. **17**(8): 1990–1996.
59. Ahir, S. V., Y. Y. Huang, and E. M. Terentjev, Polymers with aligned carbon nanotubes: Active composite materials. *Polymer*, 2008. **49**(18): 3841–3854.
60. Brooker, R. D., The morphology and toughness of thermoplastic-modified epoxy polymers with carbon nanotubes or silica nanoparticles, in *Mechanical Engineering*. 2009, Imperial College London: London.
61. Hsieh, T. H., Properties and toughening of silica nanoparticle- and carbon nanotube-modified epoxy polymers, in *Mechanical Engineering*. 2011, Imperial College London: London.
62. Halpin, J. C., and J. L. Kardos, The Halpin-Tsai equations: a review. *Polym. Eng. Sci.*, 1976. **16**(5): 344–352.

63. Reynaud, E., T. Jouen, C. Gauthier, G. Vigier, and J. Varlet, Nanofillers in polymeric matrix: a study on silica reinforced PA6. *Polymer*, 2001. **42**(21): 8759–8768.
64. Ahmed, S., and F. R. Jones, A review of particulate reinforcement theories for polymer composites. *J. Mater. Sci.*, 1990. **25**(12): 4933–4942.
65. Hull, D., *An Introduction to Composite Materials*. 1st ed. Cambridge Solid State Science Series. 1981, Cambridge: Cambridge University Press. p. 256.
66. Kinloch, A. J., and A. C. Taylor, The mechanical properties and fracture behaviour of epoxy-inorganic micro- and nano-composites. *J. Mater. Sci.*, 2006. **41**(11): 3271–3297.
67. Sheng, N., M. C. Boyce, D. M. Parks, G. C. Rutledge, J. I. Abes, and R. E. Cohen, Multiscale micromechanical modeling of polymer/clay nanocomposites and the effective clay particle. *Polymer*, 2004. **45**(2): 487–506.
68. Vanorio, T., M. Prasad, and A. Nur, Elastic properties of dry clay mineral aggregates, suspensions and sandstones. *Geophys. J. Int.*, 2003. **155**(1): 319–326.
69. Triantafyllidis, K. S., P. C. LeBaron, I. Park, and T. J. Pinnavaia, Epoxy-clay fabric film composites with unprecedented oxygen-barrier properties. *Chem. Mater.*, 2006. **18**(18): 4393–4398.
70. Chen, C., and D. Curliss, Resin matrix composites: organoclay-aerospace epoxy nanocomposites, Part II. *SAMPE J.*, 2001. **37**(5): 11–18.
71. Sorrentino, A., G. Gorrasi, M. Tortora, and V. Vittoria, Barrier properties of polymer/clay nanocomposites, in *Polymer Nanocomposites*, Y. W. Mai and Z. Z. Yu, eds. 2006, Woodhead Publishing: Cambridge. p. 594.
72. Gusev, A. A., and H. R. Lusti, Rational design of nanocomposites for barrier applications. *Adv. Mater.*, 2001. **13**(21): 1641–1643.
73. Gilman, J. W., C. L. Jackson, A. B. Morgan, R. H. Harris, E. Manias, E. P. Giannelis, M. Wuthenow, D. Hilton, and S. H. Phillips, Flammability Properties of polymer-layered-silicate nanocomposites. Polypropylene and polystyrene nanocomposites. *Chem. Mater.*, 2000. **12**: 1866–1873.
74. Gilman, J. W., Flammability and thermal stability studies of polymer layered-silicate (Clay) nanocomposites. *Appl. Clay Sci.*, 1999. **15**(1–2): 31–49.
75. Beyer, G., Carbon nanotubes as flame retardants for polymers. *Fire Mater.*, 2002. **26**(6): 291–293.

76. Gilman, J. W., R. H. Harris, and D. Hunter. Cyanate ester clay nanocomposites: synthesis and flammability studies, in *Evolving and Revolutionary Technologies for the New Millennium. 44th International SAMPE Symposium/Exhibition*. 1999. Long Beach, CA, USA: Society for the Advancement of Material and Process Engineering (SAMPE).
77. Gilman, J. W., T. Kashiwagi, M. Nyden, J. E. T. Brown, C. L. Jackson, S. Lomakin, E. P. Giannelis, and E. Manias, Flammability studies of polymer layered silicate nanocomposites: polyolefin, epoxy and vinyl ester resins, in *Chemistry and Technology of Polymer Additives*, S. Ak-Malaika, A. Golovoy, and C. A. Wilkie, eds. 1999, Blackwell Science: Malden, USA. pp. 249–265.
78. Gilman, J. W., R. H. Harris, C. L. Jackson, A. B. Morgan, L. D. Brassell, and D. L. Hunter, Phenolic cyanate ester Nanocomposites: effect of ammonium ion structure on flammability and nano-dispersion, in *Spring Meeting, Division of Polymeric Materials: Science and Engineering, American Chemical Society*. 2000. San Francisco, USA: American Chemical Society.
79. Naganuma, T., and Y. Kagawa, Effect of particle size on the optically transparent nano meter-order glass particle-dispersed epoxy matrix composites. *Compos. Sci. Technol.*, 2002. **62**(9): 1187–1189.
80. Sandler, J. K. W., J. E. Kirk, I. A. Kinloch, M. S. P. Shaffer, and A. H. Windle, Ultra-low electrical percolation threshold in carbon-nanotube-epoxy composites. *Polymer*, 2003. **44**(19): 5893–5899.
81. Song, Y. S., and J. R. Youn, Influence of dispersion states of carbon nanotubes on physical properties of epoxy nanocomposites. *Carbon*, 2005. **43**(7): 1378–1385.
82. Montanari, G. C., F. Ciani, L. Testa, F. Guastavino, G. Coletti, and R. Mulhaupt. Electric strength, space charge and surface discharge characterization of nanostructured epoxy-silicate insulating materials, in *2005 International Symposium on Electrical Insulating Materials*, Vols. 1–3. 2005. Tokyo: Inst Electr Engineers Japan.
83. Imai, T., F. Sawa, T. Nakano, T. Ozaki, T. Shimizu, M. Kozako, and T. Tanaka, Effects of nano- and micro-filler mixture on electrical insulation properties of epoxy based composites. *IEEE Trans. Dielectr. Electr. Insul.*, 2006. **13**(2): 319–326.
84. Cao, Y., P. C. Irwin, and K. Younsi, The future of nanodielectrics in the electrical power industry. *IEEE Trans. Dielectr. Electr. Insul.*, 2004. **11**(5): 797–807.

85. Rouyre, T., A. C. Taylor, M. Fu, F. Perrot, and I. James. Nano-and micro-silica modification of epoxy polymers, in *10th IEEE International Conference on Solid Dielectrics (ICSD)*. 2010. Potsdam: IEEE.
86. Guevara Morales, A., A. C. Taylor, M. Fu, A. Hyde, and F. Perrot, Nanoclay-filled epoxy composites for electrical insulation applications, in *ICPADM 2009: Proceedings of the 9th International Conference on Properties and Applications of Dielectric Materials*, Vols. 1–3, 2009: 868–871.
87. Tanaka, T., G. C. Montanari, and R. Mulhaupt, Polymer nanocomposites as dielectrics and electrical insulation-perspectives for processing technologies, material characterization and future applications. *IEEE Trans. Dielectr. Electr. Insul.*, 2004. **11**(5): 763–784.
88. Kinloch, A. J., and A. C. Taylor, Mechanical and Fracture Properties of Epoxy/Inorganic Micro- and Nano-composites. *J. Mater. Sci. Lett.*, 2003. **22**(20): 1439–1441.
89. Tarrant, A. E., *Thermoset-Acrylic/Layered-Silicate Nanocomposites: Synthesis and Structure-Property Relationships*. 2004, Imperial College London: London, UK.
90. Karger-Kocsis, J., O. Gryshchuk, J. Frohlich, and R. Mulhaupt, Interpenetrating vinylester/epoxy resins modified with organophilic layered silicates. *Compos. Sci. Technol.*, 2003. **63**(14): 2045–2054.
91. Ragosta, G., M. Abbate, P. Musto, G. Scarinzi, and L. Mascia, Epoxy-silica particulate nanocomposites: Chemical interactions, reinforcement and fracture toughness. *Polymer*, 2005. **46**(23): 10506–10516.
92. Zhang, H., Z. Zhang, K. Friedrich, and C. Eger, Property improvements of in situ epoxy nanocomposites with reduced interparticle distance at high nanosilica content. *Acta Mater.*, 2006. **54**(7): 1833–1842.
93. Rosso, P., L. Ye, K. Friedrich, and S. Sprenger, A toughened epoxy resin by silica nanoparticle reinforcement. *J. Appl. Polym. Sci.*, 2006. **100**(3): 1849–1855.
94. Kinloch, A. J., and A. C. Taylor, The toughening of cyanate-ester polymers. part I: physical modification using particles, fibres and woven-mats. *J. Mater. Sci.* 2002. **37**(3): 433–460.
95. Azimi, H. R., R. A. Pearson, and R. W. Hertzberg, Fatigue of rubber-modified epoxies: effect of particle size and volume fraction. *J. Mater. Sci.*, 1996. **31**(14): 3777–3789.
96. Liang, Y. L., and R. A. Pearson, Toughening mechanisms in epoxy-silica nanocomposites (ESNs). *Polymer*, 2009. **50**(20): 4895–4905.

97. Liang, Y. L., and R. A. Pearson, The role of nanosilica dispersion and particle size in hybrid epoxy-silica nanocomposites toughening mechanism, in *31st Annual Meeting of the Adhesion Society*. 2008. Austin, USA: Adhesion Society: Blacksburg, USA.
98. Blackman, B. R. K., A. J. Kinloch, J. Sohn Lee, A. C. Taylor, R. Agarwal, G. Schueneman, and S. Sprenger, The fracture and fatigue behaviour of nano-modified epoxy polymers. *J. Mater. Sci.*, 2007. **42**(16): 7049–7051.
99. Manjunatha, C. M., A. C. Taylor, A. J. Kinloch, and S. Sprenger, The tensile fatigue behaviour of a silica nanoparticle-modified glass fibre reinforced epoxy composite. *Compos. Sci. Technol.*, 2010. **70**(1): 193–199.
100. Zhang, W., R. C. Picu, and N. Koratkar, The effect of carbon nanotube dimensions and dispersion on the fatigue behavior of epoxy nanocomposites. *Nanotechnology*, 2008. **19**(28).
101. Sohn Lee, J., The fatigue behaviour of nano-modified epoxy adhesives, in *Mechanical Engineering*. 2009, Imperial College London: London.
102. Zhang, W., R. C. Picu, and N. Koratkar, Suppression of fatigue crack growth in carbon nanotube composites. *Appl. Phys. Lett.*, 2007. **91**(19): 193109.
103. Brooker, R. D., F. J. Guild, and A. C. Taylor, Quantifying the dispersion of carbon nanotubes in thermoplastic-toughened epoxy polymers. *J. Mater. Sci.*, 2011. **46**(9): 3108–3118.
104. Sawyer, L. C., D. T. Grubb, and G. F. Meyers, *Polymer Microscopy*. 3rd ed. 2008, New York: Springer. p. 540.
105. Guild, F. J., and J. Summerscales, Quantitative microstructural analysis for continuous fibre composites, in *Microstructural Characterisation of Fibre-Reinforced Composites*, J. Summerscales, ed. 1998, Woodhead Publishing: Cambridge.
106. Bray, D. J., S. G. Gilmour, F. J. Guild, and A. C. Taylor, Quantifying nanoparticle dispersion by using the area disorder of Delaunay triangulation. *J. R. Stat. Soc. Ser. C*, 2012. **61**(2): 253–275.
107. Khare, H. S., and D. L. Burris, A quantitative method for measuring nanocomposite dispersion. *Polymer*, 2010. **51**(3): 719–729.
108. Bray, D. J., S. G. Gilmour, F. J. Guild, T. H. Hsieh, K. Masania, and A. C. Taylor, Quantifying nanoparticle dispersion: application of the Delaunay network for objective analysis of sample micrographs. *J. Mater. Sci.*, 2011. **46**(19): 6437–6452.

109. Bakshi, S. R., R. G. Batista, and A. Agarwal, Quantification of carbon nanotube distribution and property correlation in nanocomposites. *Composites Part A*, 2009. **40**(8): 1311–1318.

Chapter 14

Modeling of the Mechanical Properties of Nanoparticle/Polymer Composites*

G. M. Odegard,[a] T. C. Clancy,[b] and T. S. Gates[c]
[a]*Michigan Technological University, Department of Mechanical Engineering – Engineering Mechanics, Houghton, MI 49931, USA*
[b]*National Institute of Aerospace, 144 Research Drive, Hampton, VA 23666, USA*
[c]*NASA Langley Research Center, MS 188E, Hampton, VA 23681, USA*
gmodegar@mtu.edu

A continuum-based elastic micromechanics model is developed for silica nanoparticle/polyimide composites with various nanoparticle/polyimide interfacial treatments. The model incorporates the molecular structures of the nanoparticle, polyimide, and interfacial regions, which are determined using a molecular modeling method that involves coarse-grained and reverse-mapping techniques. The micromechanics model includes an effective interface between the polyimide and nanoparticle with properties and dimensions that are determined using the results of molecular dynamics simulations. It is shown that the model can be used to predict the elastic properties of silica nanoparticle/polyimide composites for a large range of

*Originally published as Odegard, G. M., T. C. Clancy, and T. S. Gates, Modeling of the mechanical properties of nanoparticle/polymer composites. *Polymer*, **46**(2), 553–562 (2005). Used with permission from Elsevier.

Characterization of Nanocomposites: Technology and Industrial Applications
Edited by Frank Abdi and Mohit Garg

ISBN 978-981-4669-02-3 (Hardcover), 978-1-315-36489-6 (eBook)
www.panstanford.com

nanoparticle radii, 10 Å to 10,000 Å. For silica nanoparticle radii above 1,000 Å, the predicted properties are equal to those predicted using the standard Mori–Tanaka micromechanical approach, which does not incorporate the molecular structure. It is also shown that the specific silica nanoparticle/polyimide interface conditions have a significant effect on the composite mechanical properties for nanoparticle radii below 1,000 Å.

14.1 Introduction

In the design and development of unmanned aerial vehicles (UAVs) at the National Aeronautics and Space Administration, the primary requirements are long-duration, high-altitude flights. These requirements necessitate the use of lightweight, durable materials for most of the structural components of these UAVs. Polyimide-based composites are excellent candidates for this purpose because of their resistance to degradation under various environmental conditions. Furthermore, the use of nanostructured reinforcement in polymers has the potential to provide increases in the mechanical properties relative to larger-scale reinforcements that are currently used for aerospace applications, such as carbon fibers [1–3]. To facilitate the development of nanostructured polyimide composite materials for this purpose, constitutive relationships must be developed that predict the bulk mechanical properties of the materials as a function of the molecular structure of the polyimide and reinforcement. These constitutive relationships can be used to influence the design of these materials before they are synthesized.

In the past few years, a considerable amount of research has been conducted to examine the modeling of mechanical properties of polymer composites with nanoscale reinforcement. The majority of this work has focused on carbon nanotube-reinforced polymers [4–9]. Even though it has been shown that these materials have the potential to have excellent mechanical properties, the relatively high costs of development and manufacturing of nanotube/polymer composites has been prohibitive. A lower cost approach is the use of clays and particles in the polymer. Several efforts have focused on the modeling of mechanical properties of nanoclay-reinforced polymer composites [10] and nanoparticle-reinforced polymer composites [11, 12]. These modeling efforts have demonstrated the need for the development of a model that will predict the mechanical properties of nanoparticle/polyimide composites as a function of the nanoparticle size and volume fraction, and the molecular structure of the nanoparticle/polyimide interface.

In the present chapter, a continuum-based constitutive model was developed for silica nanoparticle/polyimide composites with four different nanoparticle/polyimide interfacial treatments. The model incorporates the molecular structure of the nanoparticle, polyimide, and interfacial region. The model was used to examine the elastic properties of the composite as a function of nanoparticle radius, ranging from 6 Å to 1 μm, and particle/matrix interfacial treatments, including untreated nanoparticles, nanoparticles with attached hydroxyl and phenoxybenzene groups, and nanoparticles attached to the polyimide via covalent bonding.

14.2 Materials

The constitutive models developed in this study are for four variations of silica (SiO_2) nanoparticle-reinforced polyimide composites. The silica had an α-quartz crystal structure, and the nanoparticles were nearly spherical in shape, with an approximate radius of 6 Å. The polymer modeled was based on a polyimide from 3,3′,4,4′-biphenyltetracarboxylic dianhydride (BPDA) and 1,3-bis(4-aminophenoxy)benzene (APB) monomers (Fig. 14.1) [13, 14]. The polymer was modeled as having an amorphous molecular structure. The first variation of the composite had a silica nanoparticle without surface treatment that was not bonded to the surrounding polyimide. The second variation had the nanoparticle surface comprised of hydroxyl groups that were bonded to silicon atoms. In this variation, there were no covalent bonds between the polyimide molecules and the nanoparticle. The third variation had phenoxybenzene groups ($-C_6H_4-O-C_6H_5$) chemically bonded to the surface of the nanoparticle, and the phenoxybenzene groups were not directly bonded to the polyimide matrix. The fourth variation had a hydroxylated surface (as in the second variation) with the

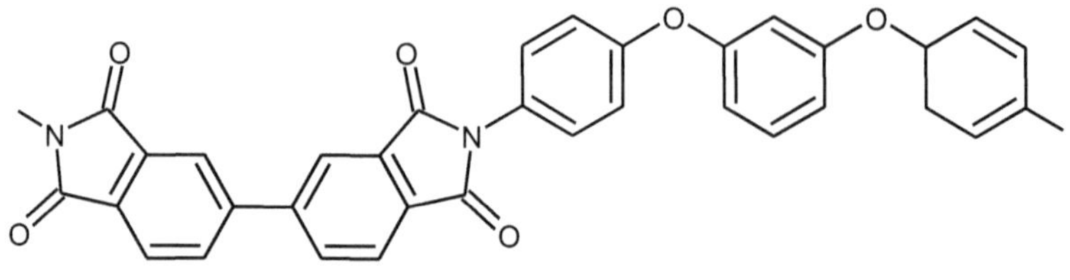

Figure 14.1 Schematic illustration of BPDA (1,3,4) APB polyimide monomer unit.

nanoparticle covalently bonded (functionalized) to the surrounding polyimide molecules. In addition to the four composite systems, the pure silica and the pure polyimide materials were examined.

14.3 Molecular Structure

The first step in establishing structure–property relationships of nanostructured materials is the determination of the molecular structure of the nanoparticle, polyimide matrix, and nanoparticle/polyimide interfacial area. The equilibrium molecular structures of representative volume elements (RVEs) of the six material systems were determined using molecular modeling techniques. Atomistic molecular modeling techniques have been in use for several decades to calculate mechanical properties of polymers [15]. Although modern computational limitations restrict the number of atoms in a typical simulated system, and hence the molecular weight of the polymer chains which can be simulated, mechanical properties consistent with experimental results can usually be obtained. The primary challenge in polymer molecular modeling is the generation of suitably equilibrated atomistic polymer molecular structures. To this end, multi-scale modeling techniques are often used to establish the molecular structures. For this study, the molecular structures were prepared with the aid of a reverse-mapping procedure from a coarse-grained model [16]. The methods are similar to those previously employed for multi-scale polymer modeling [17, 18]. A brief summary of the method is given here.

For each polyimide molecule, a linked vector model was used to represent the rigid rings that comprise the polyimide

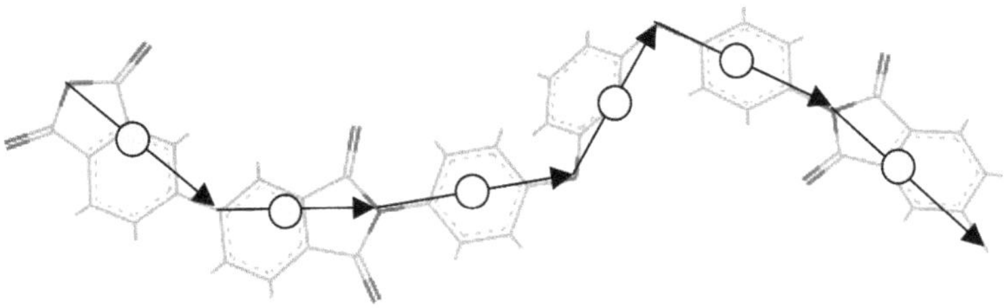

Figure 14.2 Depiction of the mapping of the atomistic polymer model to the coarse-grained linked vector model.

backbone (Fig. 14.2). The linked vectors followed the contour of the molecule. The parameters used for this model consisted of angular distributions between consecutive vectors and long-range forces between beads placed along the midpoint of each vector. These parameters were estimated from molecular dynamics (MD) simulation of the polyimide monomers with the CVFF force field [19]. The centroids of the beads placed at the midpoint of each vector were the centers for interaction forces between non-adjacent beads along the chain of the polymer and between beads on different chains. Additional details on this procedure are presented elsewhere [20].

Once the coarse-grained model for the single polyimide molecule was established, it was subsequently used to assemble the coarse-grained bulk model with multiple polymer molecules and the nanoparticle. The coarse-grained polymers were initially placed as random walk chains inside a simulation box (with periodic boundary conditions) close to their bulk density. In this initial placement, only the angular distributions between adjacent vectors along the chain were considered. Monte Carlo simulation was used to equilibrate the chains from their initial starting configuration. The nanoparticle was modeled as an effective hard sphere. As the Monte Carlo simulation proceeded, the long-range energy interactions established the excluded volume characteristics of the bulk polymer model. The repulsive interaction of the hard sphere with the polymer excluded the polymer chains from this region. The bulk polymer model consisted of seven chains of polymers each composed of ten of the repeat units shown in Fig. 14.1. This chain length is typical for MD simulations of polymers. The simulation box was a cube of side length 42 Å with a hard sphere having a diameter of 15 Å. The periodic box dimensions were chosen to allow the polymer to be close to the equilibrium bulk density. The simulation ran at 650 K until relaxation of the autocorrelation function of the end vectors was achieved and the average centers of mass were displaced a distance greater than the average radii of gyration squared. After sufficient equilibration with the coarse-grained Monte Carlo model, the chains were reverse-mapped by replacing the deleted atoms back into position along the vectors of the coarse-grained model. The nanoparticle was placed in the cavity that resulted from the effective hard sphere in the coarse-grained simulation. The resulting atomistic structures were subsequently minimized by the following procedure. A short energy minimization was applied to the structures. This was followed by constant-pressure MD simulation for 200 ps at 300 K and 1 atm. This constant-pressure MD simulation allowed the atomistic structures to relax to the equilibrium density. The employed algorithm preserved the cubic structure of the simulation box while allowing the box size to change. The final periodic boundary box size varied from 37.6 Å to 39.9 Å on a side depending on the nanoparticle involved. These equilibrated structures were then subjected to a final energy minimization with the criteria of convergence of 0.01 kcal/(mol Å). The functionalized structure was constructed by inserting an oxygen atom covalently bonded to a silicon atom in the nanoparticle and a nearby carbon atom in the polyimide. A total of ten chemical bonds were inserted between the silica particle and the polymer matrix. The structure was then subjected to a molecular mechanics-based energy minimization [21] using the CVFF force field [19]. The pure polyimide matrix was generated using the same procedure as with the composite materials without the hard sphere particle in the coarse-grained simulation. After the constant pressure MD simulation, the density of the pure polyimide matrix was 1.33 g/cm^3. This density is a reasonable value for polyimides [22].

The resulting RVEs of the molecular structures for the four composite systems are shown in Figs. 14.3–14.6. The left-hand sides of Figs. 14.3–14.6 show the silica nanoparticle, and the right-hand sides show the particle in the polyimide matrix. The various

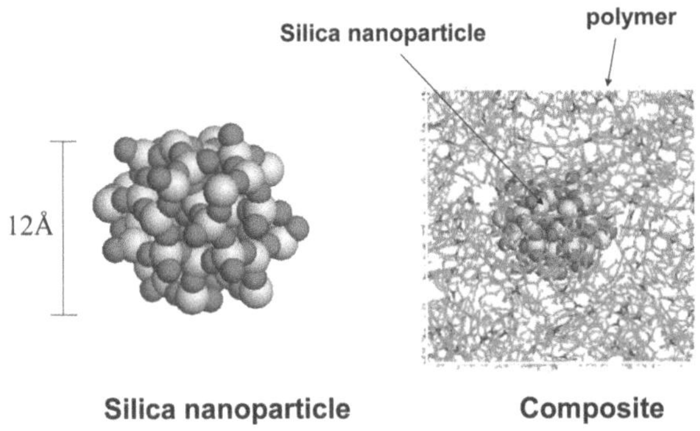

Figure 14.3 Molecular model of silica nanoparticle/polyimide composite.

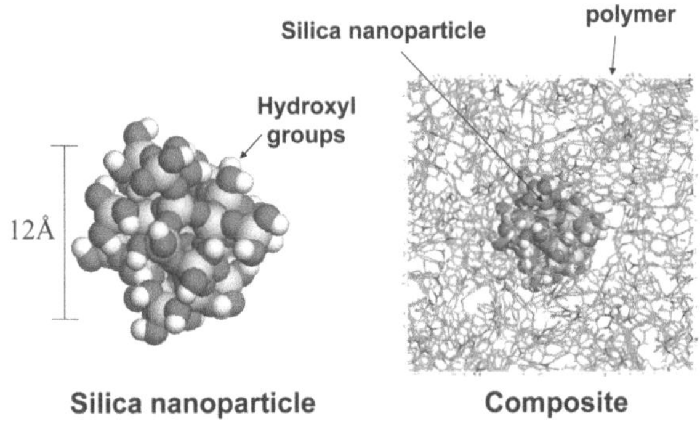

Figure 14.4 Molecular model of hydroxylated silica nanoparticle/polyimide composite.

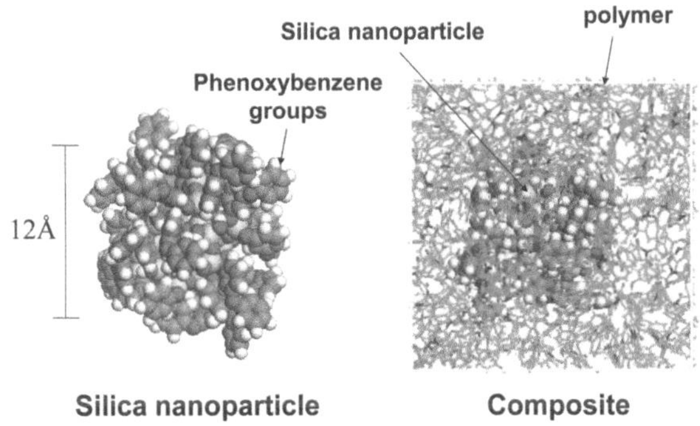

Figure 14.5 Molecular model of phenoxybenzene silica nanoparticle/polyimide composite.

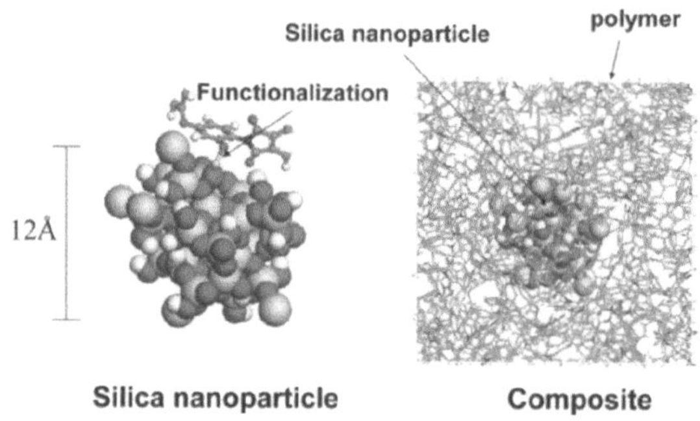

Figure 14.6 Molecular model of functionalized silica nanoparticle/polyimide composite. An example of the bonding to the matrix is illustrated at the top of the nanoparticle.

surface treatments are shown on the nanoparticle in each figure. In Fig. 14.6 an example of the chemical bonding between the nanoparticle and the matrix is shown. For each of the four composite molecular models, the radial density profiles of the nanoparticle and polymer (including the polyimide and interfacial molecules) were also determined and plotted in Fig. 14.7 as a function of the

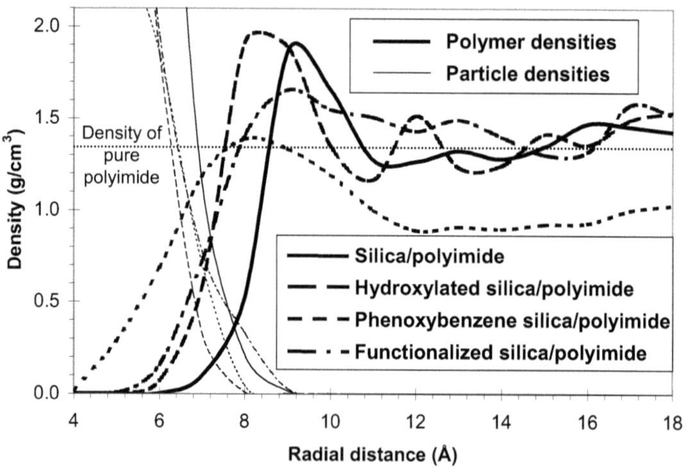

Figure 14.7 Radial density profiles of nanoparticle and polyimide. For the phenoxybenzene silica/polyimide composite, the density includes the polyimide and phenoxybenzene molecules.

radial distance from the center of the nanoparticle. In Fig. 14.7, the full magnitudes of the nanoparticle densities are not shown since only the nanoparticle density drop-off and the polymer densities are necessary in the micromechanical model development in this paper. The density of the core structure of the nanoparticle for all four materials remained at the value of the quartz structure from which it was derived (2.65 g/cm^3).

14.4 Elastic Constants

The mechanical behavior of the molecular systems was described using continuum mechanics. Since the molecular system has a discrete (not continuous) structure, the model is an equivalent-continuum model [23] in which the overall mechanical response of a RVE to an applied set of boundary conditions is equivalent to the response of the molecular system RVE subjected to the same set of boundary conditions. The equivalent-continuum constitutive equation for the materials considered herein is developed below.

It was assumed that the equivalent continuum had a linear-elastic constitutive behavior. The generalized constitutive equation of the equivalent continuum is

$$\sigma_{ij} = C_{ijkl}\varepsilon_{kl} \tag{14.1}$$

where σ_{ij} are the components of the stress tensor ($i, j = 1, 2, 3$), C_{ijkl} are the components of the linear-elastic stiffness tensor, ε_{ij} are the components of the strain tensor, and the summation convention associated with repeated subscripted indices is used. It was further assumed that the composite, pure polymer, and pure silica RVEs had isotropic material symmetry. The composite was assumed to be isotropic because of the presence of spherical reinforcement. Therefore, after expansion, Eq. 14.1 can be expressed as

$$\begin{aligned}
\sigma_{11} &= C_{11}\varepsilon_{11} + C_{12}\varepsilon_{22} + C_{12}\varepsilon_{33} \\
\sigma_{22} &= C_{12}\varepsilon_{11} + C_{11}\varepsilon_{22} + C_{12}\varepsilon_{33} \\
\sigma_{33} &= C_{12}\varepsilon_{11} + C_{12}\varepsilon_{22} + C_{11}\varepsilon_{33} \\
\sigma_{44} &= 1/2\,(C_{11} - C_{12})\,\gamma_{23} \\
\sigma_{55} &= 1/2\,(C_{11} - C_{12})\,\gamma_{13} \\
\sigma_{66} &= 1/2\,(C_{11} - C_{12})\,\gamma_{12}
\end{aligned} \tag{14.2}$$

where $\gamma_{ij} = 2\varepsilon_{ij}$, and C_{kl} are the elastic stiffness tensor components written in the usual contracted notation ($k, l = 1, \ldots, 6$). While the shear modulus of the materials, G, is simply $1/2(C_{11} - C_{12})$, the Young's modulus, E, is determined by inverting the stiffness tensor components, C_{kl}, to determine the components of the compliance tensor, S_{kl}, followed by an inversion of S_{11}.

The elastic constants in Eq. 14.2 were determined using the approach outlined by Theodorou and Suter [15], using the Materials Studio 2.2 [19] software package. With this approach, a set of small, static (i.e., no thermal motion) deformations of the RVE was performed from an equilibrated, non-deformed state. The deformed atomic coordinates were determined through an additional energy minimization step after the deformation was applied. The energy difference between the deformed and undeformed states was used to calculate one elastic constant for each deformation applied. Further details on this method can be found elsewhere [15].

14.5 Micromechanics Models

With the elastic constants of the equivalent-continuum models in hand, two continuum-based micromechanics techniques were used to predict the bulk elastic properties of composites comprised of the polyimide and silica nanoparticles for various effective particle sizes and effective interfaces. Both models are described below.

14.5.1 *Mori–Tanaka Model*

The Mori–Tanaka approach [24, 25] was used to predict the elastic properties of two-phase composites (matrix and effective particle phases) as a function of the effective particle volume fraction and geometry (Fig. 14.8). For this method, the overall elastic-stiffness tensor of the composite containing the isotropic constituents is

$$\mathbf{C} = (c^m\mathbf{C}^m + c^p\mathbf{C}^p\mathbf{T}^p)(c^m\mathbf{I} + c^p\mathbf{T}^p)^{-1} \tag{14.3}$$

where the boldface terms indicate tensor quantities, c^p and c^m are the effective particle and matrix volume fractions, respectively, $\mathbf{C}^p$ and $\mathbf{C}^m$ are the stiffness tensors of the effective particle and matrix,

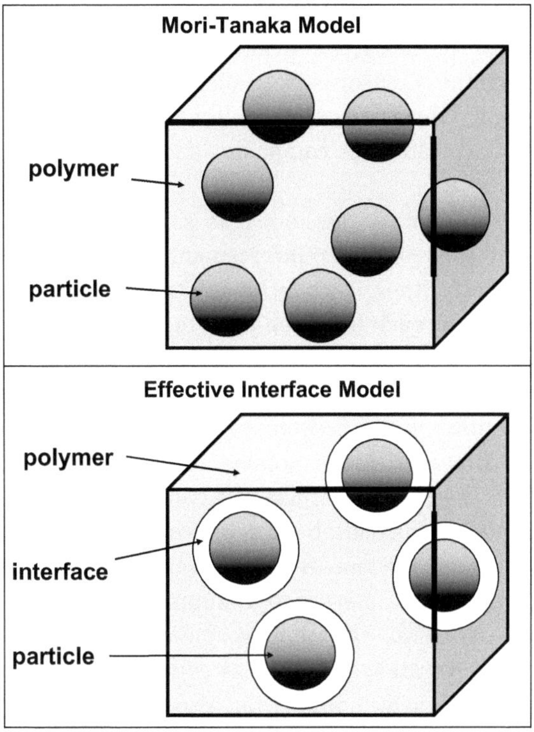

Figure 14.8 Schematic illustration of Mori–Tanaka and Effective Interface micromechanics approaches.

respectively, $\mathbf{I}$ is the identity tensor, and $\mathbf{T}^p$ is the dilute strain-concentration tensor of the effective particles, and is given by

$$\mathbf{T}^p = \left[\mathbf{I} + \mathbf{S}^p\left(\mathbf{C}^m\right)^{-1}\left(\mathbf{C}^p - \mathbf{C}^m\right)\right]^{-1} \tag{14.4}$$

where $\mathbf{S}^p$ is the Eshelby tensor [26]. For spherical effective particle and an isotropic matrix, the components of the Eshelby tensor are [27]

$$\begin{aligned}
S_{1111} &= S_{2222} = S_{3333} = \frac{7-5\nu}{15\left(1-\nu\right)} \\
S_{1122} &= S_{2233} = S_{3311} = S_{1133} = S_{2211} = S_{3322} = \frac{5\nu-1}{15\left(1-\nu\right)} \\
S_{1212} &= S_{2323} = S_{3131} = \frac{4-5\nu}{15\left(1-\nu\right)}
\end{aligned} \tag{14.5}$$

where ν is the Poisson's ratio of the matrix. For a composite with spherical particles, it is evident from Eqs. 14.3–14.5 that the composite stiffness tensor is isotropic. For composites with fibrous reinforcement, which are not considered in this study, Eq. 14.5 has a different form and the composite stiffness tensor is generally anisotropic.

In the Mori–Tanaka method, it is assumed that only the two phases exist (matrix and reinforcement), and they are perfectly bonded to each other. Through a multitude of publications, the Mori–Tanaka approach has been used to accurately predict overall properties of composites when the reinforcements are on the micrometer-scale level, or higher. At these higher length scales, the assumption of the existence of two phases is apparently acceptable. However, for nanometer-sized reinforcement, it has been shown that the molecular structure of the polymer matrix is significantly perturbed at the reinforcement/polymer interface, and this perturbed region is on a length scale that is the same at that of the nanometer-sized reinforcement [9]. Therefore, at the nanometer level, the reinforcement and adjacent polymer region is not accurately described as consisting of just two phases, thus the Mori–Tanaka model is not expected to perform well for nanostructured reinforcements.

14.5.2 *Effective Interface Model*

Because of the aforementioned drawbacks to the Mori–Tanaka approach, another modeling approach was developed. The effective interface model was used to predict the elastic properties of a composite with effective particles that have an interface of the same spherical shape as the effective particle (Fig. 14.8). The effective interface has a finite size and models the region immediately surrounding the spherical reinforcement, which is commonly referred to as an interphase or an interaction zone. The micromechanical model used for this was developed for composites with two-phase particles [28]. For this model, the bulk elastic stiffness tensor is

$$\mathbf{C} = \mathbf{C}^m + \left[\left(c^p + c^i\right)\left(\mathbf{C}^i - \mathbf{C}^m\right)\mathbf{T}^{pi} + c^p\left(\mathbf{C}^p - \mathbf{C}^i\right)\mathbf{T}^p\right] \times \left[c^m\mathbf{I} + \left(c^p + c^i\right)\mathbf{T}^{pi}\right]^{-1} \tag{14.6}$$

where c^i and $\mathbf{C}^i$ are the volume fraction and stiffness tensor for the interface, respectively, and $\mathbf{T}^p$ and $\mathbf{T}^{pi}$ are the dilute strain-concentration tensors given by

$$\mathbf{T}^p = \mathbf{I} - \mathbf{S}^p \left[\mathbf{S}^p + \left(\mathbf{C}^p - \mathbf{C}^m\right)^{-1} \mathbf{C}^m\right]^{-1}$$

$$\mathbf{T}^{pi} = \mathbf{I} - \mathbf{S}^p \left\{ \frac{c^p}{c^i + c^p} \left[\mathbf{S}^p + \left(\mathbf{C}^p - \mathbf{C}^m\right)^{-1} \mathbf{C}^m\right]^{-1} \right.$$
$$\left. + \frac{c^{i\ i}}{c} + c^p \left[\mathbf{S}^p + \left(\mathbf{C}^i - \mathbf{C}^m\right)^{-1} \mathbf{C}^m\right]^{-1} \right\} \qquad (14.7)$$

In Eq. 14.7, the Eshelby tensor, $\mathbf{S}^p$, is given by Eq. 14.5. As with the Mori–Tanaka model, it is evident with the effective interface model the composite stiffness tensor in Eq. 14.6 is isotropic for spherical particle reinforcement.

Similar to the Mori–Tanaka model, the effective interface model assumes that the phases are perfectly bonded to each other. However, the region that physically exists between the surface of the particle and the polymer matrix that has the bulk polymer molecular structure can be considered as an effective interface. The presence of the effective interface allows the region that consists of the perturbed polyimide and interfacial molecules (e.g., phenoxybenzene) to be modeled as a phase unto itself. Unlike the Mori–Tanaka model, the effective interface model should be applicable to both nanometer-sized and larger-sized reinforcement.

At this point, it is appropriate to discuss some of the assumptions in the effective-interface model. In Figs. 14.3–14.7 it is clear that the molecular structure of the interfacial region is neither continuous nor homogeneous. The use of the effective interface model assumes that the equivalent-continuum interfacial region is both continuous and homogeneous. The assumptions of continuity and homogeneity must be made to achieve the objectives of this study. The assumption of material continuity is required in order to develop structure–property relationships within a continuum mechanics-based framework. The assumption of homogeneity within the effective interface is employed here to develop an accurate and efficient equivalent-continuum model for nanostructured composites. As a consequence, the effective interface must be modeled as having a finite size with a discrete transition, even though the actual molecular structure consists of a gradual transition to the bulk molecular structure.

14.6 Results and Discussion

The calculated Young's and shear moduli of the six material systems are shown in Tables 14.1 and 14.2, respectively, determined using the molecular modeling procedure discussed in Sections 14.3 and 14.4. It is clear that the presence of the nanoparticle adversely affected the Young's moduli of all four composite systems and the shear moduli of the three non-functionalized composites systems. Therefore, in these instances, there was no elastic reinforcement of the matrix. For the functionalized composite, the shear modulus was equal to that of the pure polyimide. Furthermore, for the functionalized composite system, the Young's and shear moduli were higher than those of the other three composite systems, indicating that direct functionalization of the silica nanoparticle resulted in improved elastic properties over those obtained with non-functionalized silica nanoparticles. It is noted here that it is assumed that all of the predicted properties correspond to room temperature mechanical properties.

A micromechanics analysis was used to predict the elastic properties of the four composite systems using the Mori–Tanaka approach with the elastic properties of the pure polyimide matrix and silica materials determined with the molecular modeling. The effective nanoparticle volume fraction was chosen to be that of the molecular model RVEs (Figs. 14.3–14.6), which was determined to be 1.7%. The resulting predicted Young's and shear moduli are shown in Tables 14.1 and 14.2, respectively. Because the Mori–Tanaka approach does not consider the molecular structure of the particle/matrix interface, the properties for each of the four composite systems are the same. In contrast to the properties predicted with the molecular modeling, the properties predicted with the Mori–Tanaka approach have Young's and shear moduli that are equal to or larger than those of the polyimide matrix, as expected. Clear discrepancies exist between the values of Young's and shear moduli predicted by the molecular modeling and Mori–

Table 14.1 Young's moduli of the material systems (GPa)

	Silica	Polyimide	Silica composite	Hydroxylated silica composite	Phenoxybenzene silica composite	Functionalized silica composite
Molecular model	88.7	4.2	3.4	3.3	2.2	4.0
Mori–Tanaka	–	–	4.3	4.3	4.3	4.3
% Difference	–	–	26.5%	30.3%	95.5%	7.5%

Table 14.2 Shear moduli of the material systems (GPa)

	Silica	Polyimide	Silica composite	Hydroxylated silica composite	Phenoxybenzene silica composite	Functionalized silica composite
Molecular model	41.0	1.5	1.2	1.2	0.8	1.5
Mori–Tanaka	–	–	1.5	1.5	1.5	1.5
% Difference	–	–	25.0%	25.0%	87.5%	0.0%

Tanaka approach, as is shown in Tables 14.1 and 14.2 with the listed percentage differences. The difference between the two approaches is much smaller for the functionalized silica nanoparticle/polyimide composite.

The data presented in Fig. 14.7 may explain the discrepancies between the predicted elastic properties of the Mori–Tanaka and molecular modeling approaches. For the non-functionalized composites, that is; the silica nanoparticle/polyimide, hydroxylated silica nanoparticle/polyimide, and phenoxybenzene silica nanoparticle/polyimide systems; a 33–40% increase in the polymer density exists at a radial distance of about 8–10 Å, as measured by the peak density value relative to the density at 18 Å. For the functionalized system, only a 6% increase in density is observed. For radial distances between 10 and 15 Å, the initial rise in polymer densities is followed by a decrease in density. For the non-functionalized composites in this radial distance range, the polymer densities associated with the non-functionalized composites are below that of the bulk polymer density (1.33 g/cm^3), before they rise to the bulk density level after a radial distance of at least 15 Å. For the functionalized polymer, the initial rise in density is followed by a gradual decrease in the density that approaches (but does not consistently drop below) the bulk density level. This decrease in polymer density is most likely a result of the initial increase of the density close to the particle and conservation of mass. The decrease of polymer density below the bulk density level (with an expected decrease in stiffness with a decrease in polymer density) and the nature of the molecular interactions between the particle and polymer molecules could be the causes for the substantial decrease in the predicted properties of the polymer when the non-functionalized particles are added. This behavior at the interface agrees with other results found in the literature [12]. These local changes in molecular structure are not incorporated into the Mori–Tanaka model, where it is assumed that the matrix material has a uniform density up to the surface of the effective particle, to which the matrix is perfectly bonded. Therefore, it is expected that the Mori–Tanaka approach would predict an increase in the elastic properties of the polyimides upon reinforcement, whereas the molecular model predicts a decrease in the elastic properties. Furthermore, the closer agreement between the Mori–Tanaka and molecular models for the functionalized system is expected because of the more uniform density profile of the functionalized system, which is evident in Fig. 14.7. Because the density of the functionalized system does not consistently drop below the bulk density, a mechanical reinforcement of the polymer matrix is observed.

From the results presented thus far, continuum-based micromechanics do not accurately predict elastic properties of the silica nanoparticle/polyimide composite system when it is assumed that only two uniform phases exist, as is the case with the Mori–Tanaka approach. To improve the micromechanics formulations for nanoparticle-reinforced composite systems, the change in the molecular structure near the particle/polymer interface must be incorporated into the micromechanics model. At the same time, the micromechanics model needs to be able to accurately predict elastic properties for the composite as the effective particle radius is increased from the nanometer level up to the micrometer level.

In the proposed model, it is assumed that an effective interface existed between the surface of the effective particle and the bulk matrix material. This effective interface included the region that contained the denser polymer region and any interfacial molecules, and the schematic in Fig. 14.9 shows how this effective interface region was defined. For clarity, the graph depicted in Fig. 14.9 is the same data as in Fig. 14.7, without the data labels. The effective interface extended from the surface of the effective nanoparticle, at a radial distance of 6 Å, to a radial distance of 18 Å, which corresponded to the edge of the simulation volume in the molecular modeling. The effective interface was spherical and centered at the center of the effective particle, and had a thickness of 12 Å. In Fig. 14.7 it appears that there is a non-zero polymer density for radii below 6 Å for the phenoxybenzene silica/polyimide composite. Thus, there are small amounts of polymer in the region that is modeled as an effective particle in the effective interface model. The polymer density in this region is relatively small, and results from the non-perfect spherical shape of the nanoparticle in the molecular model. The radius of the effective particle was chosen to approximate the average radii of the nanoparticles in all of

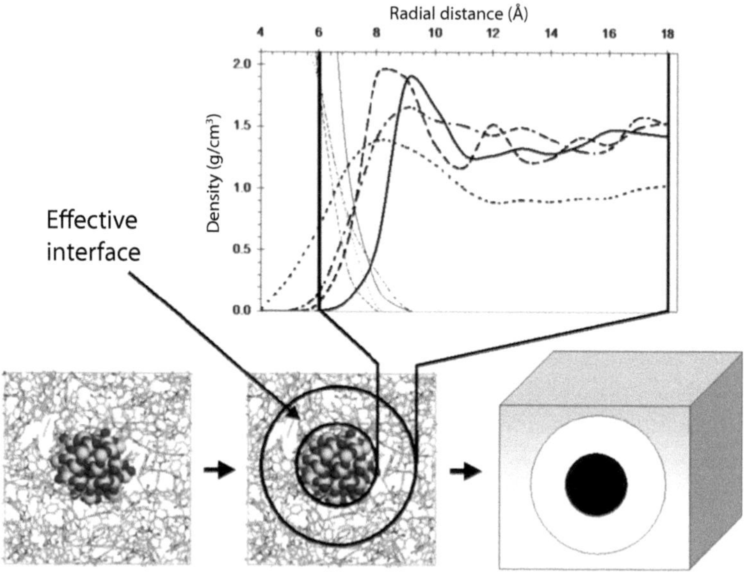

Figure 14.9 Schematic of the process used to determine effective interface (see Fig. 14.7 for graph legend).

the molecular models. It is expected that for MD simulations of phenoxybenzene silica/polyimide composite with larger simulation box sizes, the densities of the polymer would approach those of the bulk polyimide for large radial distances.

The elastic properties of the effective interface were determined using Eqs. 14.6 and 14.7 and the elastic properties of the composites, silica, and pure polyimide determined using the molecular modeling. The volume fractions of the effective particle, effective interface, and matrix were determined based on the volumes shown in Fig. 14.9, and were 1.7%, 45.2%, and 53.1%, respectively. The only unknown parameter in Eqs. 14.6 and 14.7 is the elastic stiffness tensor of the effective interface, $\mathbf{C}^i$. It was assumed that $\mathbf{C}^i$ was isotropic, and thus two independent elastic constants were required to define $\mathbf{C}^i$. Even though the molecular structure of the effective interface region (Fig. 14.7) does not necessarily posses isotropic symmetry, this assumption was made to maintain a reasonable level of simplicity in the modeling. It was also assumed that the

Table 14.3 Effective interface elastic properties (GPa)

	Silica composite	Hydroxylated silica composite	Phenoxybenzene silica composite	Functionalized silica composite
Young's modulus	2.4	2.2	0.3	3.5
Shear modulus	0.9	0.8	0.1	1.3

Poisson's ratio of the effective interface was approximately equal to that of the polyimide, 0.4. The resulting elastic properties of the effective interface are given in Table 14.3. For all four composites, the effective interface elastic properties were lower than those of the polyimide material (Tables 14.1 and 14.2), which is self-consistent with the results discussed above. The elastic properties of the effective interface for the functionalized composite were higher than those of the three non-functionalized composites, indicating that the presence of covalent bonds between the silica and polyimide allowed for an improved load transfer over those of the non-functionalized systems. The elastic properties of the phenoxybenzene silica composite were particularly low, most likely because of the nature of the interactions of the phenoxybenzene groups with the surrounding polyimide.

The Young's and shear moduli of the four composite systems were determined as a function of effective particle size (Figs. 14.10 and 14.11, respectively) using the effective interface model; the elastic properties of the silica, polyimide, and effective interface materials; an effective particle volume fraction of 5%; and a constant effective interface thickness of 12 Å. From both Figs. 14.10 and 14.11, it is clear that as the effective particle radius is increased up to 1000 Å, the Young's and shear moduli increase. Therefore, as the effective particle size increases, and thus the ratio of the volume of the interface to the volume of the effective particle decreases, the effect of the interface on the composite elastic properties diminishes. For further increases in effective particle size beyond 1000 Å, the elastic properties do not increase significantly.

Also included on Figs. 14.10 and 14.11 are the predicted elastic properties determined using the Mori–Tanaka method with the elastic properties of the silica and polyimide materials for an

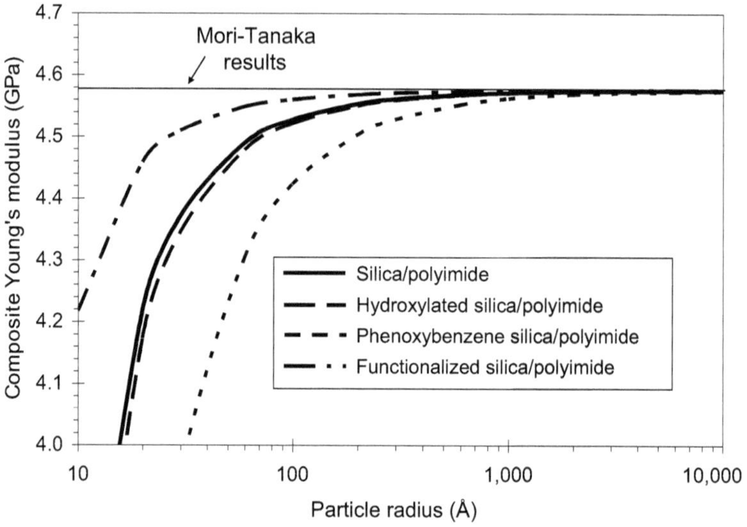

Figure 14.10 Young's modulus of the four composite systems versus effective particle radius.

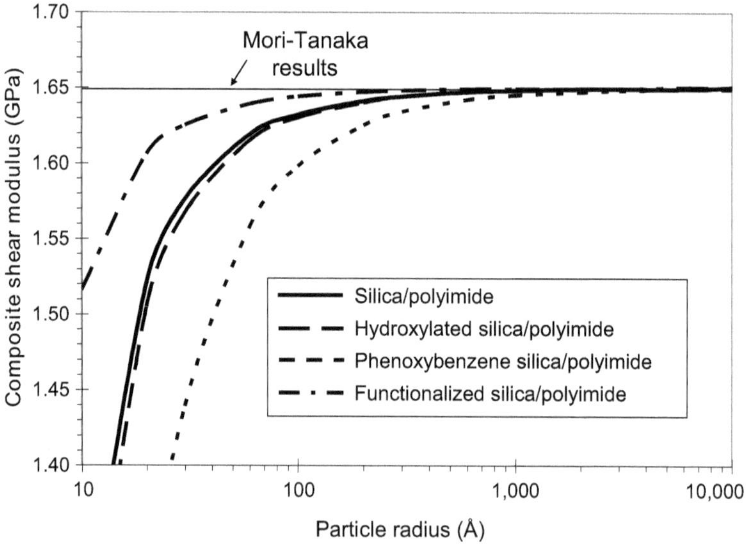

Figure 14.11 Shear modulus of the four composite systems versus effective particle radius.

effective particle volume fraction of 5%. Since the Mori–Tanaka method does not assume the existence of an effective interface, the predicted properties were independent of effective particle size and the particle/polyimide interfacial molecular structure. Comparing all of the curves determined from the two modeling approaches, it is clear that the moduli predicted from the effective interface model approach those predicted from the Mori–Tanaka model as the effective particle size increases. At effective particle radii of about 1000 Å and larger, the predicted Young's moduli and shear moduli of the two models agree. Furthermore, at effective particle radii greater than 1,000 Å, $c^i \rightarrow 0$ and it can be seen that Eqs. 14.6 and 14.7 reduce to Eqs. 14.3 and 14.4, respectively. Therefore, the two models are nearly the same for relatively large effective particle radii.

14.7 Summary and Conclusions

A continuum-based elastic micromechanics model was developed for silica nanoparticle/polyimide composites with various nanoparticle/polyimide interfacial treatments. The model incorporated the molecular structures of the nanoparticle, polyimide, and interfacial regions, which were determined using a molecular modeling method that involved coarse-grained and reverse-mapping techniques. The micromechanics model included an effective interface between the polyimide and effective nanoparticle with properties and dimensions that were determined using the results of MD simulations.

The predicted elastic properties from the effective interface model were compared to those predicted from the Mori–Tanaka method, which does not incorporate the molecular structure of the nanoparticle, polyimide, and interfacial region. The results of the models were determined for four versions of the composite with different nanoparticle/polyimide interface treatments for effective particle radii ranging from 10 Å to 10,000 Å. It was shown that the composite Young's moduli and shear moduli determined with the effective interface model increased with increasing effective particle size, and asymptotically approached the moduli predicted from the Mori–Tanaka method, which were independent of effective particle size. This increase in mechanical properties relates to the diminishing effect of the effective interface as its relative volume, with respect to the volume of the particle, decreases when the nanoparticle size becomes larger. The predicted moduli of the two models converged at an effective nanoparticle radius around 1,000 Å. Therefore, for the silica nanoparticle/polyimide composites studied, while the effective interface model can be used for any effective nanoparticle size, the Mori–Tanaka method will only predict accurate elastic properties for effective particles greater than 1,000 Å.

In addition to the observations on the limits to the micromechanics model, other conclusions can be drawn regarding the material performance. For all four composite systems studied, the elastic properties were lower than those of the polyimide matrix alone when the effective nanoparticle radii were on the order of 10 Å. Based on the predicted density profiles, this loss of properties with the addition of nanoparticles is because of the decrease in the polymer density below the bulk polymer density level and the nature of the nanoparticle/polymer molecular interactions. The loss in the elastic stiffness of the composite was significantly reduced by the introduction of chemical functionalization between the silica nanoparticle and the surrounding polyimide molecules. Clearly, for the material considered in this study, the chemical functionalization strengthened the bonding between matrix and reinforcement. As the effective particle size increased, and the size of the zone of perturbed polymer remained the same, the influence of the perturbed polymer densities diminished, and became insignificant at an effective particle diameter around 1000 Å. Therefore, for the specific materials investigated in this study, nanometer-sized reinforcement does not offer advantages over larger-scale reinforcement in terms of elastic properties of the composite. Due to the sensitivity of the macro-scale elastic properties to changes in molecular structure of the material, the benefits of nano-scale reinforcement must be considered on a case-by-case basis for each combination of nanostructured reinforcement and matrix materials.

References

1. Gomoll, AH, Bellare, A, Fitz, W, Thornhill, TS, Scott, RD, Jemian, PR, and Long, GG. A Nano-Composite Poly(Methyl-Methacrylate) Bone Cement. In: Komarneni, S., Parker, J.C., and Hahn, H., editors. Nanophase and Nanocomposite Materials III: Materials Research Society Proceedings, Vol. 581. Warrendale, PA: Materials Research Society, 2000. pp. 399–404.
2. Thostenson, ET, Ren, Z, and Chou, TW. *Composites Science and Technology* 2001; **61**(13): 1899–1912.
3. Vaia, RA, and Giannelis, EP. *MRS Bulletin* 2001; **26**(5): 394–401.
4. Frankland, SJV, Caglar, A, Brenner, DW, and Griebel, M. *Journal of Physical Chemistry* B 2002; **106**(12): 3046–3048.
5. Bradshaw, RD, Fisher, FT, and Brinson, LC. *Composites Science and Technology* 2003; **63**(11): 1705–1722.
6. Fisher, FT, Bradshaw, RD, and Brinson, LC. *Composites Science and Technology* 2003; **63**(11): 1689–1703.
7. Frankland, SJV, Harik, VM, Odegard, GM, Brenner, DW, and Gates, TS. *Composites Science and Technology* 2003; **63**(11): 1655–1661.
8. Odegard, GM, Frankland, SJV, and Gates, TS. The Effect of Chemical Functionalization on Mechanical Properties of Nanotube/Polymer Composites. 44th AIAA/ASME/ASCE/AHS Structures, Structural Dynamics, and Materials Conference. Norfolk, VA, 2003.
9. Odegard, GM, Gates, TS, Wise, KE, Park, C, and Siochi, E. *Composites Science and Technology* 2003; **63**(11): 1671–1687.
10. Sheng, N, Boyce, MC, Parks, DM, Rutledge, GC, Abes, JI, and Cohen, RE. *Polymer* 2004; **45**(2): 487–506.
11. Smith, GD, Bedrov, D, Li, L, and Byutner, O. *Journal of Chemical Physics* 2002; **117**(20): 9478–9489.
12. Brown, D, Mele, P, Marceau, S, and Alberola, ND. *Macromolecules* 2003; **36**(4): 1395–1406.
13. Srinivas, S, Caputo, FE, Graham, M, Gardner, S, Davis, RM, McGrath, JE, and Wilkes, GL. *Macromolecules* 1997; **30**(4): 1012–1022.
14. Hergenrother, PM, Watson, KA, Smith, JG, Connell, JW, and Yokota, R. *Polymer* 2002; **43**(19): 5077–5093.
15. Theodorou, DN, and Suter, UW. *Macromolecules* 1986; **19**(1): 139–154.

16. Baschnagel, J, Binder, K, Doruker, P, Gusev, AA, Hahn, O, Kremer, K, Mattice, WL, Muller-Plathe, F, Murat, M, Paul, W, Santos, S, Suter, UW, and Tries, V. *Advances in Polymer Science* 2000; **152**: 41–156.
17. Tschop, W, Kremer, K, Batoulis, J, Burger, T, and Hahn, O. *Acta Polymerica* 1998; **49**(2-3): 61–74.
18. Doruker, P, and Mattice, WL. *Macromolecular Theory and Simulations* 1999; **8**(5): 463–478.
19. Materials Studio, Ver. 2.2. Accelrys Inc., San Diego, 2002.
20. Clancy, TC, and Hinkley, J. NASA/TM-2004-213030, 2004.
21. Rappe, AK, and Casewit, CJ. *Molecular Mechanics across Chemistry*. Sausalito, CA: University Science Books, 1997.
22. Ashby, MF, and Jones, DRH. *Engineering Materials 1: An Introduction to Their Properties and Applications*. Oxford, UK: Butterworth-Heinemann, 1996.
23. Odegard, GM, Gates, TS, Nicholson, LM, and Wise, KE. *Composites Science and Technology* 2002; **62**(14): 1869–1880.
24. Mori, T, and Tanaka, K. *Acta Metallurgica* 1973; **21**(5): 571–574.
25. Benveniste, Y. *Mechanics of Materials* 1987; **6**(2): 147–157.
26. Eshelby, JD. *Proceedings of the Royal Society of London, Series A* 1957; **241**: 376–396.
27. Mura, T. *Micromechanics of Defects in Solids*. The Hague: Martinus Nijhoff, 1982.
28. Dunn, ML, and Ledbetter, H. *Journal of Applied Mechanics* 1995; **62**(4): 1023–1028.

Chapter 15

Predicting the Elastic Properties of CNF/Thermoset Polymer Composites Considering the Effect of Interphase and Fiber Waviness

Masoud Rais-Rohani and Mohammad Rouhi
Mississippi State University, Starkville, MS 39762, USA
masoud@ae.msstate.edu, mrouhi@gmail.com

The addition of nanoreinforcements, such as carbon nanofibers (CNFs), to polymer matrix materials has been shown to significantly enhance the overall elastic properties of the resulting material. One of the challenging issues in using these advanced materials in industrial design applications is the accurate modeling of their mechanical responses. These notes are intended as a short introduction to the existing micromechanical modeling approaches for capturing the effect of CNFs in polymer materials while considering the variable properties in the CNF–matrix interphase region and waviness of the nanofibers in CNF/thermoset polymer composites.

Characterization of Nanocomposites: Technology and Industrial Applications
Edited by Frank Abdi and Mohit Garg

ISBN 978-981-4669-02-3 (Hardcover), 978-1-315-36489-6 (eBook)
www.panstanford.com

15.1 Introduction

The addition of nanoreinforcements such as carbon nanofibers (CNFs) to a polymer matrix material has been shown to enhance the overall mechanical properties of the resulting material referred to as nano-enhanced matrix (NEM). In hybrid multiscale composite (HMC) materials, where conventional reinforcing fibers are combined with nano-enhanced matrix, the nanoreinforcements can enhance the interfacial shear strength properties between the conventional fibers and the enhanced matrix (Thostenson et al., 2002; Garg et al., 2008), thus improving the overall stiffness and strength properties of the hybrid composite material (Chisholm et al., 2005; Gojny et al., 2005; Zhou et al., 2008).

One of the challenging issues in using NEM and HMC materials in industrial design applications is the accurate modeling of their mechanical responses. In recent years, many models have been developed to predict the properties of NEM and HMC materials. Molecular dynamics simulation (Odegard et al., 2002, 2003) is one of the most accurate amongst them, but it cannot be used in solving large length-scale problems. On the other hand, finite element simulation with homogenized properties has been used successfully in modeling large length-scale problems, but it cannot accurately capture phenomena present at smaller length scales. Micromechanical methods, which are based on the work of Eshelby (1957), are among the modeling approaches that fall in between the previous two approaches with the capability to capture phenomena in relatively small scales as well as having sufficient computational efficiency to be used in large scale problems. Hybrid methods (Chisholm et al., 2003; Bradshaw et al., 2003) that use both finite element simulation techniques and micromechanical models at different length scales seem to have the advantages of both previously mentioned approaches in terms of both accuracy and computational performance.

Since NEM and HMC are heterogeneous materials with spatially varying properties, statistical average properties are often used to replace the real material by a homogeneous one over a large enough representative volume element (RVE) that is replicated in the domain of the real material. Different effective medium theories, such as micromechanical approaches, try to estimate those average properties. In order to obtain accurate predictions of mechanical responses of NEM and HMC materials, it is important to model the effect of CNF waviness as well as the interphase region between CNF and the matrix.

In this chapter, numerical modeling methods are used to determine the influence of CNF reinforcement on elastic (stiffness) properties of a thermoset polymer matrix (such as vinyl ester) material. Effects of CNF waviness and inhomogeneous properties of the interphase region have been included in modeling the overall elastic properties of NEM using analytical and hybrid techniques.

15.2 Modeling of Nanofiber Enhanced Matrix

For stiffness evaluation of NEM material with randomly oriented CNF, two micromechanical approaches are used. In the first approach (Garg et al., 2008), for the case of unidirectionally aligned nanofibers, the two-dimensional simplified Mori–Tanaka method (Tandon and Weng, 1984) is used to calculate the in-plane stiffness properties of the modified matrix modeled as a transversely isotropic material (material A). The effect of aspect ratio and the volume fraction of the inhomogeneities are also studied for this case. For the calculation of the same in-plane properties for randomly distributed nanofibers, the matrix (material B) is treated as an equivalent quasi-isotropic $[0/45/\text{-}45/90]_S$ laminate made of multiple layers of material A. It should be noted that the laminated composite layers in this stage are not reinforced with conventional fibers. They are reinforced by unidirectional discontinuous nanofibers. The out-of-plane stiffness properties are assumed to be the same as the in-plane properties in the quasi-isotropoic laminate. The calculated stiffness properties of the quasi-isotropic model are compared with the second approach, the general three-dimensional Mori–Tanaka method (Mori and Tanaka, 1973; Benveniste, 1987), which considers the randomness in the orientation distribution of the fibers but requires a more complex analysis.

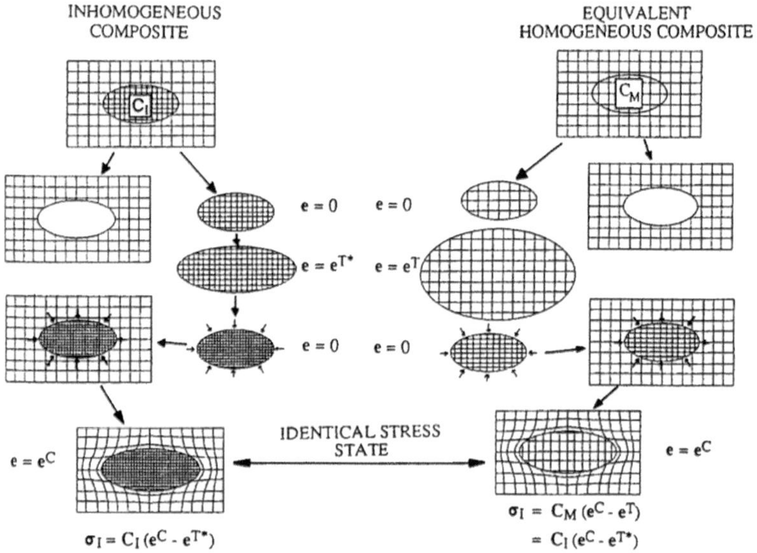

Figure 15.1

15.2.1 *Eshelby Solution*

Eshelby (1957) stated his *transformation problem* as follows:

> *A region (the "inclusion") in an infinite homogeneous isotropic elastic medium undergoes a change of shape and size which, but for the constraint imposed by its surroundings (the "matrix"), would be an arbitrary homogeneous strain. What is the elastic state of the inclusion and matrix?*

He solved the problem using a set of imaginary cutting, straining, and welding operations as shown in Fig. 15.1.

First, the region that is to undergo a transformation (shape and size change) is cut and removed from the matrix. The isolated cut region is allowed to transform in an unconstrained way. A surface traction is then applied to the isolated region to restore it to its original untransformed shape (form and size). In this form, it may be placed back in the original location to rejoin the material across the cut. In this state, stress is zero in the matrix and has a known constant value in the inclusion. It is possible to treat the applied surface traction as a layer of body force distributed over the interface between matrix and inclusion. This unwanted layer needs to be removed to complete the solution. To this end, an equal and opposite layer of body force is applied, introducing an additional elastic field, which may be found by integration from the expression for the elastic field of a point force.

So far, the inclusion may have any shape. However, Eshelby showed that if it is ellipsoidal, the stress within the inclusion is uniform. The details of these calculations can be found in the literature (Philips, 2001; Mura 1987; Khachaturyan, 1983).

Philips (2001) and Mura (1987) invoke the elastic Green function to solve the transformation problem for a single inclusion, while Khachaturyan (1983) describes the generalization of the inclusion problem to multiple inclusions. Eshelby (with his cut, strain, and weld) has shown that the transformation problem is equivalent to solving the equations of elastic equilibrium of a homogeneous body with a known body force distribution.

For the case that the ellipsoidal region in the solid domain has elastic constants different from those of its surrounding domain (if the constants are zero within the ellipsoid, it represents a cavity or void), the ellipsoidal inclusion is called an inhomogeneity and Eshelby called the problem the inhomogeneity problem. A way to solve this problem is to replace the inhomogeneity with an equivalent (hypothetical) inclusion that exerts the same action on its surrounding (matrix) as the original inhomogeneity. Eshelby, in his inclusion and inhomogeneity problems, showed that by knowing only the uniform strain inside the ellipsoid, it is possible to find the following items of physical or engineering interest:

(i) the elastic field far from the inclusion
(ii) all the stress and strain components at a point immediately outside the inclusion
(iii) the total strain energy in matrix and the inclusion
(iv) the interaction energy of the elastic field of the inclusion with another elastic field
(v) the elastic field far from an inhomogeneity

(vi) all the stress and strain components at a point immediately outside the inhomogeneity (solving the problem of stress concentration)
(vii) the interaction energy of the inhomogeneity with an elastic field
(viii) the change in the gross elastic constants of a material when a dilute dispersion of ellipsoidal inhomogeneities is introduced into it

Items (i) to (iv) can also be found for an inclusion of arbitrary shape, (i) and (iv) using simple steps, (ii) and (iii) if one can evaluate the necessary integrals. Items (v) and (viii) can only be found if the shape is ellipsoid.

Changing the aspect ratio of the ellipsoid to its geometrical limits, it is possible to utilize Eshelby solution for inclusions of different geometries such as spheres, platelets, or fibers.

As one of his major contributions, Eshelby showed that if an elastic homogeneous ellipsoidal inclusion, surrounded by an infinite linear-elastic matrix, is subjected to a uniform strain $\boldsymbol{e}^{\mathrm{T}}$ (called the "eigenstrain," "transformation strain," "unconstrained strain," or "stress-free strain"), a uniform strain $\boldsymbol{e}^{\mathrm{C}}$ is induced in the constrained inclusion, which is proportional to the induced unconstrained strain as

$$\mathbf{e}^{C} = \mathbb{S}\,\mathbf{e}^{T} \tag{15.1}$$

where $\mathbb{S}$ is the interior point Eshelby fourth-order tensor that depends solely on the geometry of the inclusion and the Poisson's ratio of the surrounding matrix (Eshelby, 1957; Mura, 1987). The Eshelby tensor $\mathbb{S}$ relates the final constrained inclusion shape to the original shape mismatch between the matrix and the inclusion (Withers et al., 1989). In Eq. 15.1, $\boldsymbol{e}^{\mathrm{T}}$ may be any kind of eigenstrain (e.g., thermal strain) that does not change the elastic constants of the inclusion and is uniform over the inclusion. For inhomogeneities embedded in a matrix, the concept of equivalent homogeneous inclusions is introduced to handle the mean field description of matrix–inhomogeneity composites. Figure 15.1 illustrates the equivalent inclusion procedure for a matrix–inhomogeneity system using Eshelby's cutting and welding exercises. The equivalent inclusion method provides a first-order approximation, resulting in an explicit outcome for the overall stiffness of a composite assuming the particles do not sense each other (the particles' interaction is neglected). This limits its validation only for very small volume fraction of particles (inclusions or inhomogeneities). A better approximation, considering a weak interaction between the particles, is the Mori–Tanaka scheme which is described in the next section.

15.2.2 *General Three-Dimensional Mori–Tanaka Homogenization Scheme*

Mori–Tanaka homogenization scheme (Mori and Tanaka, 1973; Benveniste, 1987) is a micromechanical technique used to approximate the elastic properties of a matrix–inhomogeneity composite material such as NEM. As most mean field theories in continuum micromechanics, Mori–Tanaka is also based on the work of Eshelby (1957).

As stated earlier, one of the advantages of Eshelby-solution-based approaches is that the geometry (aspect ratios) of the ellipsoidal inhomogeneity can be manipulated to obtain the elastic properties of the composite material for different types of inhomogeneities (e.g., cylindrical fibers, elliptic or circular platelets, or spherical reinforcements). It should be noted that in using the Eshelby method for calculating the elastic properties of a composite material, the interaction among inhomogeneities (equivalent inclusions) is ignored, and consequently it only applies to reinforcement volume fractions of less than a few percent ($f \ll 0.1$) called "dilute" matrix–inhomogeneity composites. For the case of relatively high volume fractions, there are several theoretical descriptions of elastic behavior of inhomogeneity-matrix composites that explicitly account for collective interactions among inhomogeneities. The Mori–Tanaka method is one of those micromechanical approaches that consider a weak interaction among the inhomogeneities. It can estimate the elastic properties of a semi-dilute composite case up to a relatively high volume fraction ($f <= 0.3$) (Friedrich, 2005).

So far, one of the challenging issues in manufacturing composite materials with CNFs or other nanoreinforcements has been dispersion and distribution of nanoreinforcements in the matrix. For example, CNFs tend to agglomerate in thermoset polymers resulting in the formation of stress concentrations in the composite material and consequently reducing strength properties, especially in tension (Gojny et al., 2005; Zhou et al., 2008). Such agglomerations limit the value of c so that after a threshold limit, the mechanical properties of the composite are degraded. Several methods (e.g., ultrasonic processor, high shear mixing) (Friedrich et al., 2005) have been used to disperse nanoreinforcements in the matrix, but one of the biggest disadvantages of these methods is that they may break the nanoreinforcements such as CNFs, thus reducing their effectiveness. Since the volume fraction of nanoreinforcements is typically low, micromechanical techniques such as Mori–Tanaka may be used for predicting effective properties of a nanoreinforced material.

For a two-phase composite material, the Mori–Tanaka method estimates the effective stiffness tensor as

$$\mathbb{L}^{MT} = \mathbb{L}_0 + f_1 \{(\mathbb{L}_1 - \mathbb{L}_0)\mathbb{T}\} [f_0\mathbb{I} + f_1 \{\mathbb{T}\}]^{-1} \tag{15.2}$$

where indices 0 and 1 represent matrix and reinforcing phase, respectively, $\mathbb{L}$ is the fourth-order stiffness tensor, f is the volume fraction, $\mathbb{I}$ is the fourth-order identity tensor, and $\mathbb{T}$ relates the uniform strain in an inhomogeneity embedded in an effective continuum to the average matrix strain. The curly brackets $\{\}$ are used to denote physical properties averaged over all possible inhomogeneity orientations. For an ellipsoidal inhomogeneity, $\mathbb{T}$ may be calculated as

$$\mathbb{T} = \left[\mathbb{I} + \mathbb{S}\mathbb{L}_0^{-1}(\mathbb{L}_1 - \mathbb{L}_0)\right]^{-1} \tag{15.3}$$

where $\mathbb{S}$ is the fourth-order interior Eshelby tensor that only depends on the geometry of the inhomogeneity and Poisson's ratio of the matrix. Having the volume fractions (f) and elastic properties ($\mathbb{L}$) of all phases as well as the geometry of the inhomogeneities ($\mathbb{S}$), the micromechanical model can be implemented on the basis of Eqs. 15.2 and 15.3.

In the case of non-aligned discontinuous fiber composites, it is necessary to average the stiffness tensor over all possible fiber orientations. Figure 15.2 shows a rotated ellipsoidal inhomogeneity and the related Euler angles necessary to average the stiffness tensor over all possible directions.

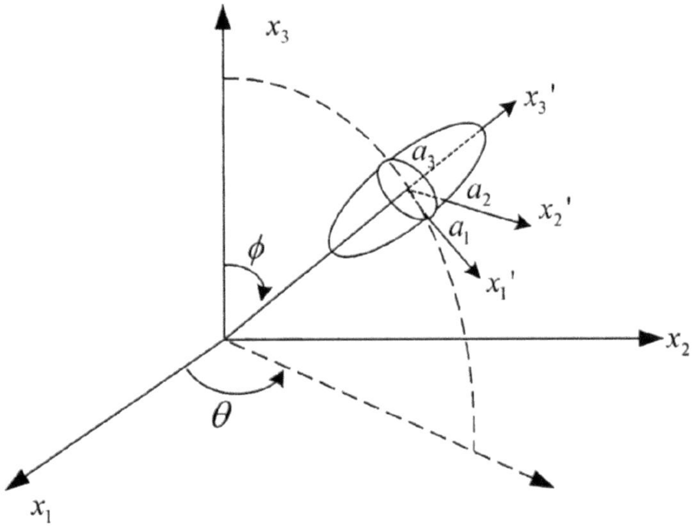

Figure 15.2 A spatially oriented ellipsoidal inhomogeneity (Huang, 2001).

To include the stiffness contribution from all orientations, $(\mathbb{L})$ has to be transformed, multiplied by an orientation distribution function, and integrated over all possible orientations as follows (Mura, 1987; Schjodt-Thomsen and Pyrz, 2001)

$$\{\mathbb{L}_{ijkl}\} = \int_0^{2\pi}\int_0^{2\pi}\int_0^{\pi} g(\phi) a_{ip} a_{jq} a_{kr} a_{ls} \mathbb{L}_{ijkl} \sin(\phi) d\phi d\theta d\beta \quad (15.4)$$

where $g(\varphi)$ is the orientation distribution function and a_{ij} are functions of Euler angles given as

$$a_{ij} = \begin{cases} a_{11} = \cos(\phi)\cos(\theta)\cos(\beta) - \sin(\theta)\sin(\beta) \\ a_{12} = -\cos(\phi)\cos(\theta)\sin(\beta) - \sin(\theta)\cos(\beta) \\ a_{13} = \sin(\phi)\cos(\theta) \\ a_{21} = \cos(\phi)\sin(\theta)\cos(\beta) + \cos(\theta)\sin(\beta) \\ a_{22} = -\cos(\phi)\sin(\theta)\sin(\beta) + \cos(\theta)\cos(\beta) \\ a_{23} = \sin(\phi)\sin(\theta) \\ a_{31} = -\sin(\phi)\cos(\beta) \\ a_{32} = \sin(\phi)\sin(\beta) \\ a_{33} = \cos(\phi) \end{cases} \quad (15.5)$$

Maekawa et al. (1989) used an incomplete beta function as the orientation distribution function, $g(\varphi)$ to account for partially aligned fibers. For the case of random fibers, $g(\varphi)$ is a constant value (Mura, 1987)

$$g(\phi) = \frac{1}{8\pi^2}$$

Let us consider the matrix and fiber properties as those of polypropylene matrix and glass fibers with the following properties: $E_0 = 1.5$ GPa, $\nu_0 = 0.32$, $E_1 = 73$ GPa, $\nu_1 = 0.22$, and the fiber volume fraction, $f_1 = 0.35$. If the glass fibers are assumed to be randomly distributed spherical particles representing an isotropic material, the 6×6 stiffness matrix is calculated as (Schjodt-Thomsen and Pyrz, 2001)

$$L^{\mathrm{MT}} = \begin{bmatrix} 4.03 & 1.67 & 1.67 & 0 & 0 & 0 \\ 1.67 & 4.03 & 1.67 & 0 & 0 & 0 \\ 1.67 & 1.67 & 4.03 & 0 & 0 & 0 \\ 0 & 0 & 0 & 1.18 & 0 & 0 \\ 0 & 0 & 0 & 0 & 1.18 & 0 \\ 0 & 0 & 0 & 0 & 0 & 1.18 \end{bmatrix} \mathrm{GPa}$$

resulting in $E^{\mathrm{MT}} = 3.05$ GPa, $\nu^{\mathrm{MT}} = 0.29$.

For the sake of verification, a computer program based on the general Mori–Tanaka approach was developed in MATLAB and the results for some cases were compared to published data as shown in Fig. 15.3–15.7. As the results reveal, for low volume fractions, the stiffness prediction based on Mori–Tanaka approach is in good agreement with experiments and finite element simulations. It should be noted that for the case of aligned fibers, the orientation averaging (curly brackets in Eqs. 15.2 and 15.4) has not been considered. As a result, the overall stiffness of the composite is as in the case of transversely isotropic materials and varies for different angles (Fig. 15.5).

15.2.3 *Quasi-Isotropic Approximation Method*

For the case of unidirectionally aligned discontinuous inhomogeneities, Tandon and Weng (1984) developed a simplified approach based on the Mori–Tanaka scheme to calculate the effective properties for an enhanced (transversely isotropic) matrix that contains aligned inhomogeneities of known aspect ratio and elastic

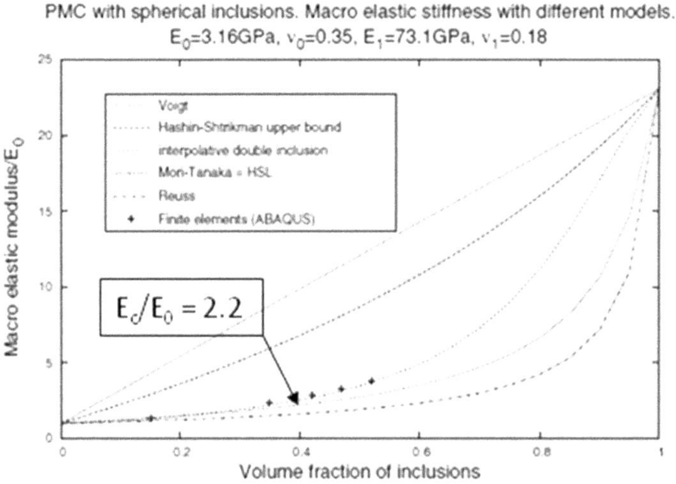

Figure 15.3 Macro elastic modulus with different models for spherical inclusions. $E_0 = 3.16$ GPa, $\nu_0 = 0.35$, $E_1 = 73.1$ GPa, $\nu_1 = 0.18$ (Pierard et al., 2004).

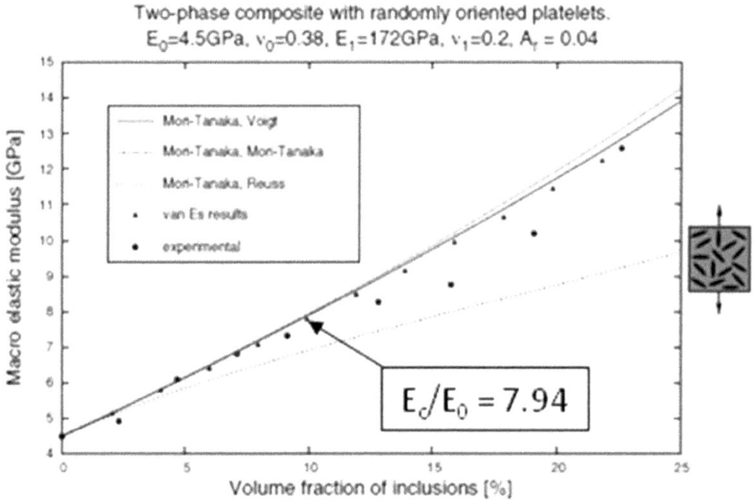

Figure 15.4 Macro elastic modulus with different models for random platelets. $E_0 = 4.5$ GPa, $\nu_0 = 0.38$, $E_1 = 172$ GPa, $\nu_1 = 0.2$, AR $= 0.04$ (Pierard et al., 2004).

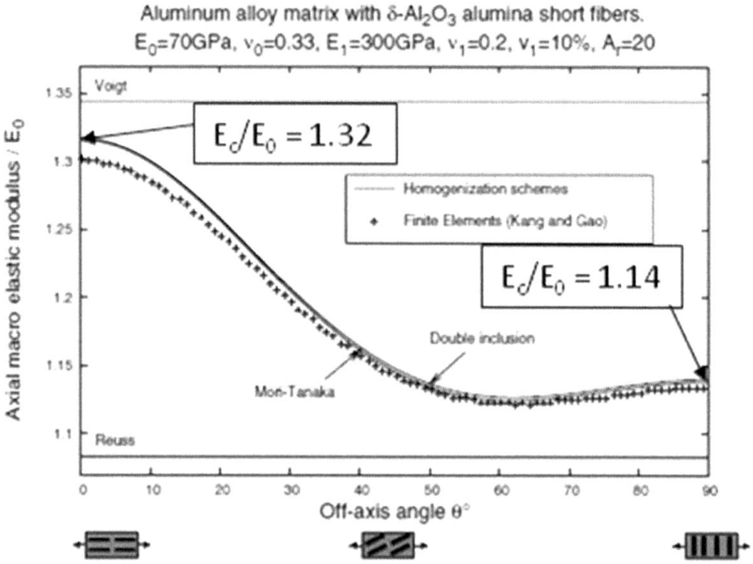

Figure 15.5 Axial macro elastic modulus with different models for short fibers. $E_0 = 70$ GPa, $\nu_0 = 0.33$, $E_1 = 3000$ GPa, $\nu_1 = 0.2$, volume fraction = 0.1, AR = 0.04 (Pierard et al., 2004).

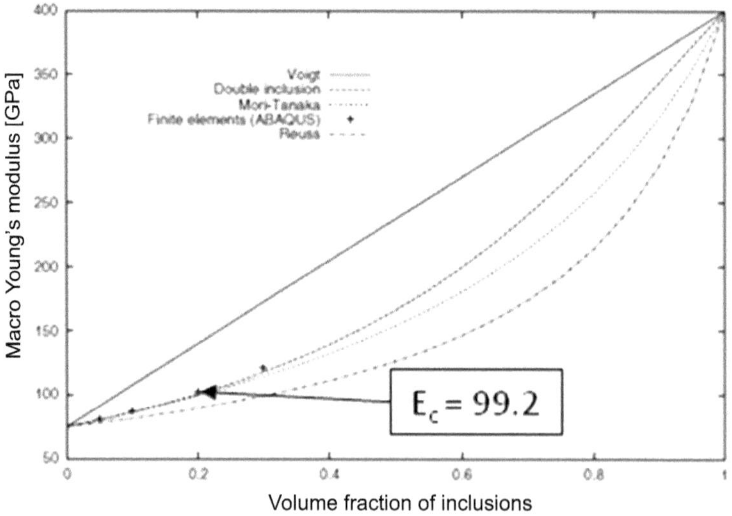

Figure 15.6 Macro elastic modulus with different models for spherical inclusions. $E_0 = 75$ GPa, $\nu_0 = 0.3$, $E_1 = 400$ GPa, $\nu_1 = 0.2$ (Doghri and Ouarr, 2003).

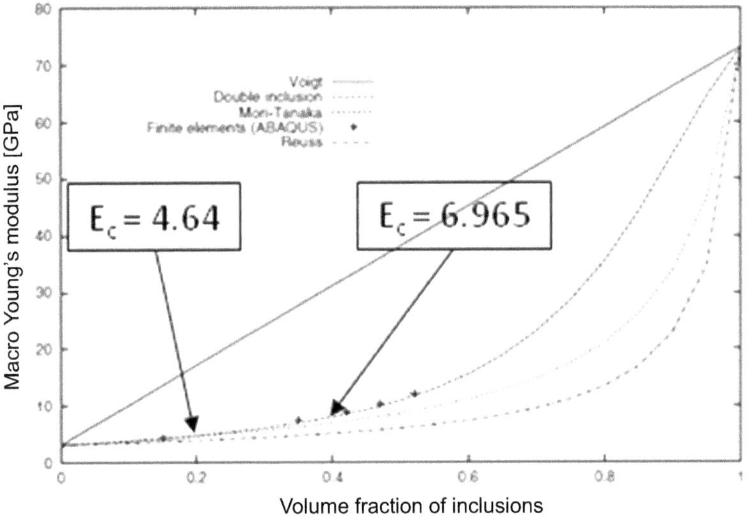

Figure 15.7 Macro elastic modulus with different models for spherical inclusions. $E_0 = 3.16$ GPa, $\nu_0 = 0.35$, $E_1 = 73.1$ GPa, $\nu_1 = 0.18$ (Doghri and Ouarr, 2003).

properties. They showed that for a given small volume fraction of inhomogeneities (CNF) aligned in one direction, the effective or homogenized elastic constants associated with the transversely isotropic composite material can be calculated analytically as (Tandon and Weng, 1984).

$$\begin{aligned}
\frac{E_{11}}{E_0} &= \frac{1}{1 + f(A_1 + 2\nu_0 A_2)/A} \\
\frac{E_{22}}{E_0} &= \frac{1}{1 + f[(-2\nu_0 A_3 + (1-\nu_0)A_4 + (1+\nu_0)A_5 A]/A} \\
\frac{G_{12}}{G_0} &= 1 + \frac{f}{\frac{G_0}{G_1 - G_0} + 2(1-f)S_{1212}} \\
\frac{G_{23}}{G_0} &= 1 + \frac{f}{\frac{G_0}{G_1 - G_0} + 2(1-f)S_{2323}}
\end{aligned} \tag{15.6}$$

where E and G represent the elastic and shear moduli, S_{ijkl}'s are Eshelby's tensor components calculated from the aspect ratio of the inhomogeneity and the Poisson's ratio of the matrix, f is the inhomogeneity volume fraction, ν_0 is the Poisson's ratio of the matrix

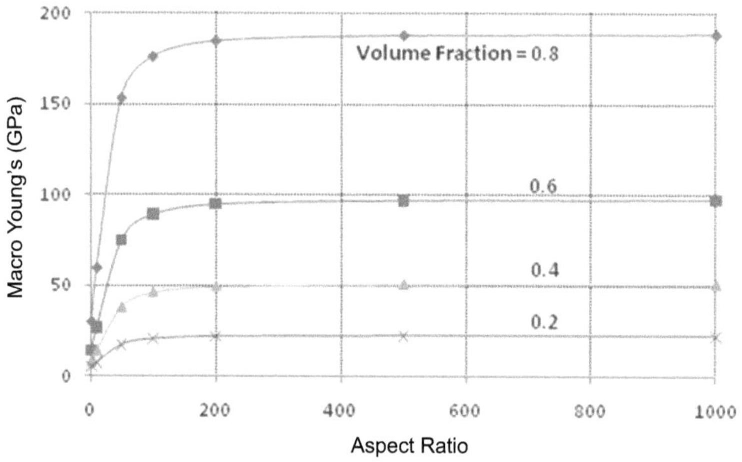

Figure 15.8 The variation of Young's modulus as a function of aspect ratio (Rouhi et al., 2010).

and A's are constants calculated from the elastic properties of the inhomogeneity and the matrix.

The above formulas can be used to calculate the stiffness properties of a thin ply made of aligned NEM material that is transversely isotropic as stated earlier. By stacking a number of such plies in the form of a quasi-isotropic laminate, Garg et al. (2008) approximated the mechanical properties of a two-dimensional randomly oriented carbon nanotube enhanced matrix. Assuming the out-of-plane elastic properties are equal to those in plane, it is possible to approximate the effect of 3-dimensional random orientation of inhomogeneities. The quasi-isotropic laminate we used here is $[0/45/\text{-}45/90]_S$.

For the case of CNF as the nanoreinforcement material and vinyl ester resin as the matrix with the following elastic properties: $E_0 = 3.5$ GPa, $\nu_0 = 0.35$, $E_1 = 450$ GPa, and $\nu_1 = 0.19$, Fig. 15.8 illustrates the variation of the Young's modulus of the NEM as a function of CNF aspect ratio for different volume fractions using the Mori–Tanaka homogenization scheme. The stiffness appears to be highly influenced by the CNF aspect ratio up to a (saturation) limit, after which there is no stiffness enhancement benefit for using CNFs with higher aspect ratios.

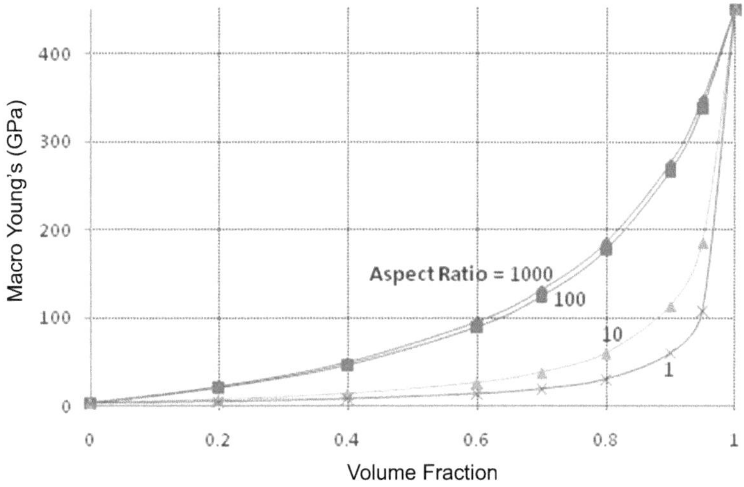

Figure 15.9 The variation of Young's modulus as a function of volume fraction (Rouhi et al., 2010).

Figure 15.9 shows the effect of CNF volume fraction on the overall stiffness of the composite with different CNF aspect ratios. At low volume fractions, the Mori–Tanaka scheme shows a nearly linear enhancement in stiffness with increasing CNF volume fraction that is much lower than that predicted by the simple rule of mixture (upper bound).

Figures 15.10–15.13 illustrate the variation of the elastic properties of unidirectionally aligned CNF composites as a function of CNF volume fraction using the simplified 2D Mori–Tanaka-based approach (Rouhi et al., 2010).

Figures 15.14–15.17 show the effect of aspect ratio variation on longitudinal, transverse, and shear moduli at volume fraction $f = 0.05$. Asymptotic value is reached in each case at aspect ratio of approximately 200.

Having the stiffness properties of the unidirectionally aligned NEM for a specific volume fraction and aspect ratio, a quasi-isotropic laminate $[0/45/\text{-}45/90]_S$ representation (Garg et al., 2008) is used to approximate the in-plane properties of NEM with randomly distributed nanoreinforcements. Assuming the out-of-plane elastic properties equal to the in-plane properties, the effect

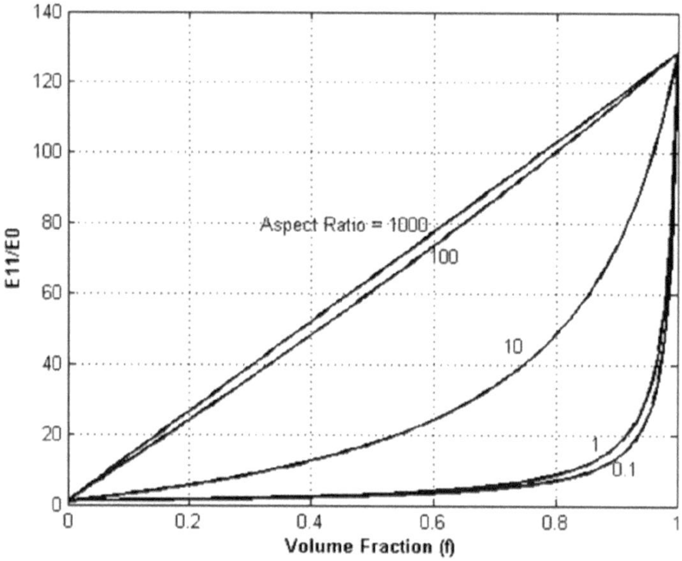

Figure 15.10 The variation of longitudinal Young's modulus as a function of volume fraction.

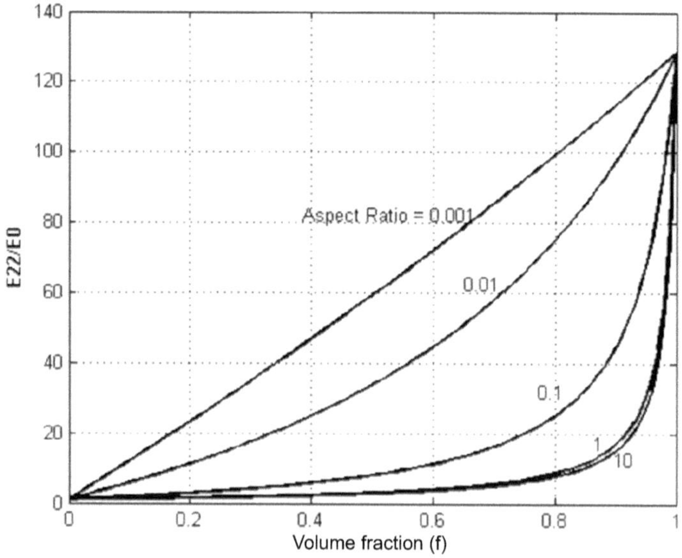

Figure 15.11 The variation of transverse Young's modulus as a function of volume fraction.

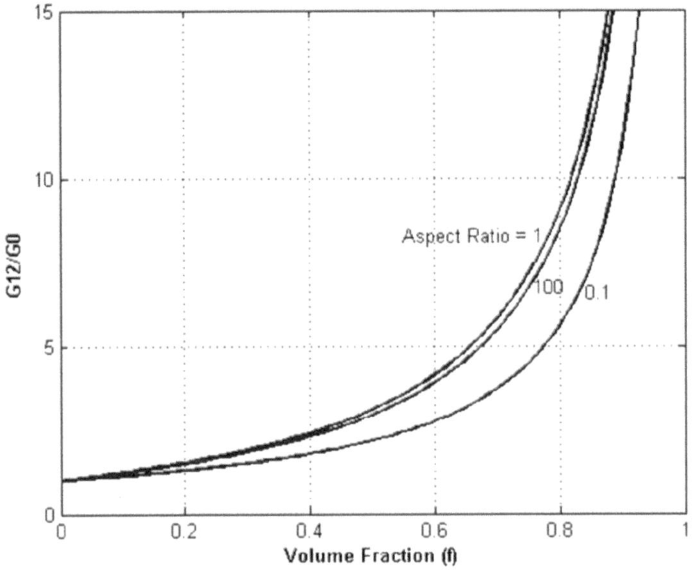

Figure 15.12 The variation of in-plane shear modulus as a function of volume fraction.

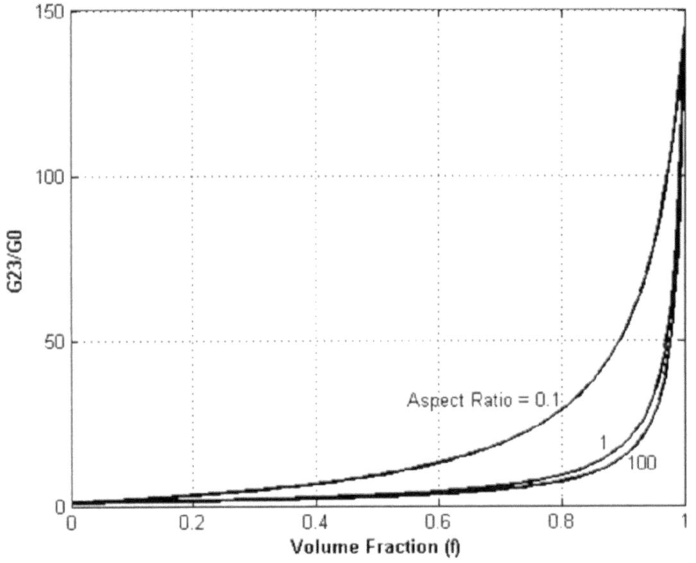

Figure 15.13 The variation of out-of-plane shear modulus as a function of volume fraction.

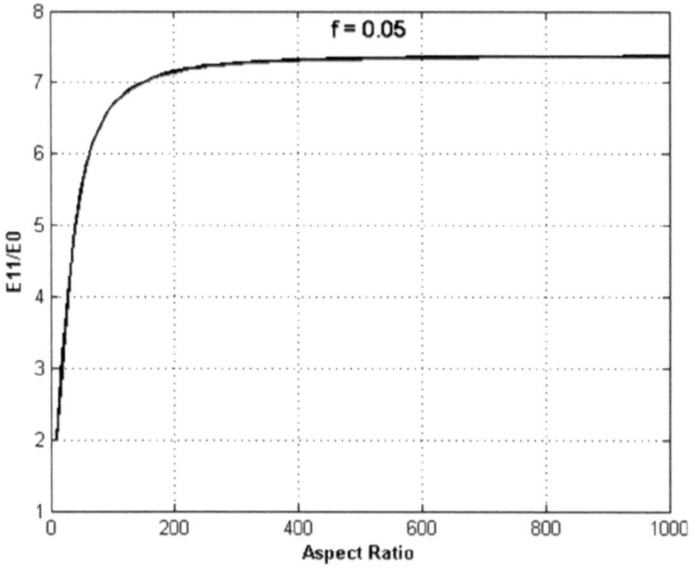

Figure 15.14 The variation of longitudinal Young's modulus as a function of aspect ratio for $f = 0.05$.

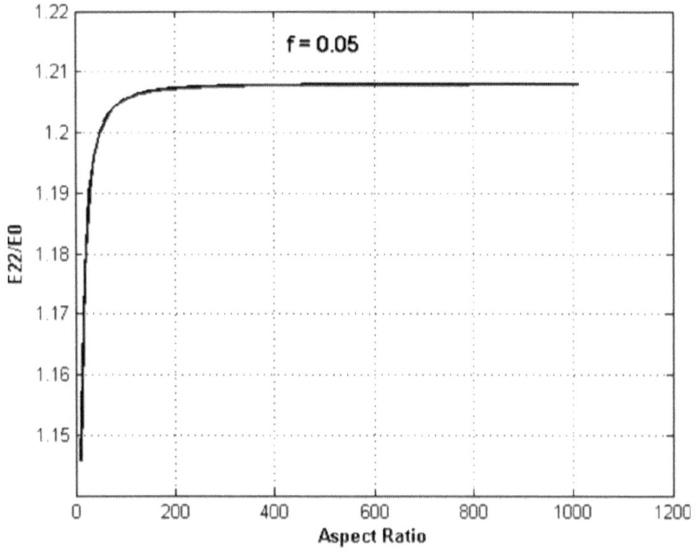

Figure 15.15 The variation of transverse Young's modulus as a function of aspect ratio for $f = 0.05$.

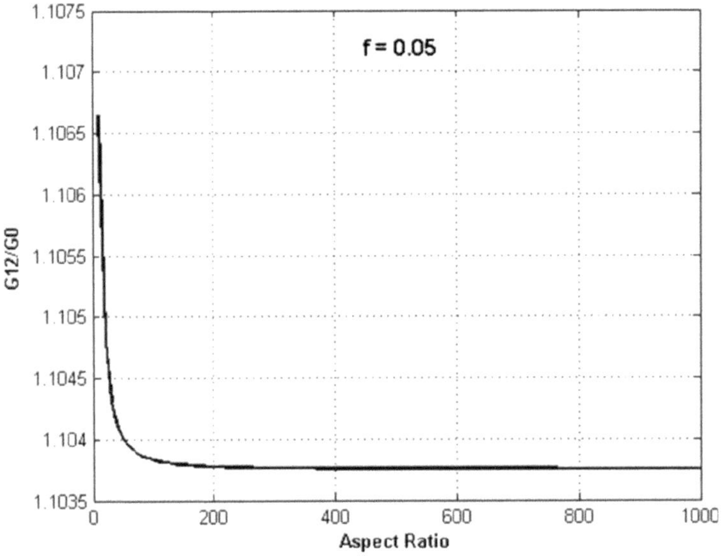

Figure 15.16 The variation of in-plane shear modulus as a function of aspect ratio for $f = 0.05$.

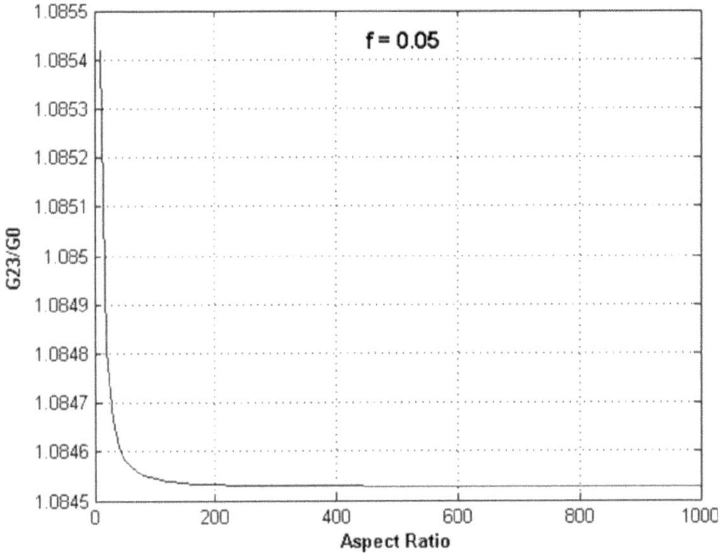

Figure 15.17 The variation of out of plane shear modulus as a function of aspect ratio for $f = 0.05$.

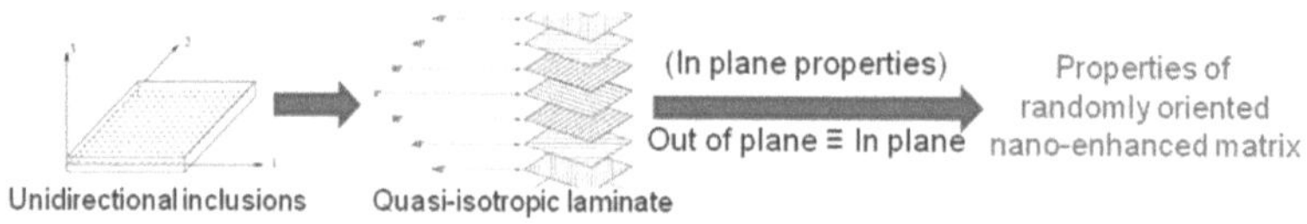

Figure 15.18 Quasi-isotropic approximation procedure.

Table 15.1 Prediction of elastic modulus for different aspect ratios and volume fractions using Mori–Tanaka (M-T) approach and quasi-isotropic (Q-I) laminate approximation (Rouhi et al., 2010)

f	0.01	0.05	0.1	0.2	0.4	AR
$E_{M\text{-}T}$ (Gpa)	3.57	3.88	4.3	5.29	8.23	1
$E_{Q\text{-}I}$ (Gpa)	3.51	3.526	3.55	3.59	3.66	
$E_{M\text{-}T}$ (Gpa)	3.67	4.36	5.32	7.55	14.09	10
$E_{Q\text{-}I}$ (Gpa)	3.88	4.97	6.466	9.91	19.69	
$E_{M\text{-}T}$ (Gpa)	4.23	7.25	11.30	20.60	46.21	100
$E_{Q\text{-}I}$ (Gpa)	4.899	10.36	17.27	31.42	61.31	
$E_{M\text{-}T}$ (Gpa)	4.29	7.56	11.95	21.99	49.46	200
$E_{Q\text{-}I}$ (Gpa)	5.02	10.94	18.37	33.4	64.37	

of three-dimensional randomness in the alignment of CNF in the matrix can be calculated. Figure 15.18 illustrates the procedure of approximating three-dimensional randomness in orientation using the quasi-isotropic laminate approach.

Table 15.1 shows the comparison between the calculated stiffness properties using a general three-dimensional Mori–Tanaka approach and the quasi-isotropic laminate approach. Only at very low volume fractions, the quasi-isotropic laminate approximation results in the Young's modulus that is in reasonable agreement with the one calculated from three-dimensional Mori–Tanaka approach.

Considering the general three-dimensional Mori–Tanaka approach as a more valid approach, one may conclude that the quasi-isotropic approximation technique is a reasonable approximation technique when the volume fraction of the reinforcing materials is very low. In other words, the validity of the quasi-isotropic laminate approach is limited to fiber volume fractions below a certain limit (depending on the nanofibers' aspect ratio) which is much lower than the validity range of Mori–Tanaka approach ($f < 0.3$).

15.3 Modeling of Nanofiber-Matrix Interphase

One of the shortcomings of the micromechanical methods, such as Mori–Tanaka, is the assumption of perfect bonding (no interphase region) between the inclusion (e.g., inhomogeneity, fiber) and the surrounding matrix material. The double-inclusion and multi-inclusion models were developed by Nemat-Nasser and Hori (1993) in an attempt to model the distinct three-dimensional interphase region whose properties may be considerably different from those of the inclusion and matrix materials. Figure 15.19 shows the double-inclusion model and Fig. 15.20 shows a multi-inclusion

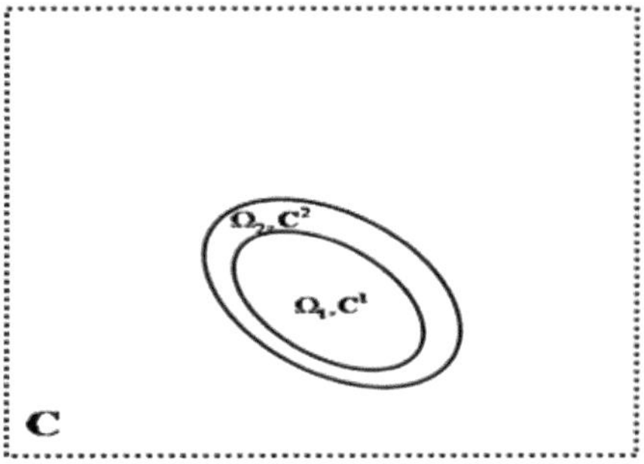

Figure 15.19 A self-similar double-inclusion model.

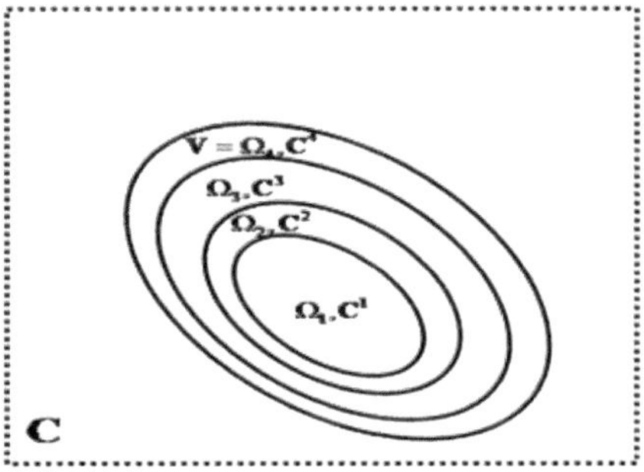

Figure 15.20 A four-phase multi-inclusion embedded in infinite domain.

representation involving four separate phases in an infinite domain of the matrix material. In both models, the overall elasticity of the multi-phase composite region is treated as the volumetric average of the material properties of the individual constituent materials.

In the multi-inclusion model, an ellipsoid V contains a nested series of ellipsoids, Ω_α ($\alpha = 1, 2, \ldots,$ m) such that $\Omega_1 \subset \Omega_2 \subset \ldots \subset \Omega_m \equiv V$, as shown in Fig. 15.20.

Multi-inclusion model estimates the overall properties of the composite as

$$\bar{\mathbb{L}}^{MP} = \mathbb{L} : \{\mathbb{I}^{(4s)} + \sum_{\alpha=1}^{n} f_\alpha(\mathbb{S}^\Omega - \mathbb{I}^{(4s)}) : (\mathbb{A}^\alpha - \mathbb{S}^\Omega)^{-1}\}$$
$$: \left\{\mathbb{I}^{(4s)} + \sum_{\alpha=1}^{n} f_\alpha \mathbb{S}^\Omega : (\mathbb{A}^\alpha - \mathbb{S}^\Omega)^{-1}\right\}^{-1} \quad \mathbb{A}^\alpha \equiv (\mathbb{L} - \mathbb{L}^\alpha)^{-1} : \mathbb{L} \tag{15.7}$$

where $\mathbf{L}$ is the fourth-order stiffness tensor, $\mathbf{I}^{(4s)}$ is the fourth-order identity tensor, f_α is the volume fraction of the α interphase region, $\mathbf{S}$ is the Eshelby tensor that depends on the geometry of the inclusion, and α is the subdivision index.

Since Eq. 15.7 gives the properties for the aligned inclusion case, the resulting stiffness needs to be averaged in all orientations for the case of randomly distributed inclusions (see Eqs. 15.4 and 15.5).

Some studies (Gao and Mader, 2002) show that the interphase region may not be homogeneous but with properties that vary as a function of distance from the vicinity of surface of the inclusions (fibers), as illustrated in Figs. 15.21 and 15.22. If the properties of the interphase region are not homogeneous, the multi-inclusion model can be used to approximate the varying properties of the interphase. As shown in Fig. 15.23, a piecewise-constant-properties approximation technique is used for the interphase region in which the properties are varying similar to functionally graded materials (Reddy, 2000; Ruhi et al., 2005).

One way to model the variation of an interphase property (e.g., Young's modulus, Poisson's ratio) from the surface of the inclusion to the outmost layer of the interphase is by using (Reddy, 2000; Ruhi et al., 2005)

$$P = P_{\text{in}} + (P_{\text{out}} - P_{\text{in}})\left(\frac{X}{N}\right)^n \tag{15.8}$$

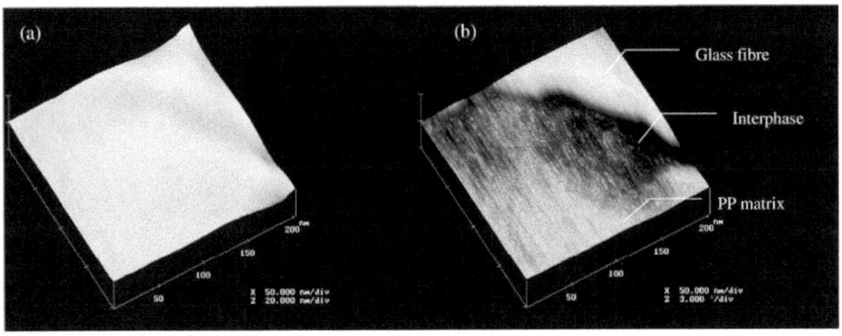

Figure 15.21 Atomic force microscopy of three-dimensional (a) topographic and (b) phase images of interphase in aminopropyltriethoxysilane/polyurethane (APS/PU) sized glass fiber/PP matrix composites (Gao and Mader, 2002).

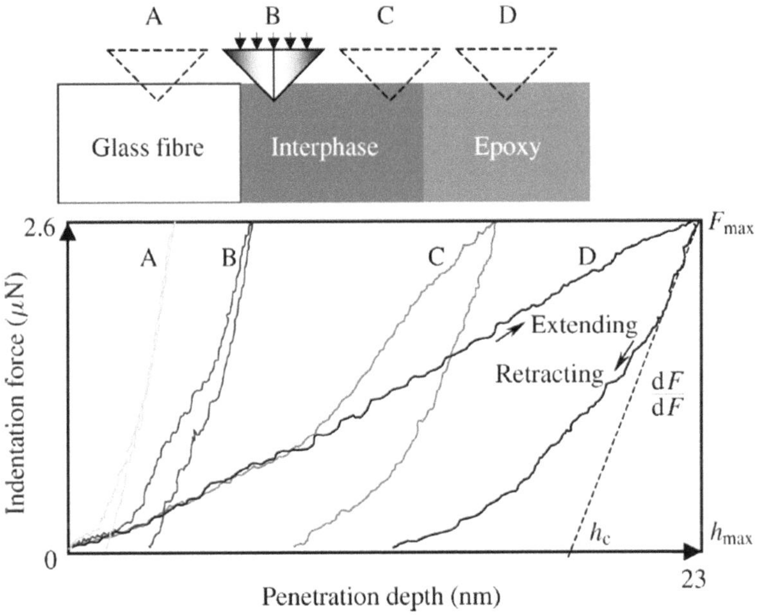

Figure 15.22 Graded characteristic of interphase region captured by indentation test (Gao and Mader, 2002).

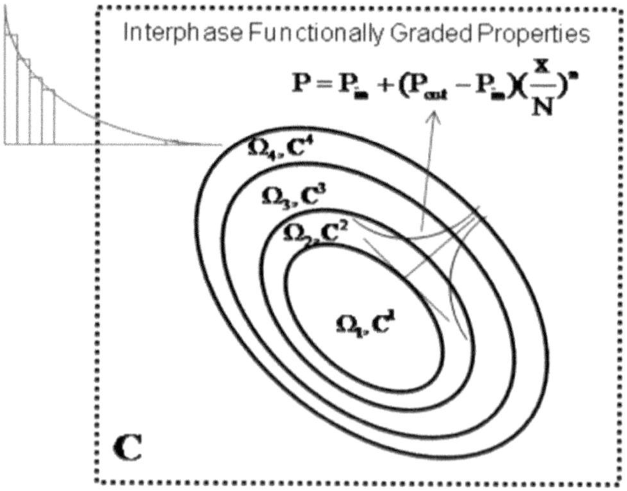

Figure 15.23 A piecewise constant model for functionally graded interphase.

where P, P_{in}, and P_{out} represent the property of interest in the interphase subdivision at a nondimensional distance x from the outer surface of the inclusion, first interphase subdivision and last or Nth interphase subdivision, respectively. The parameter n is a constant power determining the functionality of the graded property. By changing n, the rate of variation of the interphase property from P_{in} to P_{out} can be controlled in a piecewise constant fashion. For the sake of verification, one may calculate the effective properties for the case that the properties of the interphase are equal to the matrix properties in which the results have to be exactly the same as is the case of no interphase (see Figs. 15.8 and 15.9). Table 15.2 shows the results for the macroelastic modulus of a short fiber composite with different elastic properties of the interphase.

As expected, for interphases with greater elastic modulus than the matrix, the overall stiffness of the composite increases by increasing the thickness of the interphase whereas for those interphases with smaller elastic modulus than the matrix, the overall stiffness decreases by increasing the thickness of the interphase.

Table 15.2 Prediction of elastic modulus for different interphase thicknesses and elastic moduli, $E_f = 240$ GPa, $E_0 = 3.15$ GPa, AR = 100, volume fraction = 0.0382

Interphase modulus E_I (Gpa)	Composite modulus E_c (Gpa)	IPTR*
No Interphase	4.78	0.0
	4.90	0.1
100	5.72	0.5
	7.15	1.0
	4.72	0.1
2.0	4.67	0.5
	4.57	1.0

*IPTR: Interphase thickness ratio = (Interphase thickness)/(Fiber radius)

15.4 Modeling of Nanofiber Waviness

Experimental observations of polymers reinforced by carbon nanotubes (CNT) via scanning electron microscope (SEM) and transmission electron microscope (TEM) have revealed that the embedded CNT remain highly curved when dispersed in a polymer (Shaffer et al., 1999; Vigolo et al., 2000; Qian et al., 2000). This is to be expected when considering the very high aspect ratio and low bending stiffness of CNT and CNF. Hence, it is necessary to consider the effect of waviness to have a better understanding of the mechanical behavior of CNT- or CNF-reinforced composite materials. It is also necessary to have a more accurate quantitative model of the mechanical behavior of these composite materials.

Waviness has been modeled using different analytical and hybrid approaches. In analytical approaches (Fisher, 2002; Shady and Gowayed, 2010), wavy embedded fiber is isolated and its effective reinforcing modulus (E_{ERM}) is calculated. This is a representative value that accounts for the reduction in reinforcement provided by the wavy fiber compared with the reinforcement provided by a straight fiber (of modulus E_f) (Chishlom and Brinson, 2003; Bradshaw et al., 2003). Therefore, while modulus of the fiber E_f is a material property, E_{ERM} ($E_{ERM} \leq E_f$) is a material parameter influenced by the geometry of the wavy fiber.

Figure 15.24 shows how the planar sinusoidal waviness of a fiber drastically decreases E_{ERM}. This effective property can be

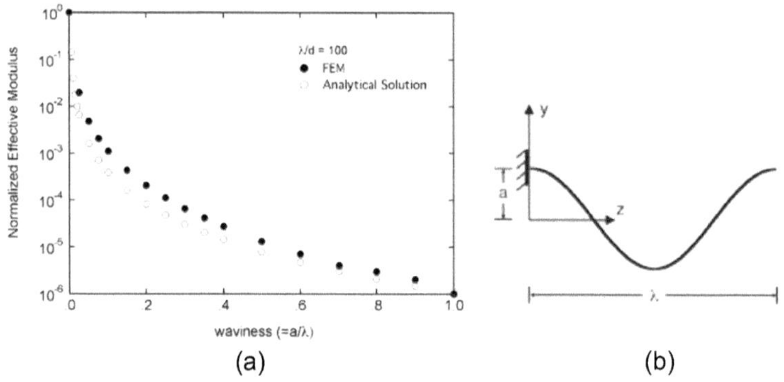

Figure 15.24 (a) Comparison of finite element and analytical solutions for the effective modulus of a free standing wavy fiber, (b) schematic of a free standing sinusoidal wavy fiber (Fisher, 2002).

used in any micromechanical model as an equivalent value for the true CNF/CNT modulus. Fisher (2002) showed that the analytical model is only useful when the modulus of the straight nanotube E_{NT} is much larger than the modulus of the matrix E_{matrix}. When this criterion is not satisfied (i.e., $E_{\text{NT}}/E_{\text{matrix}} < 1000$), a new model/analysis is needed to account for the lateral constraint (bonding between the inclusion and the matrix) imposed on the nanotube by the surrounding matrix. In a more general case of fibers with three-dimensional waviness, Shady and Gowayed (2010) assumed a helical geometry for the fibers and used the same methodology to solve for the equivalent stiffness. Figure 15.25 reveals their results for the effective stiffness compared to the one using Fisher's assumption of planar sinusoidal geometry for wavy fibers. They concluded that underprediction of the effective stiffness based on their helical assumption of the waviness is lower compared to the sinusoidal geometry assumption for the waviness.

In hybrid approaches (computational/micromechanical) (Chishlom and Brinson, 2003; Bradshaw et al., 2003) the wavy CNF/CNT is embedded in a matrix material and a three-dimensional finite element model is created to model the effectiveness of the wavy inclusion when it is perfectly bonded to the matrix. The effective modulus of the finite element cell is then related to E_{ERM} based on the volume fraction of the inclusion. The resulting E_{ERM} is used

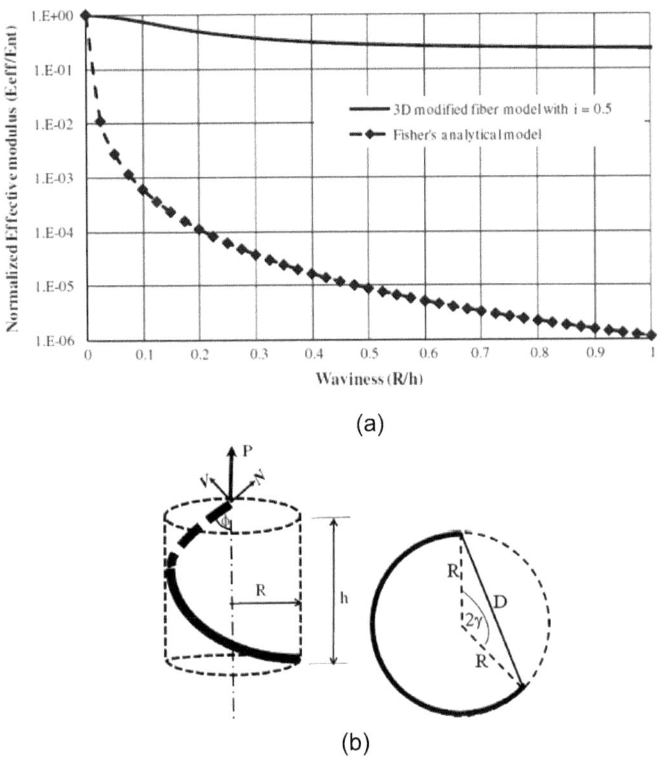

Figure 15.25 (a) The effect of curvature on the ratio of the modulus of effective fiber to the modulus of the fiber (on a semi-log scale) for Fisher's 2D analytical model (Fisher, 2002) and 3D modified fiber model (Shady and Gowayed, 2010), (b) the geometry of a helical wavy fiber in matrix, $\gamma = \frac{1}{2}\,|(2\pi i) - 2\pi|$.

in conventional micromechanical techniques such as Mori–Tanaka scheme to calculate the modulus of the composite material. Figure 15.26 shows the finite element model and the results for equivalent stiffness of a sinusoidal wavy nanotube in matrix using hybrid method (Fisher et al., 2003).

Since the hybrid method, unlike the analytical approaches, considers the effect of bonding between the wavy fiber and its surrounding matrix, it predicts the effective stiffness with less underestimating of the equivalent stiffness of the wavy fibers.

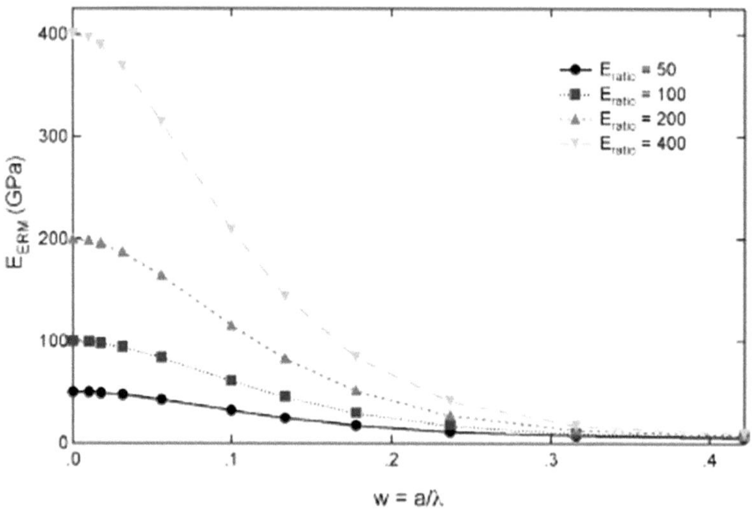

(a)

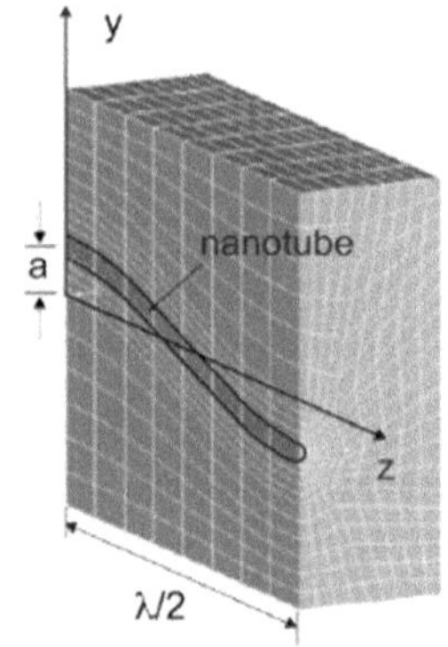

(b)

Figure 15.26 (a) E_{ERM} as a function of nanotube waviness ratio (a/λ) for different ratios of phase moduli ($\lambda/d = 100$, $E_{\text{matrix}} = 1$ GPa) and (b) finite cell model of the embedded wavy nanotube (Fisher et al., 2003).

References

Benveniste, Y. (1987). A new approach to the application of Mori–Tanaka's theory in composite materials, *Mech. Mater.*, **6**, 147–157.

Bradshaw, R. D., Fisher, F. T., and Brinson L. C. (2003). Fiber waviness in nanotube-reinforced polymer composites—II: Modelling via numerical approximation of the dilute strain concentration tensor, *Compos. Sci. Technol.*, **63**(11), 1705–1722.

Chisholm, N., and Brinson L. C. (2003). Fiber waviness in nanotube-reinforced polymer composites—I: Modulus predictions using effective nanotube properties, *Compos. Sci. Technol.*, **63**(11), 1689–1703.

Doghri, I., and Ouaar, I. (2003). Homogenization of two-phase elasto-plastic composite materials and structures study of tangent operators, cyclic plasticity and numerical algorithms, *Int. J. Solids Struct.*, **40**, 1681–1712.

Eshelby, J. D. (1957). The determination of the elastic field of an ellipsoidal inclusion, and related problems, *Proc. R. Soc. London A*, **241**, 376–396.

Fisher, F. (2002). Nanomechanics and viscoelastic behavior of carbon nanotube reinforced polymers, PhD Dissertation, Northwestern University.

Friedrich, K., Fakirov S., and Zhang, Z., eds. (2005). *Polymer Composites: From Nano- to Macro-scale*, Springer, New York.

Gao, S. L., and Mäder, E. (2002). Characterisation of interphase nanoscale property variations in glass fibre reinforced polypropylene and epoxy resin composites, *Composites Part A*, **33**, 559–576.

Gojny, F. H., Wichmann, M. H. G., Fiedler, B., Bauhofer, W., and Schulte, K. (2005a). Influence of nano-modification on the mechanical and electrical properties of conventional fibre-reinforced composites, *Composites Part A*, **36**(11), 1525–1535.

Garg, M., Abdi, F. and McHugh, S. (2008). Analyzing interlaminar shear strength of multi-scale composite via combined finite element and progressive failure analysis approach, SAMPE Fall Technical Conference on Multifunctional Materials, Memphis, TN, USA, September.

Huang, J. H. (2001). Some closed-form solutions for effective moduli of composites containing randomly oriented short fibers, *Mater. Sci. Eng. A*, **315**, 11–20.

Maekawa, Z-I., Hamada, H., and Yokoyama, A. (1989). Lamination theory of composite material with complex fiber orientation distribution, *Proc. Int. Conf. on Conceptual Structures (ICCS)*, **701–714**, 24–26.

Mori, T., and Tanaka, K. (1973). Average stress in matrix and average elastic energy of materials with misfitting inclusions, *Acta Metall.*, **21**, 571–574.

Mura, T. (1987). *Micromechanics of Defects in Solids*, Martinus Nijhoff Publishers.

Nemat-Nasser, S., and Hori, M. (1993). *Micromechanics: Overall Properties of Heterogeneous Solids*, Elsevier, Amsterdam.

Odegard, G. M., Gates, T. S., Nicholson, L. M., and Wise, K. E. (2002) Equivalent-continuum modeling of nano-structured materials, *Compos. Sci. Technol.*, **62**(14), 1869–1880.

Odegard, G. M., Harik, V. M., Wise, K. E., and Gates, T. S. (2003). Constitutive modeling of nanotube-reinforced polymer composites, *Compos. Sci. Technol.*, **63**, 1671–1687.

Pierard, O., Friebel, C., and Doghri, I. (2004). Mean-field homogenization of multi-phase thermo-elastic composites: a general framework and its validation, *Compos. Sci. Technol.*, **64**, 1587–1603.

Qian, D., Dickey, E. C., Andrews, R. and Rantell, T. (2000). Load transfer and deformation mechanisms in carbon nanotube-polystyrene composites, *Appl. Phys. Lett.*, **76**(20), 2868–2870.

Reddy, J. N. (2000). Analysis of functionally graded plates, *Int. J. Numer. Methods Eng.*, **47**, 663–684.

Rouhi, M., Rais-Rohani, M., Lacy, T. E., Garg, M., and Abdi, F. (2010). Mechanical characterization of nanofiber enhanced polymer with application to composite crush tubes, Proceedings of the 51st AIAA/ASME/ASCE/AHS/ASC Structures, Structural Dynamics, and Materials Conference, Orlando, FL, USA.

Ruhi, M., Angoshtari, A., and Naghdabadi, R. (2005). Thermoelastic analysis of thick-walled finite-length cylinders of functionally graded materials, *J. Therm. Stresses*, **28**, 391–408.

Schjodt-Thomsen, J., and Pyrz, R. (2001). The Mori–Tanaka stiffness tensor: diagonal symmetry, complex fiber orientations and non-dilute volume fractions, *Compos. Sci. Technol.*, **33**, 531–544.

Shady, E. and Gowayed, Y. (2010). Effect of nanotube geometry on the elastic properties of nanocomposites, *Compos. Sci. Technol.*, **70**, 1476–1481.

Shaffer, M. S. P. and Windle, A. H. (1999). Fabrication and characterization of carbon nanotube/poly(vinyl alcohol) composites, *Adv. Mater.*, **11**(11), 937–941.

Tandon, G. P. and Weng, G. J. (1984). The effect of aspect ratio of inclusions on the elastic properties of unidirectionally alligned composites, *Polym. Compos.*, **5**, 327–333.

Thostenson, E. T., Li, W. Z., Wang, D. Z., Ren, Z. F., and Chou, T. W. (2002). Carbon nanotube/carbon fiber hybrid multiscale composites, *J. Appl. Phys.*, **91**(9), 6034–6037.

Vigolo, B., Penicaud, A. P., Couloun, C., Sauder, C., Pailler, R., Journet, C., et al. (2000). Macroscopic fibers and ribbons of oriented carbon nanotubes, *Science*, **290**, 1331–1334.

Withers, P. J., Stobbs, W. M., and Pedersen, O. B. (1989). The application of the eshelby method of internal stress determination to short fiber metal matrix composites, *Acta Metall.*, **37**(11), 3061–3084.

Zhou, Y., Pervin, F., Jeelani, S., and Mallick P. K. (2008). Improvement in mechanical properties of carbon fiber–epoxy composite using carbon nanofibers, *J. Mater. Process. Technol.*, **198**(1–3), 445–453.

Chapter 16

Part 1
Multiscale Nanocomposite Fatigue Life Determination*

Kamran Nikbin and Anthony J. Kinloch
Imperial College London, South Kensington Campus, London SW7 2AZ, UK
k.nikbin@imperial.ac.uk, a.kinloch@imperial.ac.uk

This chapter describes an approach to predict the effect of material and manufacturing uncertainty on nanocomposite fatigue life. A multiscale micromechanics technique has been proposed to characterize the nanocomposite material constituting of matrix reinforced/infused with silica spherical shaped nano-size particles. The technique was used to first predict in situ material properties of constituents (linear elastic silica nanoparticles and nonlinear elastic matrix) from macro-level test data available from literature. Second, the in situ constituent properties were used to predict the fatigue properties of epoxy reinforced with 10 wt% of well-dispersed silica nanoparticles under tension fatigue loading. The predictions were validated with test and were found to be in good agreement within ±4% error. The nanomechanics fatigue durability simulation

*SAMPE 2011, October 17–20, 2011, Fort Worth, Texas.

Characterization of Nanocomposites: Technology and Industrial Applications
Edited by Frank Abdi and Mohit Garg

ISBN 978-981-4669-02-3 (Hardcover), 978-1-315-36489-6 (eBook)
www.panstanford.com

utilized closed-form solution from Mori–Tanaka, classical laminate theory, multiscale multiphysics progressive failure analysis of a unit cell. At the end probabilistic analysis was utilized to determine the sensitivity effect of nanomanufacturing (i.e., void volume fraction, interphase effect, and uniform versus nonuniform distribution) and matrix and nanoparticle material properties parameters on the fatigue life.

16.1 Introduction

Advanced fiber-reinforced plastics (FRP) composite materials are being used increasingly for aerospace and automotive applications because of their higher strength-to-weight ratio compared to the metals. However, the composite materials are weaker in the transverse direction (transverse to the fibers) than in the longitudinal direction, thus making them prone to delamination due to matrix failure. Test data in the literature [1] indicates that longitudinal composite compressive properties are also highly dependent on the stiffness and strength of the matrix material. This is mainly because a stiffer matrix material resists the buckling of the fibers under axially compressive load. The majority of engineering composite materials consist of a thermosetting epoxy matrix reinforced by continuous glass or carbon fibers. The epoxy, when polymerized, is an amorphous and highly cross-linked material. This cross-linked microstructure results in many useful properties such as a high modulus and failure strength, low creep, etc. However, it also leads to an undesirable property, whereby the polymer is relatively brittle, exhibits void content, and has a poor resistance to crack initiation and growth, which may affect the overall fatigue and fracture performance of FRP composite. Two ways to enhance the mechanical properties of FRPs are (i) creating a strong interphase between the conventional fibers and matrix with randomly oriented nanoparticles, as shown in Fig. 16.1a, and (ii) enhancing matrix properties by infusing it with small weight fraction of nanoparticles, as shown in Fig. 16.1b. The first approach is expensive than the second.

An example of the first approach is growing carbon nanotubes around the carbon fibers and then preparing the prepregs for use

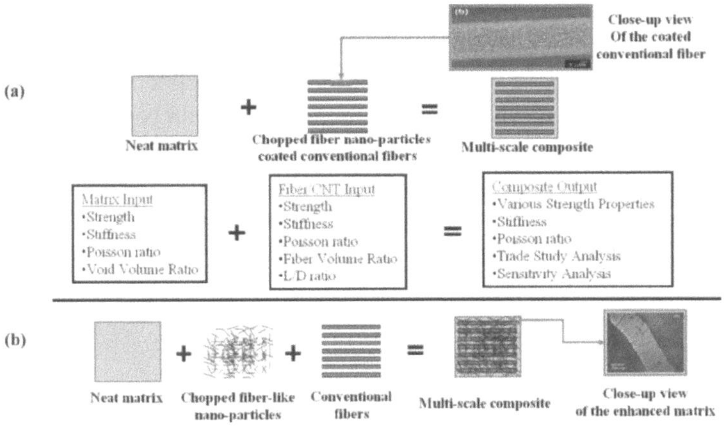

Figure 16.1 Micrographs of (a) coated conventional fiber [3] and (b) enhanced matrix [2].

in composite laminates [2, 3]. The interphase between conventional carbon fibers and neat matrix can be aligned or randomly oriented carbon nanotubes depending on the length and manufacturing process [2, 3]. The random and aligned interphase here refers to the phase where carbon nanotubes are grown with random or aligned orientation in the matrix, respectively. An example of the second approach is neat epoxy reinforced with randomly oriented carbon nanotubes and then used with regular conventional fibers (Fig. 16.1b). Another example would be infusing Nanopox F400 (SiO_2) spherical particles in diglycidyl ether of bisphenol A (DGEBA) epoxy [1, 5–8]. The nanoparticle-reinforced matrix can then be used with conventional glass or carbon fibers for improved transverse tensile, compressive, shear, and longitudinal compressive failure strengths. In this chapter, the multiscale composite refers to conventional fibers, which are used with an enhanced matrix to improve the properties of the epoxy matrix by incorporating second phase fillers into the resin.

Manufacturing these advanced nanocomposite material system is quite challenging: uniformly dispersing the nanoparticles in the matrix, improving the interface between the matrix and the particles for proper load transfer from matrix to the nanoparticles, and removing voids among others.

In this chapter the in situ silica (Nanopox F400) nanoparticles and the DGEBA matrix (assuming uniformly dispersed nanoparticles) strength and stiffness material properties have been characterized and predicted. The limited test data is obtained from the literature [5–8]. The term *in situ* is used to represent the effective fiber and matrix properties, as they vary when the constituents are cured together to form a nanocomposite material system versus their bulk state properties.

16.2 Experiment [5]

Manjunatha et al. (2010) presented their experimental process [5, 6] to create nanocomposite and multiscale composite material system that consisted of silica nanoparticle infused epoxy and E-glass fibers reinforcing silica nanoparticle infused epoxy, respectively. The epoxy used in their experimental work was a standard DGEBA. The nanoparticles consisted of nano-size spherical shape (~20 nm diameter) silica particles (Nanopox F400). According to Qian et al. [5] and Manjunatha et al. [6], the nanoparticles were uniformly distributed in the matrix and constituted 10 wt% in the end product (epoxy/silica). Figure 16.2 shows the transmission electron

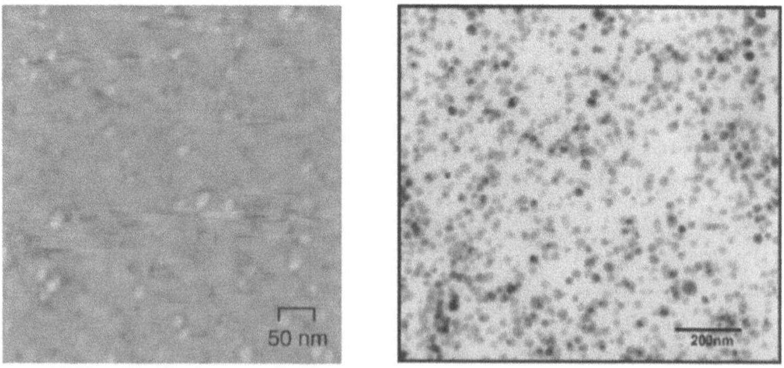

(a) 10 wt.% silica/epoxy resin [5] (b) 15 wt.% silica/epoxy resin [1]

Figure 16.2 (a) Atomic force microscope (AFM) and (b) transmission electron microscope (TEM) image.

Table 16.1 Stiffness and strength properties for DGEBA epoxy infused with and without 10 wt% Nanopox F400 silica (SiC) nanoparticles from test [5, 8]

Matrix material properties	Symbol	Control DGEBA	Nanocomposite DGEBA + Nanopox F400	Units	Reference
Desnsity	Rhom	1.22E-09	1.28E-09	[tonne/mm^3]	2
Young's modulus	Em	2.62	3.07	[GPa]	1
Poisson's ratio	vm	-	-	[-]	
Tension strength	SmT	73.3	86.9	[MPa]	1
Compression Strength	SmC	231.5	298.95	[MPa]	2
Shear strength	SmS	-	-	MPa	

microscope image of typical Nanopox F400 particles uniformly dispersed in epoxy [1, 5].

Table 16.1 lists the control (neat epoxy) and nanocomposite (DGEBA + Nanopox F400) test properties extracted from the experimental work from literature [5].

16.3 Methodology

The proposed methodology for fatigue durability simulation of nanocomposite material system consists of first utilizing the closed-form solution from Mori–Tanaka, classical laminate theory (CLT), and multiscale multiphysics progressive failure analysis of a unit cell or a representative volume element (RVE) and then applying probabilistic analysis to determine the sensitivity of the fatigue life in nanocomposites to the manufacturing (i.e., void volume fraction, interphase effect, and uniform versus nonuniform distribution) and material (matrix and nanoparticle modulus and strength) random variables.

16.3.1 *Nanocomposite: Closed-Form Analytical Solution*

First, effective nanocomposite (or enhanced matrix) material properties, where randomly dispersed Nanopox F400 (SiO_2) of a known weight fraction in neat matrix, are determined analytically. The analysis approach uses well-known Mori–Tanaka formulation

[9, 10] for calculating the anisotropic nanocomposite properties from isotropic matrix and in situ isotropic/anisotropic nanoparticle properties (stiffness, aspect ratio, and volume fraction). For a composite material reinforced with aligned nanoparticles, the Tandon and Weng (1984) prediction of the moduli E_{11} (aligned particle direction), E_{22} (transverse to the aligned particle direction), the in-plane shear modulus G_{12}, and the out-of-plane shear modulus G_{23} of the composite are

$$E_{11} = {}^{E_\mathrm{m}}\!/\left(1 + f_\mathrm{p}\left(A_1 + 2\upsilon_\mathrm{m} A_2\right)/A\right) \tag{16.1}$$

$$E_{22} = {}^{E_\mathrm{m}}\!/\left(1 + f_\mathrm{p}\left(-2\upsilon_\mathrm{m} A_3 + \left(1 - \upsilon_\mathrm{m}\right) A_4 + \left(1 + \upsilon_\mathrm{m}\right) A_5^{*} A\right)/\left(2A\right)\right) \tag{16.2}$$

$$G_{12} = {}^{G_\mathrm{m}}\!/\left(1 + f_\mathrm{p}\left(\mu_\mathrm{m}/\left(\mu_\mathrm{p} - \mu_\mathrm{m}\right)\right) + 2\left(1 - f_\mathrm{p}\right) H_{1212}\right) \tag{16.3}$$

$$G_{23} = {}^{G_\mathrm{m}}\!/\left(1 + f_\mathrm{p}\left(\mu_\mathrm{m}/\left(\mu_\mathrm{p} - \mu_\mathrm{m}\right)\right) + 2\left(1 - f_\mathrm{p}\right) H_{2323}\right) \tag{16.4}$$

where A and A_i are constants depending on the components of the Eshelby tensor and the matrix/nanoparticles properties, and H_{ijkl} are the Cartesian components of the Eshelby tensor [11].

In this chapter, a closed-form analytical solution for the complete set of anisotropic elastic properties of the nanocomposite derived by Tandon and Weng (1984) by combining the Eshelby theory [11] and the Mori–Tanaka model [9], as shown in Eqs. 16.1–16.4 has been developed.

The in situ strength properties of the nanocomposites are calculated based on stress and strain-based micromechanics theories developed earlier in the literature [12, 13]. At this stage, we have calculated the effective material properties of the enhanced matrix where the nanoparticles are randomly dispersed. Note that for fibrous nano particles the above methodology calculates aligned anisotropic properties. However, for spherical-shaped nanoparticles, subject of present investigation, the predicted properties are random isotropic in three-dimensional spaces. In order to predict random isotropic properties for fibrous nanoparticles composite, we follow the process described in Section 3.2.

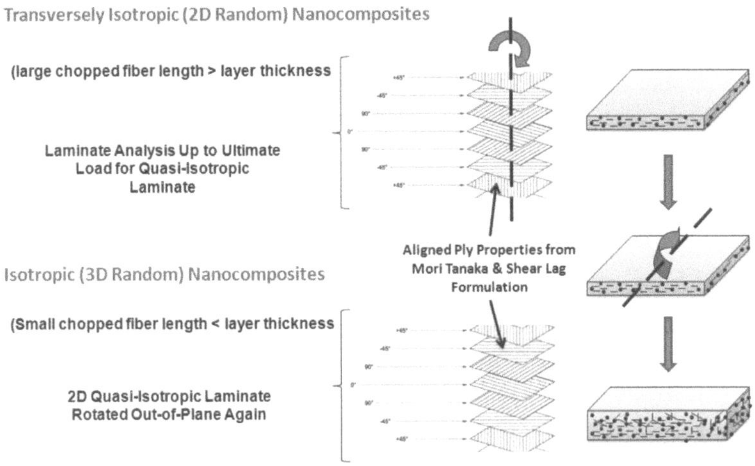

Figure 16.3 Classical laminate theory and multiscale multiphysics progressive failure analysis approach to determine 2D and 3D random nanocomposite properties.

16.3.2 *Conventional Composite: Multiscale Micromechanics*

A unit cell based coupon model is then subjected to multiscale multi-physics progressive failure analysis (PFA). The unit cell with quasi-isotropic $[45/0/\text{-}45/90]_{2S}$ lay-up is subjected to five in-plane loading conditions: longitudinal tension and compression, transverse tension and compression, and in-plane shear. Then modulus and strength values are recorded from each simulation. Each ply is assigned unidirectional enhanced matrix properties that were previously derived analytically using the closed-form Mori–Tanaka approach. The combined unit cell and PFA simulation of a quasi-isotropic ply lay-up gives equivalent effective 2D and 3D random (transversely isotropic and isotropic) stiffness, strength and Poisson's ratio of the enhanced matrix (nanocomposite), as shown in Fig. 16.3.

The constituent material properties can be estimated using limited test data by reverse engineering the constituent material properties. MCQ-Nano, a commercial nanocomposite material characterization and qualification software [14], was used to predict the constituent properties. The module uses the above-mentioned methodology to predict the aligned, transversely isotropic and isotropic nanocomposite thermoelectromechanical material properties. The formulation in the module also accounts for the manufacturing variables such as particle and void volume ratio. The following process was adopted to predict and validate the material properties using the limited test data [5, 8].

1. Obtain results for three tests: tension, compression, and shear.
2. Obtain the particle and void volume ratio for the nanocomposite. The void volume ratio was assumed to be 0% for calibration, as the information is unknown from the test data [5].
3. Obtain approximate bulk constituent (particle/matrix) properties from literature [5, 8] or manufacturer.
4. Reverse-engineer/predict in situ constitute material properties using above methodology, such that the predictions match the tests listed under Step 1. Note that if three tests are not available, then the manufacturer's properties are assumed to be the constituent properties.
5. Once the constituent material properties are known, progressive failure analysis (PFA) of unit cell is used to predict the five tests for validity.

16.3.3 *Progressive Failure Analysis (PFA)*

Progressive failure analysis is the preferred approach for failure analysis in composite materials. In this approach, representative volume element (RVE) independent of finite element model (FEM) failure in each ply constituting the laminate, under investigation, is assessed based on the ply limit strength/strain properties, as shown in Fig. 16.3. Some failure criteria are maximum stress- or strain-based and some are interactive. The damaged plies are then either removed from the unit cell model or given a negligible stiffness to simulate a broken ply and study its effect on overall laminate. Later, once all the plies in the unit cell are damaged, it is assumed to be completely broken. The stresses in individual plies are obtained using classical laminate theory, which are then compared against

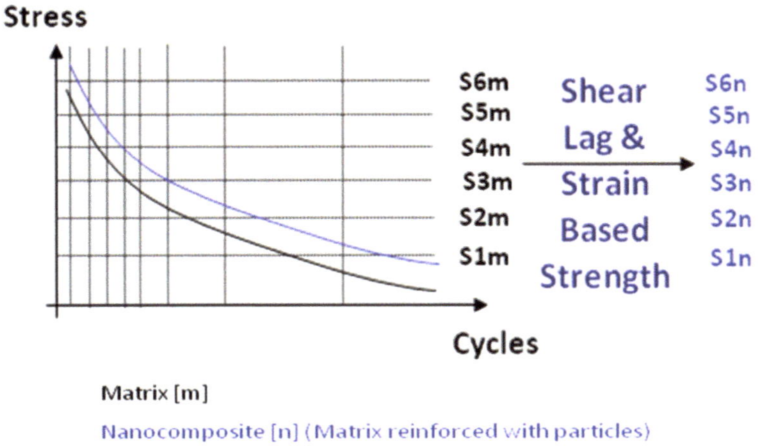

Figure 16.4 Methodology to predict S-N curve for nanocomposites from S-N curve of control matrix as input (m = matrix, n = nanocomposite, S = strength, number = location on S-N curve, black = control matrix from test input, blue = nanocomposite).

the limit stresses or strains parameters using default failure criteria [15]. The predictions from progressive failure analysis then give the laminate modulus, Poisson's ratio, and strength in addition to other thermal and electrical properties that consists of randomly oriented particles in 2D or 3D space, are shown in Fig. 16.3.

16.3.4 *Fatigue S-N Curve*

1. Predict S-N curve (blue) for particulate-reinforced matrix using neat matrix S-N curve (black) as input. New strength is calculated for epoxy containing random oriented particles at each increment to predict the overall S-N curve of the reinforced matrix (Fig. 16.4).
2. Predict S-N curve for reinforced matrix for stress ratio similar to the input S-N curve for neat matrix.
3. Apply Goodman's or similar rules to shift the predicted S-N curve for different stress ratio.

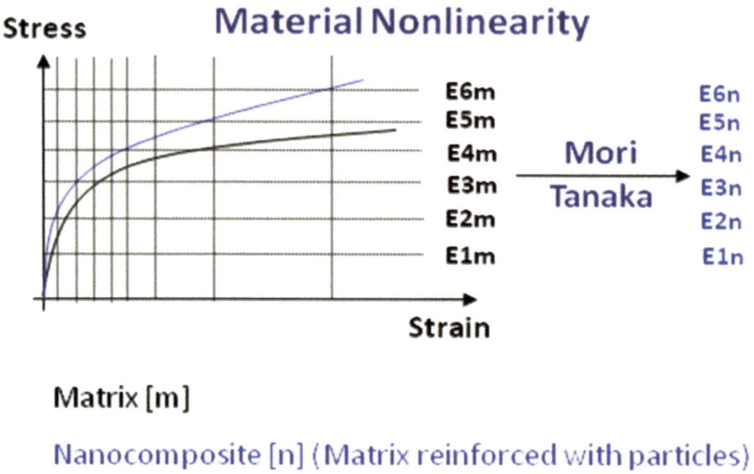

Figure 16.5 Methodology to predict stress–strain curve for nanocomposites from S-S curve of control matrix as input (m = matrix, n = nanocomposite, S = strength, number = location on S-S curve, black = control matrix from test input, blue = nanocomposite).

16.3.5 *Stress–Strain Curve*

1. Predict stress–strain (S-S) curve (blue) for particulate-reinforced matrix using neat matrix S-S curve (black) as input. The updated tangential modulus is calculated for epoxy containing random oriented particles at each increment to predict the overall S-S curve of reinforced matrix (Fig. 16.5).
2. Predict S-S curve for reinforced matrix for strain rate similar to the input S-S curve for neat matrix.
3. The prediction is limited to tangential S-S curve at present and cannot capture the drop in the S-S curve beyond yield stress as some materials show.

16.3.6 *Probabilistic Material Uncertainty Analysis*

With the direct coupling of composite nano- and micromechanics, classical laminate theory, and probabilistic methods, it is possible to simulate material and manufacturing/fabrication uncertainties in all inherent scales of nanocomposites, from constituent materials level to macro-level for specified loading conditions. The evaluation process starts with the identification of the primitive variables at the nano, micro, and macro scales. These variables are then selectively perturbed assuming normal or any other distribution in order to generate a database in the form of probability density function (PDF) and cumulative density function (CDF) for determining the relationships between the desired materials behavior and/or structural fatigue response and the primitive variables using first-order reliability method (FORM). FORM is a semi-probabilistic reliability analysis method devised to evaluate the reliability of the system [14]. The generated probabilistic data is then used to perform sensitivity analysis.

16.3.7 *Material Modeling Assumptions*

Because of lack of availability of the three basic tests (under tension, compression, and shear) listed earlier, some assumptions have been made to characterize the particle/matrix material properties and are listed below:

1. Isotropic and linear elastic material properties for matrix and particles.
2. The void content was assumed to be 0% during the material characterization process.
3. No interphase during the material characterization process. However, for material uncertainty analysis for fatigue properties, the following properties were assumed for the interphase (Table 16.2 and Fig. 16.6) between the matrix and the particle. Note that the Young's modulus for the interphase is assumed to be 1/10th of the control matrix. The compression strength is assumed to be same as yield strength from literature [8]. The shear yield strength is dropped by the same amount as the compressive yield strength. The tensile yield strength is assumed to be same as the rupture strength based on linear elastic behavior in tension for the epoxy.

Table 16.2 In situ interphase properties (thickness = 1 nm)

Matrix material properties	Symbol	Effective	Units
Desnsity	Rhom	1.221E-09	[tonne/mm^3]
Young's modulus	Em	262	[MPa]
Poisson's ratio	vm	0.35	[-]
Tension strength	SmT	73.3	[MPa]
Compression strength	SmC	131.5	[MPa]
Shear strength	SmS	150	[MPa]

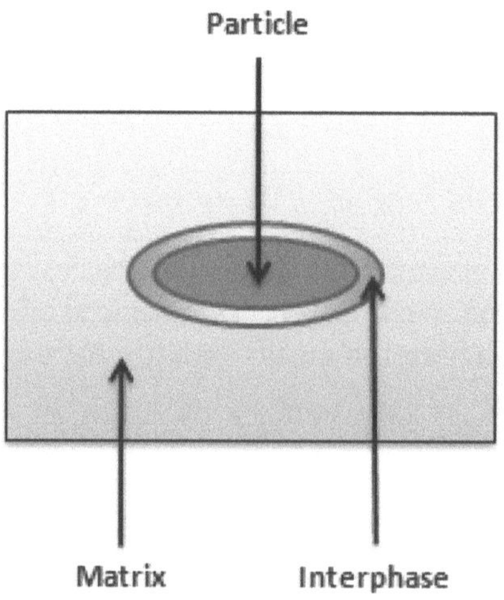

Figure 16.6 Modeling interphase effects for progressive failure analysis.

16.4 Results and Discussion

16.4.1 *Static Property Prediction/Test Validation of Control and Nanocomposite*

Using the above methodology material characterization and qualification Nano (MCQ-Nano) tool and the limited test data listed under Table 16.1, the constituent (Nanopox F400 nanoparticles and DGEBA epoxy) in situ material properties were predicted and are listed in Tables 16.3 and 16.4. The in situ properties were found using nanoparticle weight fraction of 10 wt% and void volume ratio

Table 16.3 In situ Nanopox F400 (SiO_2) nanoparticles reverse-engineered from nanocomposite test data [5, 8] (particle diameter = 20 nm; % weight fraction = 10%)

Matrix material properties	Symbol	Effective	Units
Desnsity	Rhop	1.8E-09	[tonne/mm^3]
Young's modulus	Ep	70	[GPa]
Poisson's ratio	vp	0.19	[-]
Tension strength	SpT	5.30E+04	[MPa]
Compression strength	SpC	1.03E+05	[MPa]

Table 16.4 In situ epoxy (ly556) reverse-engineered from nanocomposite test data [5, 8] (particle diameter = 20 nm; % weight fraction = 10%)

Matrix material properties	Symbol	Effective	Units
Desnsity	Rhom	1.221E-09	[tonne/mm^3]
Young's modulus	Em	2.62	[GPa]
Poisson's ratio	vm	0.35	[-]
Tension strength	SmT	73.3	[MPa]
Compression strength	SmC	231.5	[MPa]
Shear strength	SmS	250	[MPa]

of 0.0%. The in situ nanoparticle modulus of 70.0 GPa is similar to tested data in literature [16].

16.4.2 *Fatigue Property Prediction and Test Validation*

The fatigue behavior of bulk epoxy, load-controlled, constant-amplitude, tension–tension cyclic- fatigue test results obtained for the neat and silica nanoparticle modified bulk epoxy polymers are shown in Fig. 16.7. It may be clearly seen that both predicted and tested overall fatigue life, for a given maximum cyclic stress, of the nanoparticle-modified epoxy is greater than that of the neat epoxy. The fatigue life enhancement of about three to four times in the nanoparticle-modified epoxy is observed over the entire range of stress levels predicted in this chapter and tested. The predicted and tested S-N curve is in good agreement overall range of the S-N curve.

Table 16.5 Comparison between predicted and test data for nanocomposite containing randomly oriented nanoparticles [5, 8] (particle diameter = 20 nm; weight fraction = 10%)

SJO2/LY556 (Nanocomposite): FVR = 7.01%; VVR = 0.0%				
Material		**SiO_2/LY556**		
Property	**Units**	**Test**	**MCQ nanocomposite**	**% Error**
Rhoc	[tonne/mm^3]	-	1.26E-09	-
Ec	[GPa]	3.07	3.028	−1.37
vc	[-]	0.35	0.3434	−1.89
ScT	[MPa]	86.9	91.9	5.75
ScC	[MPa]	300	286.5	−4.50
ScS	[MPa]	-	153.3	-

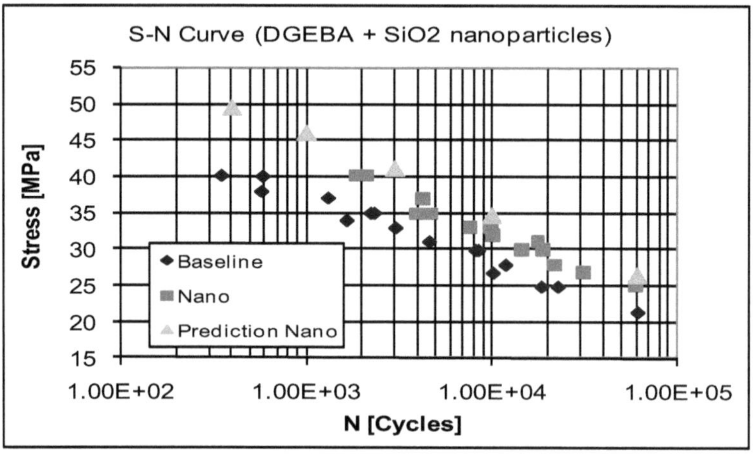

Figure 16.7 Comparison between tested and predicted data [5]. Stress ratio (R) = 0.1; max load = 25 kN.

16.5 Material Uncertainty Analysis due to Manufacturing and Design Parameters on Fatigue Life

The effect of manufacturing (void, interphase, and distribution of nanoparticles in the matrix), and constituent properties (nanoparticles/matrix) on the tensile fatigue behavior was evaluated.

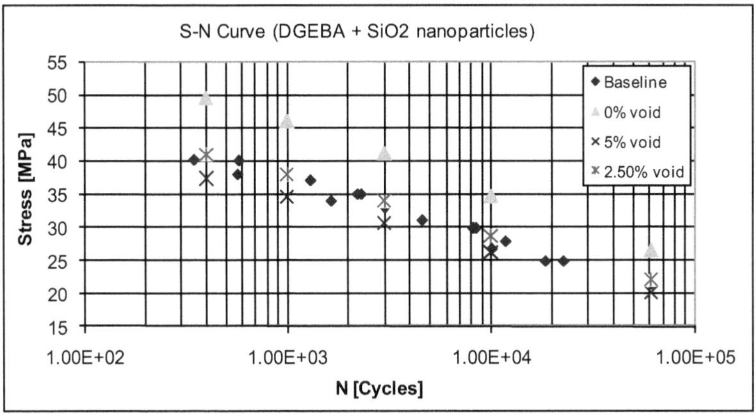

Figure 16.8 Effect of voids on fatigue life of nanocomposites. Stress ratio $(R) = 0.1$; max load = 25 kN.

Figure 16.8 shows the effect of voids on the fatigue life of nanocomposite compared against the test control matrix and 0% void nanocomposite from actual test data. Figure 16.8 shows that a 5% void volume ratio can significantly reduce the contributions from nanoparticles to below the control matrix. On the other hand, a 2.5% void volume ratio nullifies the contributions from the addition of Nanopox F400.

Figure 16.9 shows the effect of variation in Nanopox F400 particle weight fraction (pvr) on the fatigue life of the nanocomposites. It is clear from Fig. 16.9 that the fatigue life is insensitive to slight variation (COV = ±20%) in nanoparticle weight fraction.

Figures 16.10–16.13 show the effect of variations in Nanopox F400 particle modulus (E_p) and strength (S_pT) and control matrix modulus (E_m) and strength (S_mT) on the fatigue life of nanocomposites. Figures 16.10–16.13 clearly show that the fatigue life is insensitive to slight variation (COV $= \pm 11\%$) in the nanoparticle modulus. Note that the COV for particle strength was chosen to be ±18%.

Figure 16.14 shows the effect of an interphase on the fatigue life of nanocomposites. The interphase was assumed to be in between the particle and matrix. The interphase is assumed to be plastically yielded due to thermal mismatch between the nanoparticles and

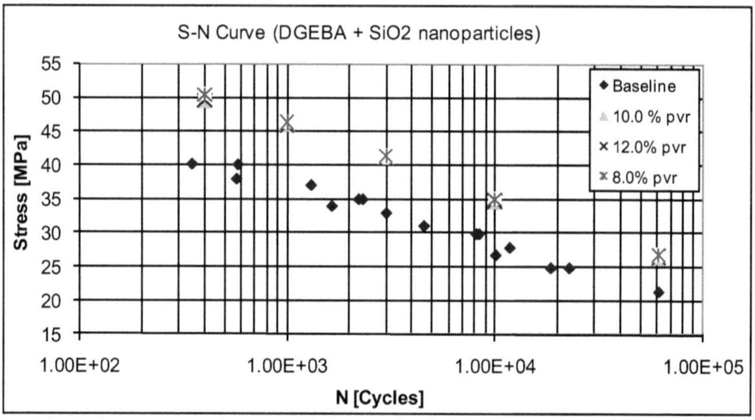

Figure 16.9 Effect of particle weight fraction on fatigue life of nanocomposites. Stress ratio $(R) = 0.1$; max load = 25 kN; pvr = particle weight fraction.

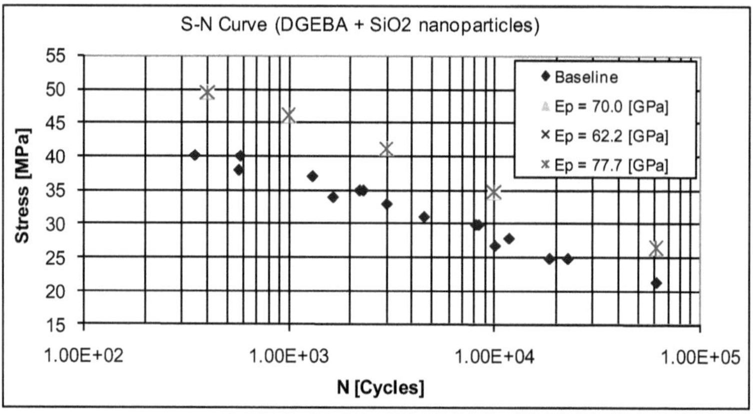

Figure 16.10 Effect of nanoparticle modulus on fatigue life of nanocomposites. Stress ratio $(R) = 0.1$; max load = 25 kN.

matrix. The assumed properties are listed in Table 16.2 for the interphase. The interphase is also assumed to be perfectly bonded to the particle and the matrix.

Figure 16.15 shows the effect of uniform and nonuniform nanoparticle dispersion on the fatigue life of the nanocomposites. For nonuniform dispersion of nanoparticles, the authors assumed

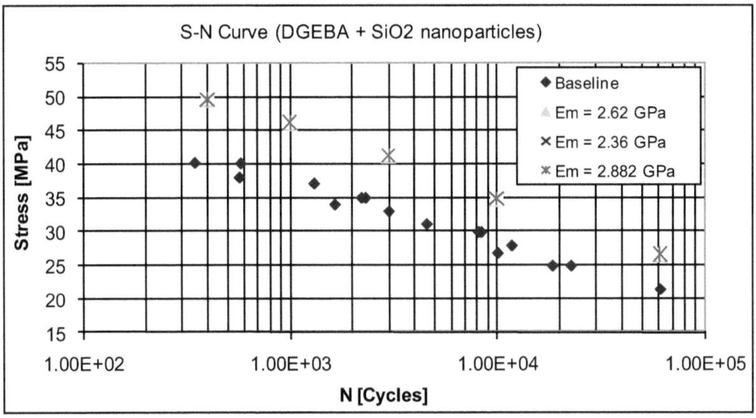

Figure 16.11 Effect of matrix modulus on fatigue life of nanocomposites. Stress ratio $(R) = 0.1$; max load = 25 kN.

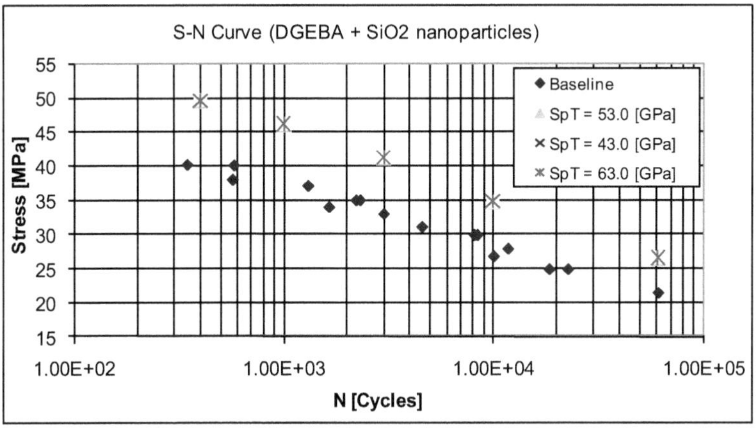

Figure 16.12 Effect of particle tensile strength on fatigue life of nanocomposites. Stress ratio $(R) = 0.1$; max load = 25 kN.

large spherical shape particles consisting of agglomerates of nanoparticles in neat matrix. The authors assumed the ratio of agglomerates to the nanocomposite volume to be 0.5 and that of nanoparticles to agglomerates to be 1.0. All other variables were same as the baseline case (Fig. 16.7). Figure 16.15 also shows a

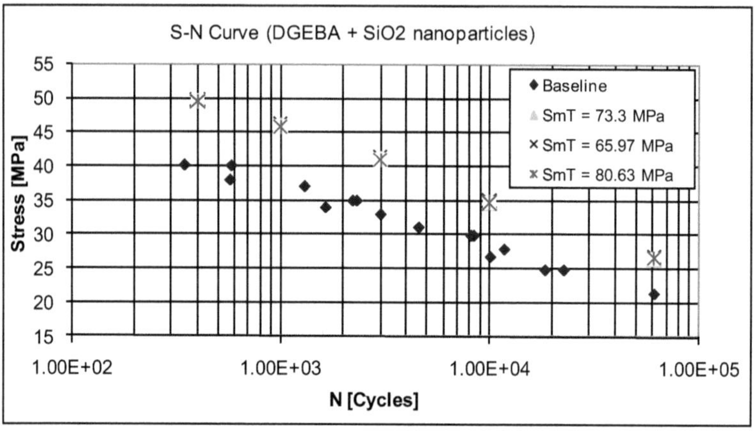

Figure 16.13 Effect of matrix tensile strength on fatigue life of nanocomposites. Stress ratio $(R) = 0.1$; max load $= 25$ kN.

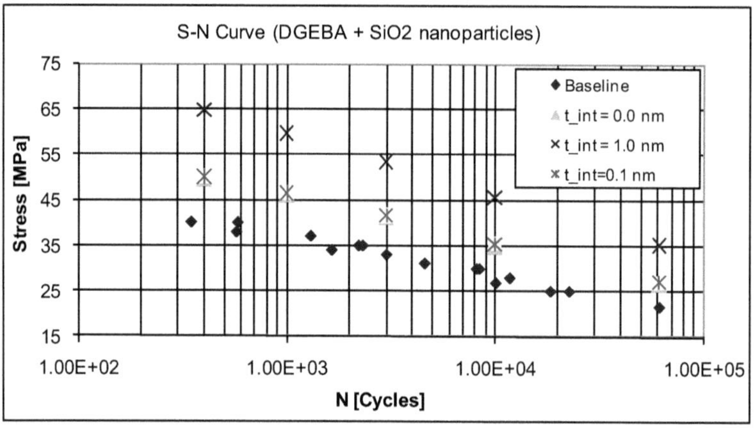

Figure 16.14 Effect of Interphase and thickness on fatigue life of nanocomposites. Stress ratio $(R) = 0.1$; max load $= 25$ kN.

significant effect of the dispersion of particles on the fatigue life of the nanocomposite, i.e., less than the original matrix fatigue life.

In summary, the results above (Figs. 16.8–16.15) and probabilistic sensitivity analysis (Figs. 16.16 and 16.17) indicate that the fatigue properties of the nanocomposite (DGEBA epoxy reinforced with Nanopox F400 nanoparticles) show a high degree of sensitivity

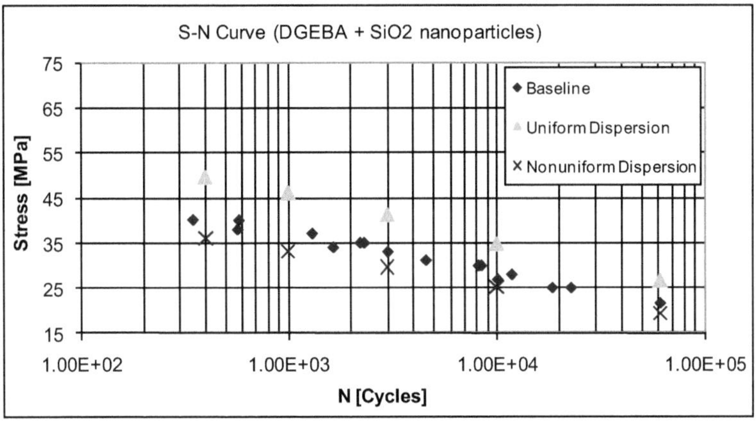

Figure 16.15 Effect of nanoparticle dispersion on fatigue life of nanocomposites. Stress ratio (R) = 0.1; max load = 25 kN.

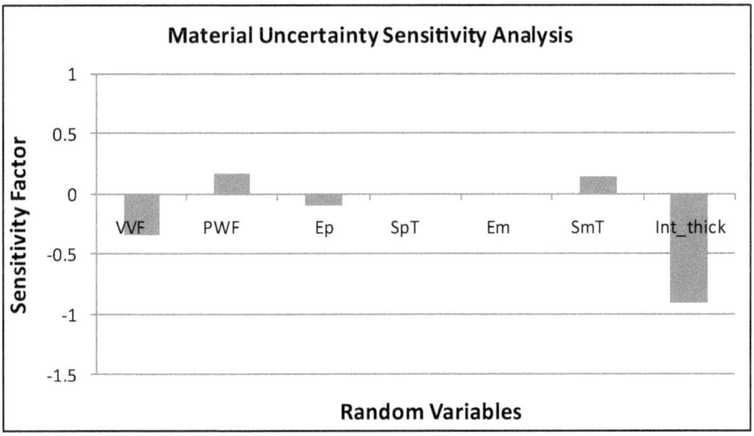

Figure 16.16 Material uncertainty analysis considering effect of material and manufacturing variables on fatigue life of nanocomposites at 3000 cycles in Figs. 16.7–16.14. Stress ratio (R) = 0.1; max load = 25 kN. VVF = void volume factor; PWF = particle weight fraction; E_p = particle modulus; S_pT = particle tensile strength; E_m = matrix modulus; S_mT = matrix tensile strength; Int_thick = interphase thickness (assuming on an average ±10% variation, with normal distribution).

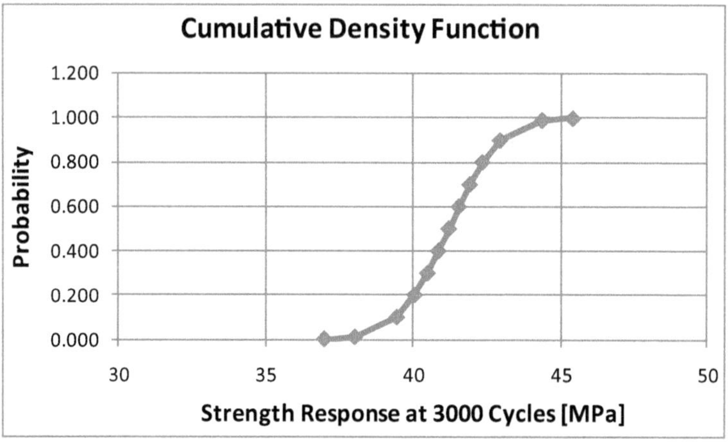

Figure 16.17 Cumulative density function from material uncertainty analysis considering effect of material and manufacturing variables on fatigue life of nanocomposites at 3000 cycles in Figs. 16.7–16.14. Stress ratio $(R) = 0.1$; max load = 25 kN (assuming on an average ±10% variation, with normal distribution).

to void concentration, interphase thickness, and dispersion of the nanoparticles in the epoxy, and least sensitivity to particle and matrix modulus and strength properties (assuming on an average ±10% variation, with normal distribution).

16.6 Conclusions

The authors have described and applied a novel computational simulation approach to predict the static properties (strength, modulus, and Poisson's ratio), and fatigue life degradation of control and nanocomposite polymeric materials used in aerospace and automobile applications. The predictions were made using limited macro-level (coupon) test data [5, 8]. The predicted in situ constituent (particle and matrix) properties were then used to predict the fatigue life and perform probabilistic material uncertainty sensitivity analyses. The uncertainty analysis determined the effect of material (particle and matrix modulus and strength) and manufacturing parameters (void and particle volume fraction, dispersion, and interphase) affecting the tensile fatigue life for thermoset epoxy and their nanocomposites consisting of silica nanoparticles under tension–tension fatigue loading. The overall static and fatigue predictions matched the test data reasonably well within ±4% error.

From the work and research performed to date, the authors conclude the following:

1. The scatter inherent in the material data can be reproduced analytically by combining various methodologies (analytical, classical laminate theory, progressive failure analysis, probabilistic analysis) together and after characterizing in situ constituent material properties.
2. In general, a high degree of sensitivity to void concentration, interphase thickness, and dispersion of the nanoparticles in the epoxy was identified.
3. Sensitivity analysis can be used to reduce scatter in the material response by identifying the more sensitive material (nanoparticle and matrix modulus and strength) and manufacturing (voids, interphase, dispersion, etc.) random variables.
4. The validation with experimental results suggests the methodology is robust and may possibly be used for other nanocomposite material combination.

References

1. Uddin, M. F., and Sun, C. T. (2008). Strength of unidirectional glass/epoxy composite with silica nanoparticle-enhanced matrix. *Compos. Sci. Technol.* **68**:1637–1643.
2. Garg, M., Abdi, F., and McHugh, S. (2008). Analyzing interlaminar shear strength of multi-scale composites via combined finite element and progressive failure analysis approach. SAMPE 2008, Memphis Tennessee, September 2008.
3. Thostenson, E. T., Li, W. Z., Ren, Z. F., and Chow, T. W. (2002). Carbon nanotube/carbon fiber hybrid mutli-scale composites. *J. Appl. Phys.* **91**(9):6034–6037.

4. Qian, D., Dikey, E. C., Andrews, R., and Rantell, T. (2000). Load transfer and deformation mechanisms in carbon nanotube-polystyrene composites. *Appl. Phys. Lett.* **76**:2868–2870.
5. Manjunatha, C. M., Taylor, A. C., Kinloch, A. J., and Sprenger, S. (2010). The tensile fatigue behaviour of a silica nanoparticle-modified glass fibre reinforced epoxy composite. *Compos. Sci. Technol.* **70**(2010):193–199.
6. Manjunatha, C.M., Sprenger, S., Kinloch, A. J., and Taylor, A. C. (2010). The tensile fatigue behavior of a glass-fibre reinforced plastic composite using a hybrid-toughened epoxy matrix. *J. Compos. Mater.* **44**(17):2095–2109.
7. Hsieh, T. H., Kinloch, A.J., Masania, K., Lee, J. S., Taylor, A. C., and Sprenger, S. (2010). The toughened epoxy polymers and fibre composite modified with rubber microparticles and silica nanoparticles. *J. Mater. Sci.* **45**:1193–1210.
8. Jumahat, A., Soutis, C., Jones, R. F., Hodzic, A. (2010) Effect of silica nanoparticles on compressive properties of an epoxy polymer. *J. Mater. Sci.* **45**(21):5973–5983.
9. Mori, T., and Tanaka, K. (1973). Average stress in matrix and average elastic energy of materials with misfitting inclusions. *Acta Metall.* **21**(5):571–574.
10. Tandon, G. P., and Weng, G. J. (1984). The effect of aspect ratio of inclusions on the elastic properties of unidirectionally aligned composites. *Polym. Compos.* **5**(4):327–333.
11. Eshelby, J. D. (1957). The determination of the elastic field of an ellipsoidal inclusion and related problems. *Proc. R. Soc. London A* **241**(1226):376–396.
12. Courtney, T. H. (2000). *Mechanical Behavior of Materials* (2nd ed.). New York, McGraw-Hill.
13. Murthy, P. L. N., and Chamis, C. C. (1986). Integrated composites analysis (ICAN). Technical Paper 2515, Cleveland, Ohio, NASA.
14. *GENOA User Manual*. www.ascgenoa.com
15. Garg, M., Abumeri, G., and Abdi, F. (2010). Material properties characterization for multi-scale composites: glass/epoxy/silica nanoparticles. *J. Mater. Eval.* **68**(5):2010.
16. Tetard, L., Passian, A., Lynch, R. M., Voy, B. H., Shekhawat, G., Dravid, V., and Thundat, T. (2008). Elastic phase response of silica nanoparticles buried in soft matter. *Appl. Phys. Lett.* **93**(13), 133113-1–13113-3.

Chapter 17

Part 2
Multiscale Nanocomposite Fatigue Life Determination

Kamran Nikbin and Anthony J. Kinloch
Imperial College London, South Kensington Campus, London SW7 2AZ, UK
k.nikbin@imperial.ac.uk, a.kinloch@imperial.ac.uk

This chapter describes an approach to predict the effect of material and manufacturing uncertainty on the fatigue life of multiscale composites (epoxy reinforced with E-glass fibers and silica nanoparticles). A multiscale micromechanics technique was proposed to characterize the nanocomposite material consisting of matrix reinforced/infused with spherical-shaped nano-size silica particles in Part 1. The technique was used to first predict in situ material properties of constituents (linear elastic silica nanoparticles and nonlinear elastic matrix) from macro-level test data available from the literature. In this chapter, additional constituent in situ material properties of E-glass fibers are predicted to predict the fatigue properties of epoxy reinforced with 10 wt% of well-dispersed silica nanoparticles and E-glass fibers

Characterization of Nanocomposites: Technology and Industrial Applications
Edited by Frank Abdi and Mohit Garg

ISBN 978-981-4669-02-3 (Hardcover), 978-1-315-36489-6 (eBook)
www.panstanford.com

under tension fatigue loading. The predictions were validated with test and were found to be in good agreement (−6% error). The micromechanics fatigue durability simulation utilized closed-form solutions from the ideas of Mori–Tanaka, classical laminate theory, and multiscale multiphysics progressive failure analysis of a unit cell. At the end, probabilistic analysis was utilized to determine the sensitivity effects of manufacturing (i.e., void, silica particle and fiber volume fraction, interphase effect, and uniform versus nonuniform distribution) and matrix, nanoparticle and E-glass material properties parameters on the fatigue life of E-glass composites reinforced with silica nanoparticles.

17.1 Introduction

Advanced fiber-reinforced plastic (FRP) composite materials are being used increasingly for aerospace and automotive applications because of their higher strength-to-weight ratio compared to metals. However, the composite materials are weaker in the transverse direction (transverse to the fibers) than in the longitudinal direction, thus making them prone to delamination due to matrix failure. Test data in the literature [1, 2] indicate that the longitudinal composite compressive properties are highly dependent on the stiffness and strength of the matrix material. This is mainly because a stiffer matrix material resists the buckling of the fibers under an axially compressive load. The majority of engineering composite materials consist of a thermosetting epoxy matrix reinforced by continuous glass or carbon fibers. The epoxy, when polymerized, is an amorphous and highly cross-linked material. This cross-linked microstructure results in many useful properties such as a high modulus and failure strength, low creep, etc. However, it also leads to an undesirable property, whereby the polymer is relatively brittle and has a poor resistance to crack initiation and growth which may affect the overall fatigue and fracture performance of FRP composite. Two ways to enhance the mechanical properties of FRPs are (i) creating a strong interphase between the conventional fibers and matrix with randomly oriented nanoparticles, as shown in Fig. 17.1a, and (ii) enhancing the matrix properties by infusing it with a small weight fraction of nanoparticles, as shown in Fig. 17.1b. The first approach is relatively expensive than the second.

An example of the first approach is growing carbon nanotubes around the carbon fibers and then preparing the prepregs for use in the composite laminates [3, 4]. The interphase between conventional carbon fibers and neat matrix can consist of aligned or randomly oriented carbon nanotubes depending on the length and manufacturing process [3, 4]. The random and aligned interphase here refers to the phase where carbon nanotubes are grown with random or aligned orientation in the matrix, respectively. An example of the second approach is neat epoxy reinforced with randomly oriented carbon nanotubes and then used with regular conventional fibers (Fig. 17.1b). Another example would be infusing Nanopox F400 (SiO_2) spherical particles in diglycidyl ether of bisphenol A (DGEBA) epoxy [1, 5–8] into the fiber lay-up. The nanoparticle-reinforced matrix can then be used with conventional glass or carbon fibers for improved transverse tensile, compressive, shear and longitudinal compressive failure strengths. In the present chapter, the multiscale composite refers to conventional fibers, which are used with an enhanced matrix to improve the properties of the epoxy matrix by incorporating second phase fillers into the resin.

Manufacturing these advanced nanocomposite material system is quite challenging, for example: uniformly dispersing the nanoparticles in the matrix, improving the interface between the matrix and the particles for proper load transfer from matrix to the nanoparticles, and removing voids among others.

In Part 1, the in situ silica (Nanopox F400) nanoparticles and the DGEBA matrix (assuming uniformly dispersed nanoparticles) strength and stiffness material properties were characterized and predicted. In the present chapter, i.e., Part 2, the in situ E-glass fibers were characterized and used with the previously characterized nanoparticles and matrix details from Part 1. The limited test data was obtained from the literature [5–8]. The term *in situ* is used to represent the effective fiber, nanoparticles, and matrix properties, as they may vary when the constituents are cured together to

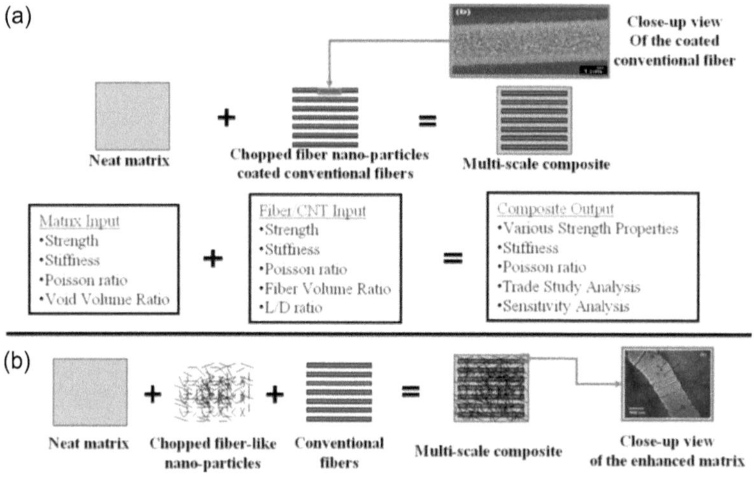

Figure 17.1 Micrographs of (a) coated conventional fiber [3] and (b) enhanced matrix [2].

form a multiscale composite material system versus their bulk state properties.

17.2 Experiment [5]

Manjunatha et al. (2010) presented their experimental process [5, 6] to create nanocomposite and multiscale composite material system that consisted of silica nanoparticle epoxy and E-glass fibers reinforcing infused with silica nanoparticle epoxy, respectively. The epoxy used in their experimental work was a standard DGEBA. The nanoparticles consisted of spherical-shaped nano-size (~20 nm diameter) silica particles (Nanopox F400). According to Qian et al. [5] and Manjunatha et al. [6], the nanoparticles were uniformly distributed in the matrix and constituted 10 wt% in the end product (epoxy/silica). Figure 17.2 shows the transmission electron microscope image of typical Nanopox F400 particles uniformly dispersed in epoxy [1, 5]. Later, this nanocomposite (epoxy/silica) was reinforced again with ~57% E-glass fibers (GFRP: glass fiber-reinforced plastic). Another sample was prepared that consisted of neat epoxy and E-glass fibers only (i.e., to provide a base line). More

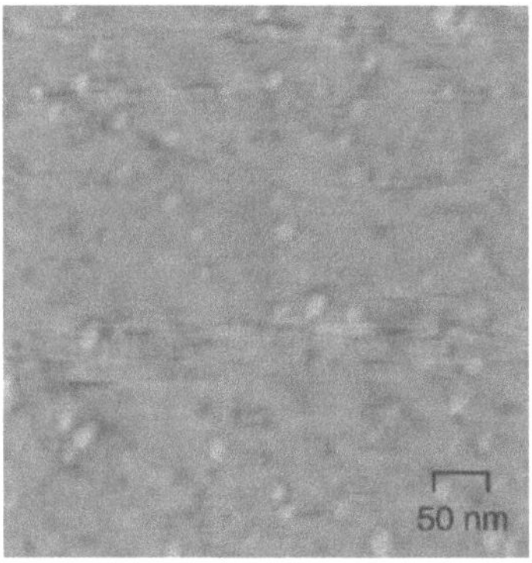

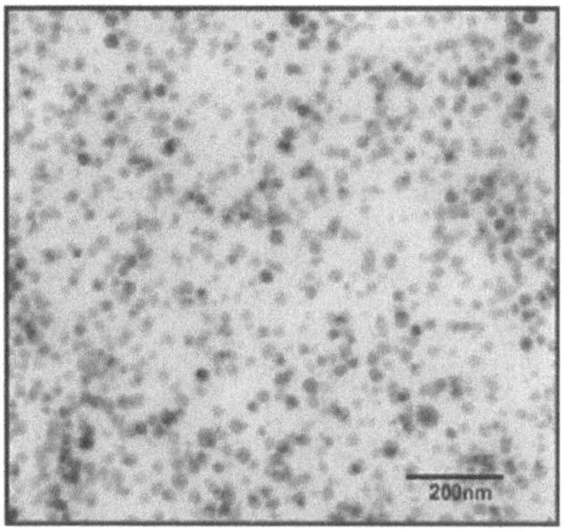

(a) 10 wt.% silica/epoxy resin [5] **(b) 15 wt.% silica/epoxy resin [1]**

Figure 17.2 (a) Atomic force microscope (AFM) and (b) transmission electron microscope (TEM) image.

Table 17.1 Stiffness and strength properties for DGEBA epoxy with and without 10 wt% Nanopox F400 silica (SiO_2) nanoparticles from test [1, 5, 6, 8]

Matrix material properties	Symbol	Control DGEBA	Nanocomposite DGEBA + Nanopox F400	Units	Ref.
Desnsity	Rhom	1.22E-09	1.28E-09	[tonne/mm^3]	2
Young's modulus	Em	2.62	3.1	[GPa]	1
Poisson's ratio	vm	–	–	[-]	
Tension strength	SmT	73.3	86.9	[MPa]	1
Compression strength	SmC	231.5	300.0	[MPa]	8
Shear strength	SmS	–	–	[MPa]	

details about the manufacturing process can be found elsewhere [5, 6].

Table 17.1 lists the control (neat epoxy) and nanocomposite (DGEBA + Nanopox F400) test properties extracted from the experimental work from literature [5]. Table 17.2 lists the modulus and strength of the GFRP composite with and without nanoparticles (Nanopox F400).

Table 17.2 Modulus and strength properties for GFRP laminate; Lay-up: $[(45/-45/0/90)_S]_2$. Epoxy with and without 10 wt% Nanopox F400 silica (SiO_2) nanoparticles from test [5, 6]; S11T corresponds to axial tension strength

GFRP properties	Symbol	Neat epoxy	Nanocomposite (10 wt.%) DGEBA + Nanopox F400	Units	Ref.
Modulus	E	17.5±0.6	18.8±0.7	[tonne/mm^3]	5, 6
Ultimate tensile strength	S11T	365±A 13	382±12	[GPa]	5, 6

17.3 Methodology

In Part 1, the authors proposed a methodology for simulating the fatigue durability of nanocomposite material system which consisted of first utilizing the closed-form solution from Mori–Tanaka, classical laminate theory (CLT), and Multiscale multiphysics progressive failure analysis of a unit cell or a representative volume element (RVE) and then applying a probabilistic analysis to determine the sensitivity of the fatigue life in the nanocomposites to the manufacturing (i.e. void volume fraction, interphase effect, and uniform versus nonuniform distribution) and material (matrix and nanoparticle modulus and strength) random variables.

Here, in Part 2, the authors use only the classical laminate theory (CLT) and multiscale multiphysics progressive failure analysis of a unit cell or a representative volume element (RVE) to predict the fatigue behavior of the GFRP laminates. Later, as in Part 1, the authors use the probabilistic analysis to determine the sensitivity of the fatigue life in the GFRP composites with and without nanoparticles considering the manufacturing and material random variables. The details of the closed-form analytical solution, multiscale micromechanics, multiscale progressive failure analysis, fatigue S-N curve, and probabilistic material uncertainty analysis are given in Part 1 of this chapter [17]. Part 1 [17] also lists other material modeling assumptions. The authors used MCQ-NANO and MCQ-Composites tools which incorporate the above-mentioned methodologies.

Table 17.3 In situ Nanopox F400 (SiO_2) nanoparticles reverse-engineered from nanocomposite test data [5, 8] (particle diameter = 20 nm; % weight fraction = 10%)

Particle material properties	Symbol	Effective	Units
Density	Rhop	1.8E-09	[tonne/mm^3]
Young's modulus	Ep	70	[GPa]
Poisson's ratio	vp	0.19	[-]
Tension strength	SpT	5.30E+04	[MPa]
Compression strength	SpC	1.03E+05	[MPa]

17.4 Results and Discussion

17.4.1 *Static Property Prediction/Test Validation of Control and Nanocomposite and GFRP*

Using the above methodology, and the Material Characterization and Qualification Nano (MCQ-Nano) tool and the limited test data listed in Table 17.1, the constituent (i.e., the Nanopox F400 nanoparticles and the DGEBA epoxy) in situ material properties were predicted and are listed in Tables 17.3 and 17.4 from Part 1 [17]. The in situ properties were found using a nanoparticle weight fraction of 10 wt% and void volume ratio of 0.0%. The authors modified the shear strength of the neat and predicted nanocomposite matrix (different from Part 1) in order to match the quasi-isotropic lay-up static strength test data from Refs. [5] and [6]. The modification in shear strength of the matrix will not affect the findings in Part 1 and

Table 17.4 In situ epoxy (LY556) reverse-engineered from nanocomposite test data [5, 8] (particle diameter = 20 nm; % weight fraction = 10%)

Matrix material properties	Symbol	Effective	Units
Desnsity	Rhom	1.22E-09	[tonne/mm^3]
Young's modulus	Em	2.6	[GPa]
Poisson's ratio	vm	0.35	[-]
Tension strength	SmT	73.3	[MPa]
Compression strength	SmC	231.5	[MPa]
Shear strength	SmS	116.8	[MPa]

Table 17.5 Comparison between predicted and test data for nanocomposite containing randomly oriented nanoparticles [5, 8] (particle diameter = 20 nm; weight fraction = 10%)

SiO_2/LY556 (Nanocomposite): FVR = 7.01%; VVR = 0.0%				
Material			SiO_2/LY556	
Property	**Units**	**Test**	**MCQ Nanocomposite**	**% Error**
Rhoc	[tonne/mm^3]	-	1.26E-09	-
Ec	[GPa]	3.07	3.0	−2.28
vc	[-]	0.35	0.34	−2.86
ScT	[MPa]	86.9	91.9	5.75
ScC	[MPa]	300	286.5	−4.50
ScS	[MPa]	-	130.3	-

was chosen arbitrarily there. The in situ nanoparticle modulus of 70 GPa is similar to the experimental data reported in the literature [16]. Table 17.6 lists the in situ (effective) E-glass fiber material properties, reverse-engineered using test data in Table 17.2 and from the literature [18]. Table 17.7, on the other hand, shows predicted unidirectional-ply properties using reverse-engineered epoxy, nanocomposite, and E-glass fiber properties in Tables 17.4, 17.5, and 17.6, respectively. Table 17.7, in addition, shows tape ply properties from test in Ref. [18] for higher fiber volume fraction of 60%.

Table 17.6 In situ fiber (E-glass) reverse-engineered from GFRP composite with neat epoxy test data [5, 8]: reverse-engineered assuming fiber and void volume fraction of 56% and 1.98%, respectively

Fiber material properties	Symbol	Effective	Units
Longitudinal young's modulus	Ef11	63.0	[GPa]
Transverse young's modulus	Ef22	63.0	[GPa]
Poisson's ratio	vf12	0.2	[-]
Poisson's ratio	vf23	0.2	[-]
Shear modulus	Gf12	26.3	[GPa]
Shear modulus	Gf23	26.3	[GPa]
Longitudinal tension strength	Sf11T	1640.0	[MPa]
Longitudinal compression strength	Sf11C	1205.0	[MPa]

Table 17.7 Predicted GFRP unidirectional-ply properties using in situ matrix, nanocomposite, and fiber properties from Tables 17.4, 17.5, and 17.6: reverse-engineered assuming fiber and void volume fraction of 56% and 1.98%, respectively

E-Glass/LY556 (Tape): FVR=56.0%; VVR=1.98%				
Material		**EGLASS/LY556**		
Property	**Units**	**Ref. test***	**Neat**	**Nano**
E11	[GPa]	45.6	35.7	35.9
E22	[GPa]	16.2	9.0	10.3
E33	[GPa]	-	9.0	10.3
G12	[GPa]	5.8	3.4	3.9
G13	[GPa]	-	3.4	3.9
G23	[GPa]	-	2.1	2.4
v12	[-]	0.28	0.27	0.26
v13	[-]	-	0.27	0.26
v23	[-]	-	0.38	0.37
S11T	[MPa]	1280.0	941.2	949.3
S11C	[MPa]	800.0	761.3	785.0
S22T	[MPa]	40.0	45.7	57.4
S22C	[MPa]	145.0	144.3	178.9
S33T	[MPa]	-	45.7	57.4
S33C	[MPa]	-	144.3	178.9
S12S	[MPa]	73.0	72.7	81.3
S13S	[MPa]	-	72.7	81.3
S23S	[MPa]	-	54.0	60.8

***Ref:** Fiber volume ratio = 60.1%; void volume ratio ≈ 1.98% [18].

17.4.2 *Fatigue Property Prediction and Test Validation*

The fatigue behavior of bulk epoxy, load-controlled, constant-amplitude, tension–tension cyclic-fatigue test results obtained for the neat and silica nanoparticle modified bulk epoxy polymers are shown in Fig. 17.3. It may be clearly seen that both predicted and tested overall fatigue life, for a given maximum cyclic stress, of the nanoparticle-modified epoxy is greater than that of the neat epoxy (i.e., the baseline values given in Fig. 17.3). The fatigue life enhancement of about three to four times in the nanoparticle-

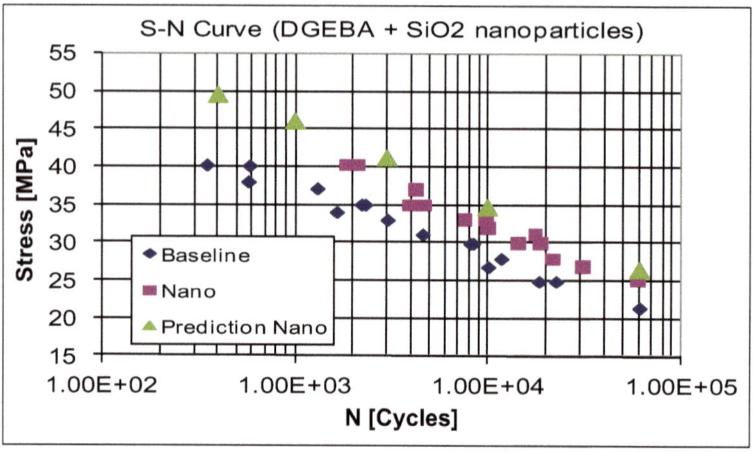

Figure 17.3 Comparison between tested and predicted data [5]. Stress ratio $(R) = 0.1$; max load = 25 kN.

modified epoxy is observed over the entire range of stress levels predicted in this chapter and tested. The predicted S-N curve is in good agreement for the whole range of the experimental results for the S-N curve [5].

Next, the authors reverse-engineered the S-N curve for the fiber degradation with cycles using test data from the literature [5] for the GFRP consisting of neat (control) epoxy (i.e., the blue dots in Fig. 17.4). The authors used the Material Characterization and Qualification (MCQ) Composites tool for this [14]. Figure 17.4 shows comparison between predicted and tested S-N curves for GFRP with neat epoxy using reverse-engineered S-N curves for E-glass fiber (Fig. 17.5) and neat epoxy (the blue curve in Fig. 17.3). Figure 17.4 also shows the predicted S-N curve for the GFRP composite with nanoparticles in the matrix. The prediction though not showing significant improvement is within the acceptable range. Note that the fiber S-N curve only represents the failure of the fiber tows with fatigue cycles, as usually fibers are considered to have no degradation with cycles. The uncertainty analysis in the next section will shed more light on this variation in the prediction compared to the test data for the nanocomposite matrix in the GFRP.

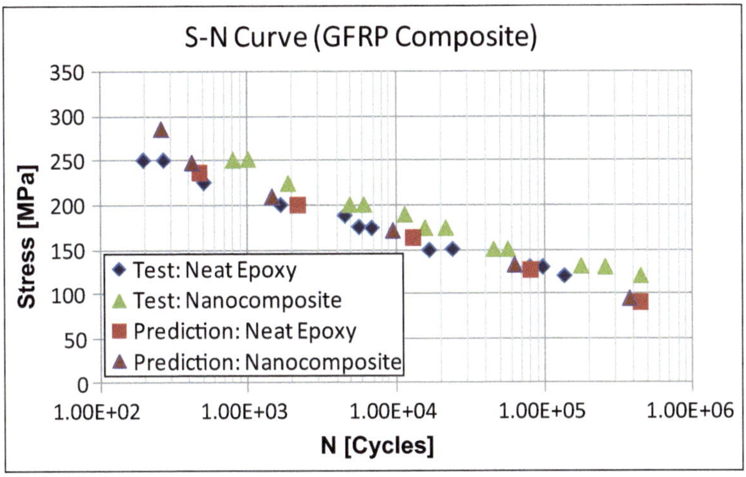

Figure 17.4 Comparison between tested and predicted data [5]. Stress ratio $(R) = 0.1$; max load = 25 kN.

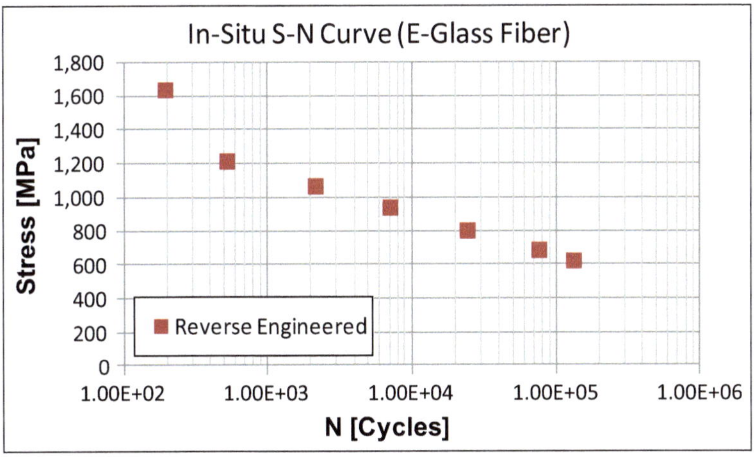

Figure 17.5 Reverse-engineered in situ E-glass fiber S-N curve. Stress ratio $(R) = 0.1$; max load = 25 kN. *Note*: Fiber S-N curve is reverse-engineered from Fig. 17.4 test data for neat epoxy.

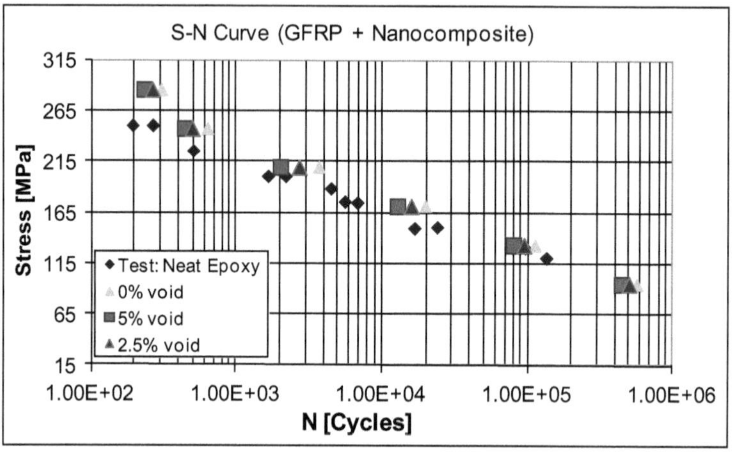

Figure 17.6 Effect of voids on fatigue life of GFRP composites. Stress ratio $(R) = 0.1$; max load = 25 kN; maximum 5% variation. Lay-up: $[(45/\text{-}45/0/90)_S]_2$.

17.4.3 *Material Uncertainty Analysis due to Manufacturing and Design Parameters on Fatigue Life*

In Part 1 the authors presented the effect of manufacturing and material uncertainty on the nanocomposite containing only silica nanoparticles in the DGEBA epoxy. In Part 2, the authors studied the effect of manufacturing and material uncertainty on the GFRP composites considering the variables from Part 1 and new ones that include aspects of the E-glass fiber manufacturing (i.e., fiber and void volume ratio, and fiber angle) and material (i.e., fiber modulus and strength) variables.

The effect of manufacturing (void, particle and fiber volume, interphase, and distribution of nanoparticles in the matrix) and constituent properties (nanoparticles/matrix/fibers) on the tensile fatigue behavior was evaluated. Figure 17.6 shows the effect of voids on the fatigue life of GRFP composite containing silica nanoparticles compared against the GFRP composite containing the control matrix from actual test data. In Part 1, the authors showed that 5% void volume ratio can significantly reduce the contributions from nanoparticles to below the control matrix. On the other hand, a 2.5%

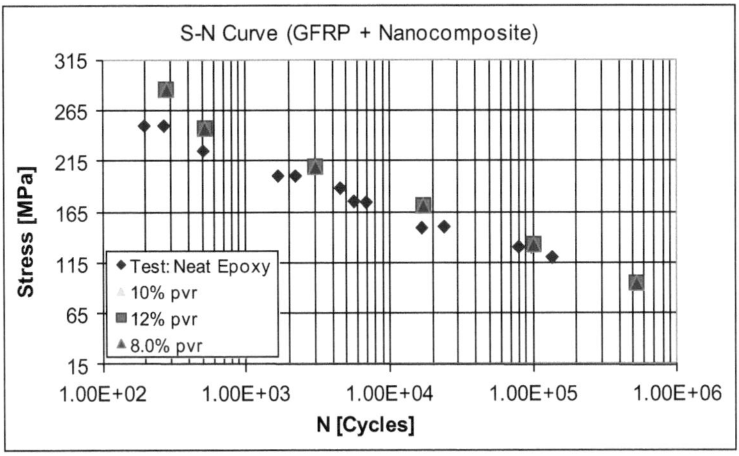

Figure 17.7 Effect of particle weight fraction on fatigue life of GFRP composites. Stress ratio $(R) = 0.1$; max load = 25 kN; pvr = particle weight fraction; maximum ± 20% variation. Lay-up: $[(45/\text{-}45/0/90)_S]_2$.

void volume ratio nullifies the contributions from the addition of Nanopox F400. For the GFRP, however, 5% variation in voids does not show a significant effect on the fatigue life.

Figure 17.7 shows the effect of variation in the Nanopox F400 particle weight fraction (pvr) on the fatigue life of the GFRP composites. It is clear from Fig. 17.7 that the fatigue life is insensitive to a slight variation (COV = ± 20%) in nanoparticle weight fraction.

Figures 17.8–17.9 show the effect of variations in the Nanopox F400 particle modulus (E_p) and strength (S_pT) and the control matrix modulus (E_m) and strength (S_mT) on the fatigue life of the nanocomposites matrices in the GFRP material. Figures 17.8–17.9 clearly show that the fatigue life is insensitive to a slight variation (COV = ± 11%) in the nanoparticle modulus. Note that the COV for particle strength was chosen to be ± 18%.

Figure 17.12 shows the effect of the interphase on the fatigue life of GFRP composites. Note that the interphase was assumed to be in between the nanoparticles and the epoxy of the GFRP matrix. The interphase is assumed to be plastically yielded due to thermal mismatch between the nanoparticles and matrix. The assumed properties for the interphase can be found in Part 1 of

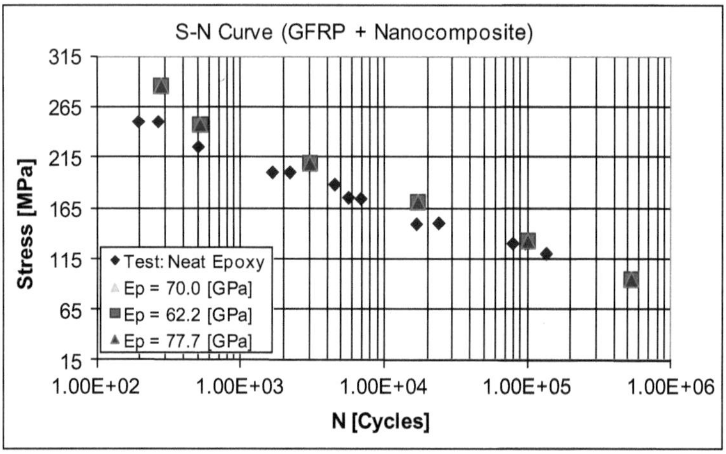

Figure 17.8 Effect of nanoparticle modulus on fatigue life of GFRP composites. Stress ratio $(R) = 0.1$; max load = 25 kN; maximum ±11.1% variation. Lay-up: $[(45/\text{-}45/0/90)_S]_2$.

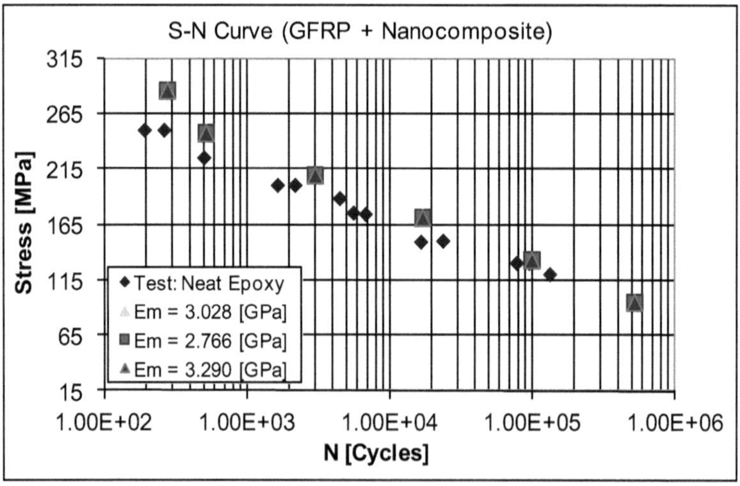

Figure 17.9 Effect of nanocomposite modulus on fatigue life of GFRP composites. Stress ratio $(R) = 0.1$; max load = 25 kN; maximum ±8.65% variation. Lay-up: $[(45/\text{-}45/0/90)_S]_2$.

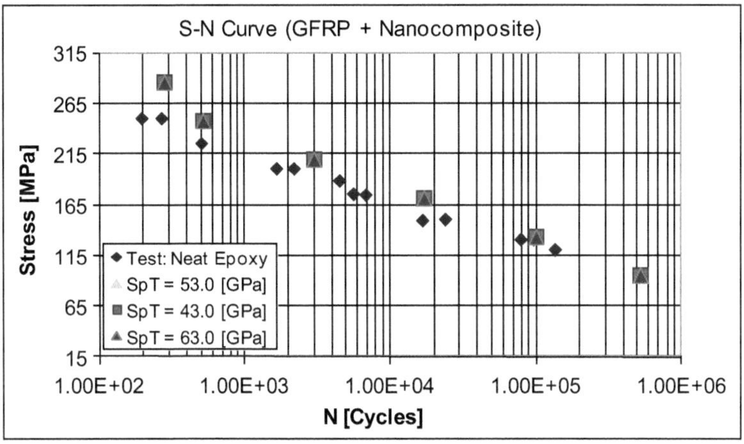

Figure 17.10 Effect of particle tensile strength on fatigue life of GFRP composites. Stress ratio $(R) = 0.1$; max load = 25 kN; maximum $\pm 18.8\%$ variation. Lay-up: $[(45/\text{-}45/0/90)_S]_2$.

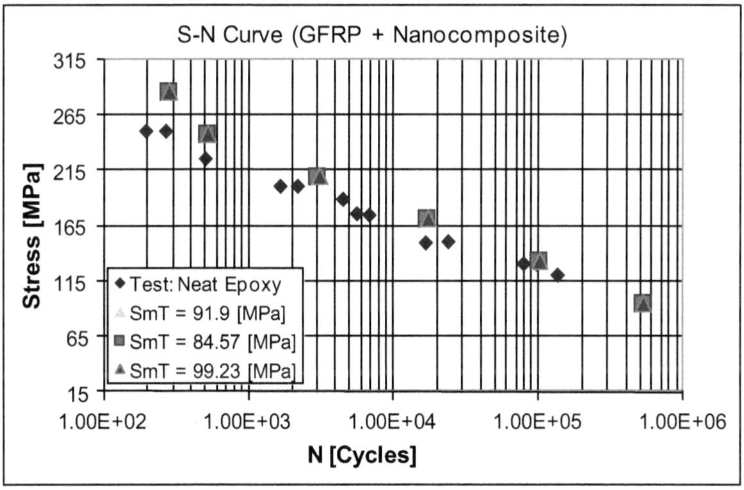

Figure 17.11 Effect of nanocomposite tensile strength on fatigue life of GFRP composites. Stress ratio $(R) = 0.1$; max load = 25 kN; maximum $\pm 8.0\%$ variation. Lay-up: $[(45/\text{-}45/0/90)_S]_2$.

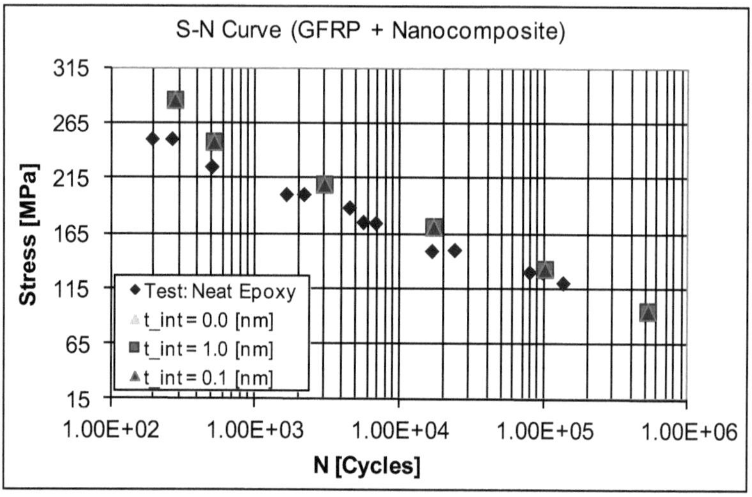

Figure 17.12 Effect of interphase and thickness on fatigue life of GFRP composites. Stress ratio (R) = 0.1; max load = 25 kN. Lay-up: $[(45/-45/0/90)_S]_2$.

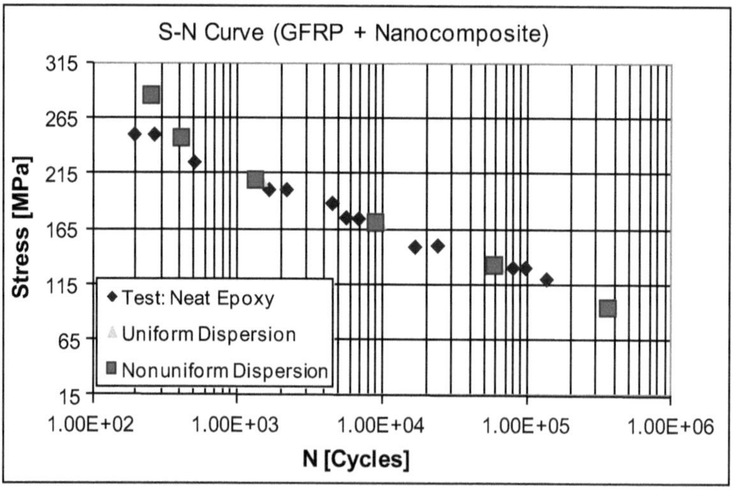

Figure 17.13 Effect of nanoparticle dispersion on fatigue life of GFRP composites. Stress ratio (R) = 0.1; max load = 25 kN. Lay-up: $[(45/-45/0/90)_S]_2$.

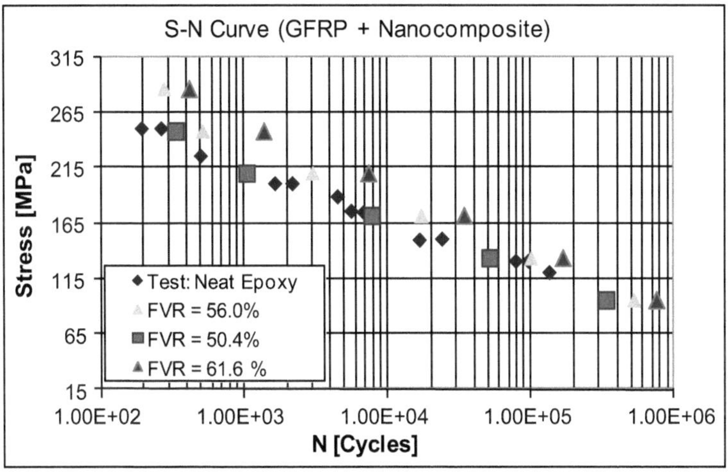

Figure 17.14 Effect of E-glass fiber volume fraction on fatigue life of GFRP composites. Stress ratio (R) = 0.1; max load = 25 kN; maximum ±10% variation. Lay-up: $[(45/\text{-}45/0/90)_S]_2$.

this publication [17]. The interphase is also assumed to be perfectly bonded to the particle and the matrix.

Figure 17.13 shows the effect of a uniform and non-uniform nanoparticle dispersion on the fatigue life of the GFRP composites. For non-uniform dispersion of nanoparticles, the authors assumed large spherical shape particles consisting of agglomerates of nanoparticles in neat matrix. The authors assumed the ratio of agglomerates to the nanocomposite volume to be 0.5 and that of nanoparticles to agglomerates to be 1.0. All other variables were same as the baseline case (Fig. 17.6). Figure 17.13 also shows a negligible effect of the dispersion of particles on the fatigue life of the GFRP composites. This is consistent with the results given in Figs. 17.8–17.9, as usually the non-uniform dispersion effects the overall modulus and strength of the nanocomposites based on results presented in Part 1 [17].

Figures 17.14–17.17 focus on a variation in material and manufacturing properties related to E-glass fibers. Figures 17.14–17.16 show the effect on fatigue life of GFRP composites containing nanocomposites as matrix material with a variation in the fiber volume fraction, fiber modulus, and fiber strength, respectively. Clearly,

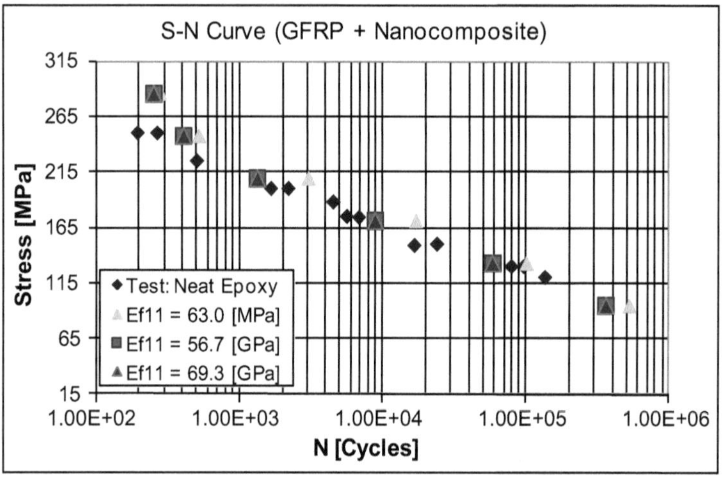

Figure 17.15 Effect of E-glass fiber modulus on fatigue life of GFRP composites. Stress ratio $(R) = 0.1$; max load = 25 kN; maximum ±10% variation. Lay-up: $[(45/\text{-}45/0/90)_S]_2$.

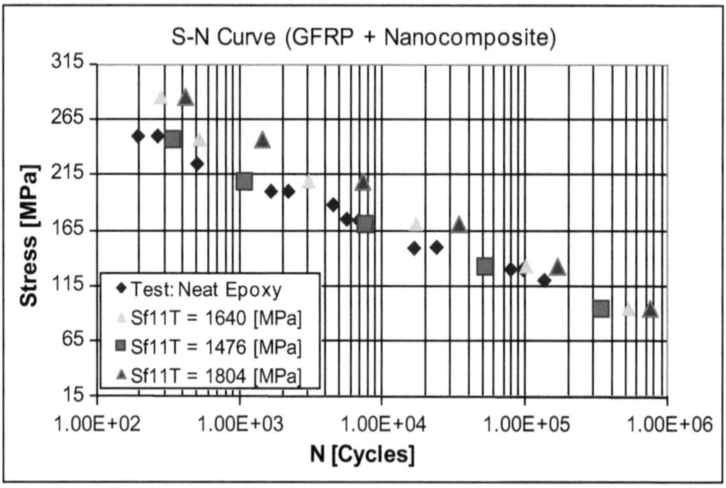

Figure 17.16 Effect of E-glass fiber tensile strength on fatigue life of GFRP composites. Stress ratio $(R) = 0.1$; max load = 25 kN; maximum ±10% variation. Lay-up: $[(45/\text{-}45/0/90)_S]_2$.

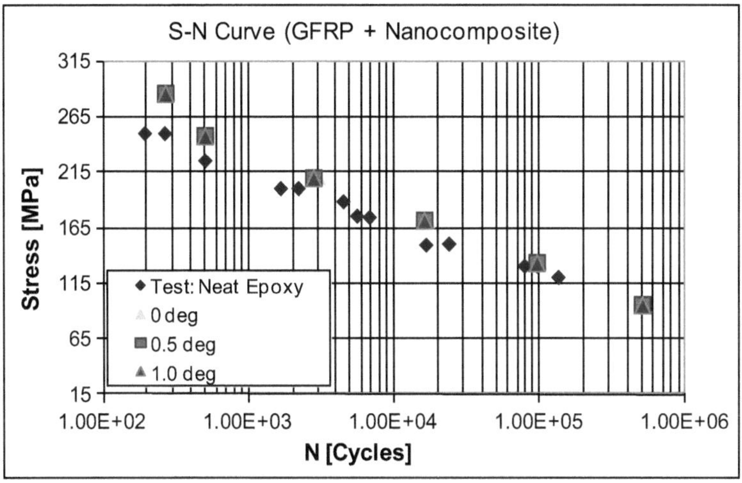

Figure 17.17 Effect of fiber misalignment on fatigue life of GFRP composites. Stress ratio $(R) = 0.1$; max load = 25 kN; maximum 1.1% variation. Lay-up: $[(45/\text{-}45/0/90)_S]_2$.

the fatigue life of the GFRP composites is affected significantly with a variation in the E-glass fiber–related properties relative to the matrix related properties shown in Figures 17.7–17.13 earlier.

As shown in Figs. 17.14–17.16, the variables of the fiber volume ratio, fiber modulus, and fiber tensile strength, all shift the S-N curve of the GFRP composites and brings the S-N curve in close agreement with the reference test data [5]. Figure 17.17 indicates that slight fiber misalignment (for the 0 degree plies only) does not affect the S-N curve of the GFRP composites significantly. It is important to note that the variation in figure misalignment selected here is typical in GFRP composites.

In summary, the results shown above (Figs. 17.6–17.17) and the probabilistic sensitivity analysis (Figs. 17.18 and 17.19) indicate that the fatigue properties of the GFRP composites reinforced with Nanopox F400 nanoparticles show a high degree of sensitivity to glass fiber–related material and manufacturing (except the fiber misalignment) variables, and least sensitivity to nanoparticles and matrix material and manufacturing variables (assuming on an average ±10% variation, with normal distribution) and quasi-

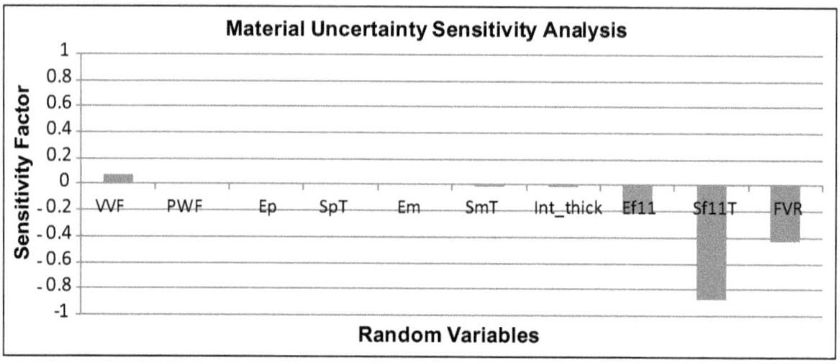

Figure 17.18 Material uncertainty analysis considering effect of material and manufacturing variables on fatigue life of nanocomposites at 3000 cycles in Figs. 17.6–17.17. Stress ratio $(R) = 0.1$; max load = 25 kN. VVF = void volume factor; PWF = particle weight fraction; E_p = particle modulus; S_pT = particle tensile strength; E_m = matrix modulus; S_mT = matrix tensile strength; Int_thick = interphase thickness (assuming on an average ± 10% variation, with normal distribution); E_{f11} = fiber tensile modulus; $S_{f11}T$ = fiber tensile strength; FVR = fiber volume ratio. Lay-up: $[(45/\text{-}45/0/90)_S]_2$.

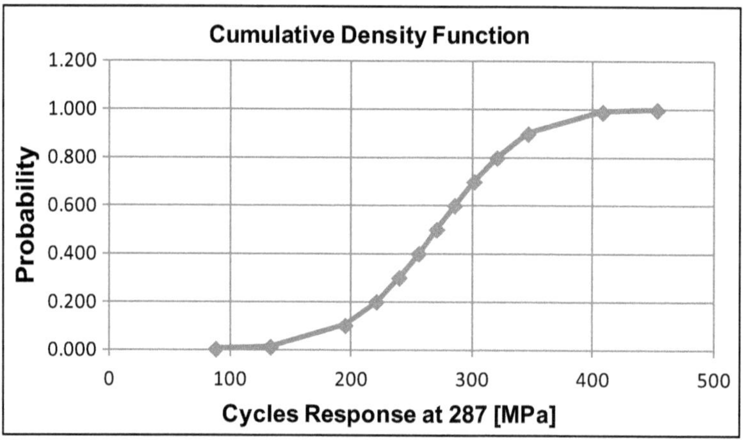

Figure 17.19 Cumulative density function from material uncertainty analysis considering effect of material and manufacturing variables on fatigue life of GFRP composites at 287 MPa in Figs. 17.6–17.17. Stress ratio $(R) = 0.1$; max load = 25 kN (assuming on an average ±10% variation, with normal distribution). Lay-up: $[(45/\text{-}45/0/90)_S]_2$.

isotropic lay-up. Figure 17.19 shows that the scatter can produce higher cycles to failure for a maximum applied load of 287 MPa and reduce the prediction error compared to the reported test data [5]. Note that changing the composite lay-up may change the sensitivity results; for instance, a lay-up that is dominated by matrix-related failure (e.g., with more shear or transverse plies) may show more dependence on the presence of the nanoparticles and matrix-related properties than for the fiber-related properties.

17.5 Conclusions

The authors described and applied a computational simulation approach to predict the static properties (strength, modulus, and Poisson's ratio), and fatigue life degradation of GFRP composites reinforced with silica nanoparticles (Nanopox F400). The predictions were made using limited macro-level (coupon) test data [5, 8, 18]. The predicted in situ constituent (fiber, nanoparticle, and matrix) properties were then used to predict the fatigue life and perform probabilistic material uncertainty sensitivity analyses for GFRP composites reinforced with nanoparticles in the epoxy matrix. The uncertainty analysis determined the effect of material (fiber, nanoparticle, and matrix modulus and strength) and manufacturing parameters (void volume fraction, nanoparticle and glass fiber volume fraction, dispersion, and interphase) affecting the tensile fatigue life for the GFRP composites containing silica nanoparticles under tension–tension fatigue loading. The overall static and fatigue predictions matched the test data reasonably well with a −6% error and a −10% error for GFRP composites with and without nanoparticles, respectively.

From the work and research performed and presented above, the authors conclude the following:

1. The scatter inherent in the material data can be reproduced analytically by combining various methodologies (analytical, classical laminate theory, progressive failure analysis, and probabilistic analysis) together and after characterizing in situ constituent (fiber, nanoparticle, and matrix) material properties.

2. In general, a high degree of sensitivity to fiber modulus, tension strength, and volume fraction in the epoxy was identified for quasi-isotropic lay-up and not significantly on nanoparticles and epoxy properties.
3. A sensitivity analysis can be used to reduce scatter in the material response (fatigue cycles) by identifying the more sensitive material (fiber, nanoparticle, and matrix modulus and strength) and manufacturing (voids, interphase, dispersion, etc.) random variables.
4. The validation with experimental results suggests that the methodology is robust and may possibly be used for other nanocomposite-and fiber-reinforced composite material combinations.

References

1. Uddin, M. F., and Sun, C. T. (2008). Strength of unidirectional glass/epoxy composite with silica nanoparticle-enhanced matrix. *Compos. Sci. Technol.* **68**:1637–1643.
2. Garg, M., Abdi, F., and McHugh, S., (2008). Analyzing interlaminar shear strength of multi-scale composites via combined finite element and progressive failure analysis approach. SAMPE 2008, Memphis Tennessee, September 2008.
3. Thostenson, E. T., Li, W. Z., Ren, Z. F., and Chow, T. W. (2002). Carbon Nanotube/carbon fiber hybrid mutli-scale composites. *J. Appl. Phys.* **91**(9):6034–6037.
4. Qian, D., Dikey, E. C., Andrews, R., and Rantell, T. (2000). Load transfer and deformation mechanisms in carbon nanotube-polystyrene composites. *Appl. Phys. Lett.* **76**:2868.
5. Manjunatha, C. M., Taylor, A. C., Kinloch, A. J., and Sprenger, S. (2010). The tensile fatigue behaviour of a silica nanoparticle-modified glass fibre reinforced epoxy composite. *Compos. Sci. Technol.* **70**(2010):193–199.
6. Manjunatha, C. M., Sprenger, S., Kinloch, A. J., and Taylor, A. C. (2010). The tensile fatigue behavior of a glass-fibre reinforced plastic composite using a hybrid-toughened epoxy matrix. *J. Compos. Mater.* **44**(17):2095–2109.

7. Hsieh, T. H., Kinloch, A. J., Masania, K., Lee, J. S., Taylor, A. C., and Sprenger, S. (2010). The toughened epoxy polymers and fibre composite modified with rubber microparticles and silica nanoparticles. *J. Mater. Sci.*, **45**:1193–1210.
8. Jumahat, A., Soutis, C., Jones, R. F., and Hodzic, A. (2010). Effect of silica nanoparticles on compressive properties of an epoxy polymer. *J. Mater. Sci.* **45**(21):5973–5983.
9. Mori, T., and Tanaka, K. (1973). Average stress in matrix and average elastic energy of materials with misfitting inclusions. *Acta Metall.* **21**(5):571–574.
10. Tandon, G. P., and Weng, G. J. (1984). The effect of aspect ratio of inclusions on the elastic properties of unidirectionally aligned composites. *Polym. Compos.* **5**(4):327–333.
11. Eshelby, J. D. (1957). The determination of the elastic field of an ellipsoidal inclusion and related problems. *Proc. R. Soc. London A* **241**(1226):376–396.
12. Courtney, T. H. (2000). *Mechanical Behavior of Materials* (2nd ed.). New York, McGraw-Hill.
13. Murthy, P. L. N., and Chamis, C. C. (1986). Integrated composites analysis (ICAN). Technical Paper 2515, Cleveland, Ohio, NASA, 1986.
14. *GENOA User Manual*. www.ascgenoa.com
15. Garg, M., Abumeri, G., and Abdi, F. (2010). Material properties characterization for multi-scale composites: glass/epoxy/silica nanoparticles. *J. Mater. Eval.* **68**(5): 542–547.
16. Tetard, L., Passian, A., Lynch, R. M., Voy, B. H., Shekhawat, G., Dravid, V., and Thundat, T. (2008). Elastic phase response of silica nanoparticles buried in soft matter. *Appl. Phys. Lett.* **93**(13):133113-1–13113-3.
17. Garg, M., Abdi, F., Kinloch, A. J., and Nikbin, K. (2011). Part 1: Multi-scale nanocomposite fatigue life determination. SAMPE 2011, Fort Worth, Texas, October, 2011.
18. Hinton, M. J., Kaddour, A. S., and Soden, P. D. (2004). Predictive capabilities of nineteen failure theories and design methodologies for polymer composite laminates. Part B: Comparison with experiments. *Failure Criteria in Fiber Reinforced Polymer Composites: The World-Wide Failure Exercise*, ed. Hilton, M. J., Kaddour, A. S., and Soden, P. D. Elsevier, Kidlington, Oxford, UK, pp. 1073–1221.

Chapter 18

Stress Analysis and Fracture in Nanolaminate Composites

Christos C. Chamis*
Former Senior Aerospace Scientist, NASA Glenn Research Center, Cleveland, OH 44135, USA
christos.c.chamis@aol.com

A nanolaminate subjected to bending is analyzed with respect to stresses in a nanoelement. A composite mechanics computer code based on constituent properties and nanoelement formulation is used for evaluation of the nanoelement stresses. The results indicated that the computer code is sufficient for the analysis. The results also show that nanoelement stresses exceed their corresponding matrix-dominated strengths and the nanofiber fracture when a stress concentration factor is used.

18.1 Introduction

A nanocomposite is analyzed which is subjected to a bending moment. The reasons for this type of composite are that (i) the

*Retired.

Characterization of Nanocomposites: Technology and Industrial Applications
Edited by Frank Abdi and Mohit Garg

ISBN 978-981-4669-02-3 (Hardcover), 978-1-315-36489-6 (eBook)
www.panstanford.com

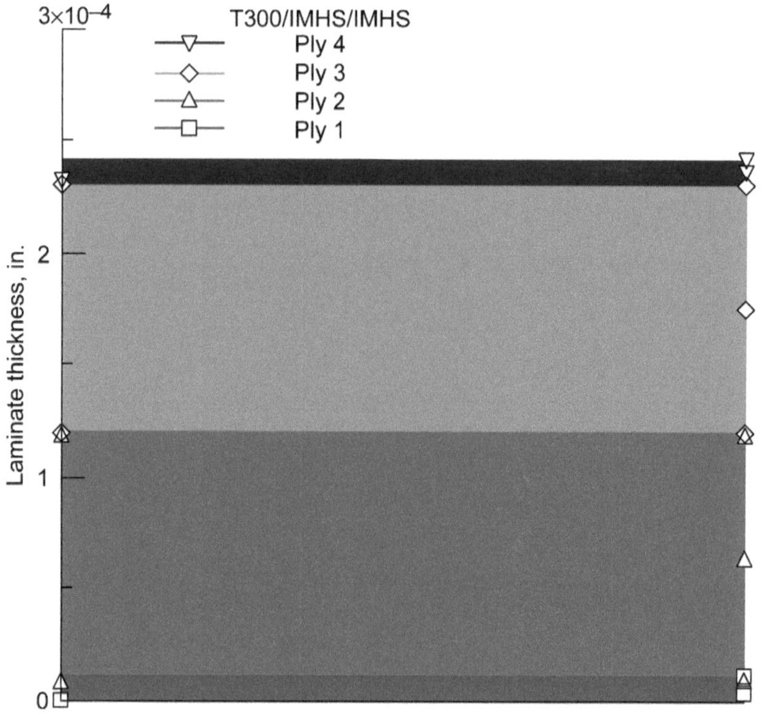

Figure 18.1 Nanolaminate though-thickness schematic (1 in. = 2.54 cm).

bottom ply and the top ply can be divided into subplies and the magnitude of the stresses can be evaluated, and (ii) there is not a comparable analysis in the literature. Both plies are assumed to be made from a single nanofiber composite, where as the core is made from 10 aligned nanofibers' composite. A schematic of the nanocomposite section is shown in Fig. 18.1, while that for the sliced subplies is shown in Fig. 18.2. Note that the two plies are subsliced to 10 slices each to obtain a detailed stress distribution of the maximum bending stresses in the section. The stress analysis includes the stress distribution through laminate thickness and the stresses in the two substructured plies. In addition, the corresponding strains are also computed for completeness. The objective of this chapter is to describe the stresses in a nanoelement

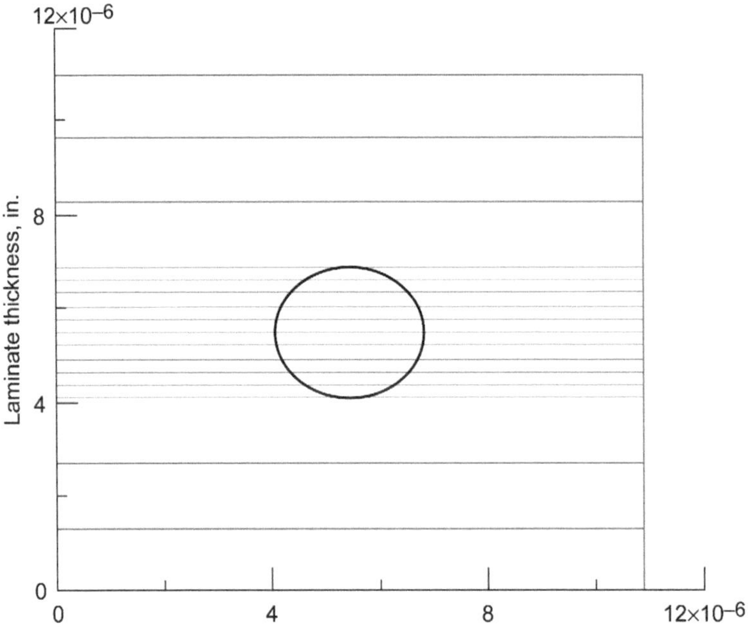

Figure 18.2 Mono-nanofiber ply (in.) (1 in. = 2.54 cm).

of a unidirectional nanolaminate subjected to a bending moment along the x direction.

18.1.1 *Nanocomposite Simulation Properties*

The constituent properties of the fiber and the matrix are given in Table 18.1 for fibers and Table 18.2 for matrix. Note that the nanofibers areassumed to be made of Thornell (T300) with modulus of 6.9×10^9 Pa (1 billion psi) and a tensile strength of 0.552 GPa (800,000) psi and a fiber diameter of (110 nm) (2.756×10^{-6}) in. The matrix is assumed to be an intermediate-modulus high tensile strength matrix. The main reason for these types of fibers and matrix is that the properties for these two are available as seen in Tables 18.1 and 18.2. The nanocomposite is usually made with relatively low fiber volume ratio (about 0.05) [1]. Entering this value and the values in Tables 18.1 and 18.2 in the composite mechanics computer code (ICAN/JAVA) [2], the following properties are predicted:

Table 18.1 T300 graphite fiber properties

Description	Symbols	Value	Units
Number of fibers per end	Nf	1.0	number
Filament equivalent diameter	df	2.756×10^{-6}	in.
Weight density	Rhof	0.064	lb/in.**3
Normal moduli (11)	Ef11	1.0×10^{9}	psi
Normal moduli (22)	Ef22	7.0×10^{7}	psi
Poisson's ratio (12)	Nuf12	0.2	Nondimensional
Poisson's ratio (23)	Nuf23	0.25	Nondimensional
Shear moduli (12)	Gf12	5.0×10^{7}	psi
Shear moduli (23)	Gf23	3.5×10^{7}	psi
Thermal expansion coefficient (11)	A1faf11	-5.5×10^{-7}	in./in./°F
Thermal expansion coefficient (22)	Alfaf22	5.6×10^{-6}	in./in./°F
Heat conductivity (11)	Kf11	444.0	Btu/hr/ft^2/°F/in.
Heat conductivity (22)	Kf22	4.0	Btu/hr/ft^2/°F/in.
Heat capacity	Cf	0.22	Btu/lb/°F
Dielectric strength (11)	KeF11	0.0	V/in.
Dielectric strength (22)	Kef22	0.0	V/in.
Dielectric constant (11)	Gamma11	0.0	in./V
Dielectric constant (22)	Gamma22	0.0	in./V
Capacitance	Cef	0.0	V
Resistivity	Ref.	0.0	Ω-in.
Tensile strength	SfT	8.0×10^{5}	psi
Compressive strength	SiC	6.0×10^{5}	psi
Shear strength	SfS	4.0×10^{5}	psi
Normal damping capacity (11)	psi11f	0.38	%Energy
Normal damping capacity (22)	psi22f	6.3	%Energy
Shear damping capacity (12)	psi12f	3.34	%Energy
Shear damping capacity (23)	psi23f	6.3	%Energy
Melting temperature	TMf	6000.0	°F

Notation: Multiply: inches by 2.54 cm; psi by 6.9 Pa; in./in./°F by 9/5 cm/cm/°C (1 btu = 1055 J; 1 lb = 4.448 N).

(i) The geometric properties of the nanocomposite fibers' volume ratio, void volume ratio, and matrix volume ratio are shown in Fig. 18.3. (ii) The laminate configuration thickness ply configuration is shown in Fig. 18.4 (iii) The laminate thermal expansion coefficient are shown in Fig. 18.5. (iv) The laminate fiber–dominated modules $E_{\ell 11}$ is shown in Fig. 18.6. (v) The matrix dominated moduli ($E_{\ell 22}$, $E_{\ell 33}$) are shown in Fig. 18.7. (vi) The corresponding shear moduli are shown in Fig. 18.8, and (vii) the corresponding Poisson's ratios

Table 18.2 Polymer matrix properties

Description	Symbols	Value	Units
Weight density	Rhom	0.044	lb/in.**3
Normal modulus	Em	500000.0	psi
Poisson's ratio	Num	0.35	Nondimensional
Thermal expansion coefficient	Alfa m	3.6×10^{-5}	in./in./°F
Heat conductivity	Km	0.008681	Btu/hr/ft^2/°F/in.
Heat capacity	Cm	0.25	Btu/lb/°F
Dielectric strength	Kem	0.0	V/in.
Dielectric constant	Gammam	0.0	in./V
Capacitance	Cem	0.0	V
Resistivity	Rem	0.0	Ω-in.
Moisture expansion coefficient	Betam	0.0033	in./in./%moisture
Diffusivity	Dm	2.16×10^{-7}	in.**2/hr
Saturation	Mm	0.0	%moisture
Tensile strength	SmT	15000.0	psi
Compressive strength	SmC	35000.0	psi
Shear strength	SmS	13000.0	psi
Allowable tensile strain	eps mT	0.02	in./in.
Allowable compression strain	eps mC	0.05	in./in.
Allowable shear strain	eps mS	0.035	in./in.
Allowable torsional strain	eps mTOR	0.035	in./in.
Normal damping capacity	psiNM	6.6	%energy
Shear damping capacity	psiSm	6.9	%energy
Void heat conductivity	Kv	0.0012	Btu/hr/in./°F
Glass transition temperature	Tgdr	420.0	°F
Melting temperature	TMm	0.0	°F

See Table 18.1 for conversion factors.

are shown in Fig. 18.9. The significance of the laminate properties shown in Figs. 18.6–18.9 are that (i) the spikes shown on the two substructure plies represent the thickness averages (these two plies have 1/10 the average thickness) and (ii) they illustrate the discontinuity that exists between the core and the two end plies. Note that in Fig. 18.5, the matrix-dominated thermal expansion coefficients $\alpha_{\ell 22}$ and $\alpha_{\ell 33}$ is about 1/4 of the $\alpha_{\ell 11}$ (fibers dominated). This type of discontinuity is also apparent in Figs. 18.6–18.8. The discontinuity effects are somewhat different in Fig. 18.9. In this figure the fiber-dominated Poisson's ratio $\nu_{\ell 21}$ has opposite effect than the Poisson's ratio $\nu_{\ell 23}$, which is matrix dominated.

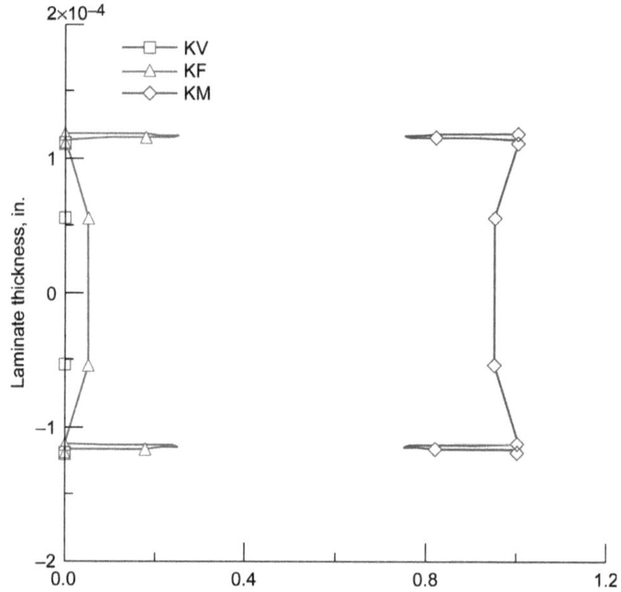

Figure 18.3 Fabrication variables (ratios).

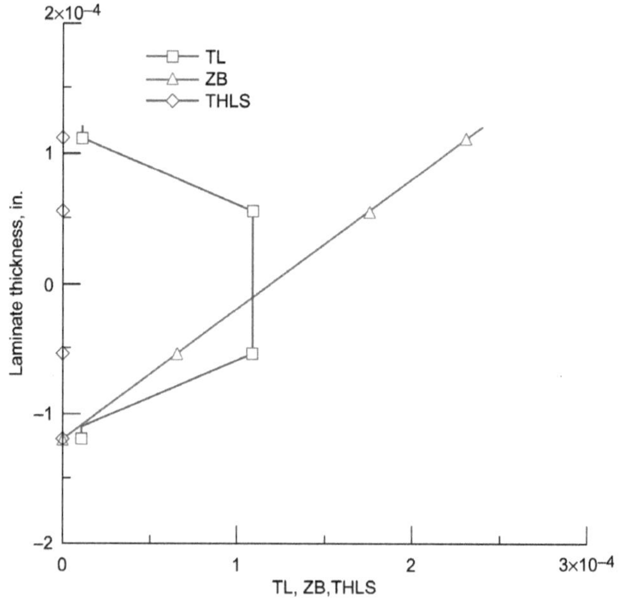

Figure 18.4 Nanofiber ply geometry (in. = 2.54 cm).

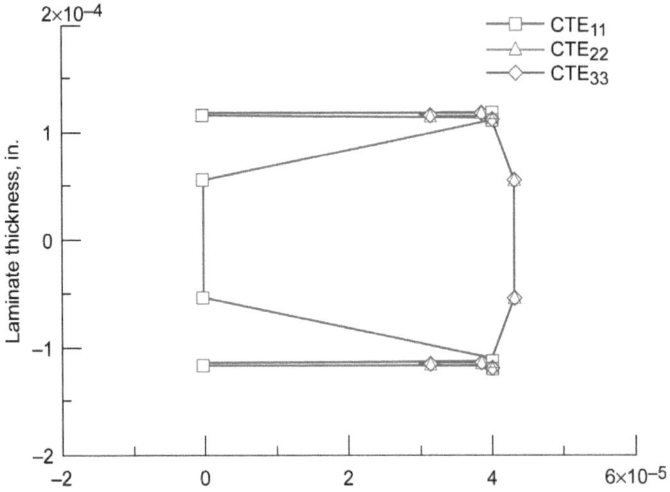

Figure 18.5 Thermal expansion coefficients (in./in./°F) (in./in./°F = 9/5 cm/cm/°C; 1 in. = 2.54 cm).

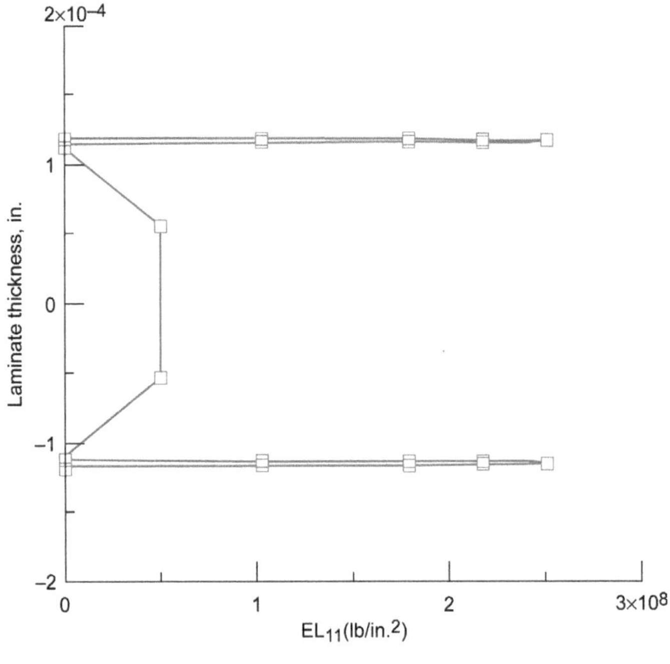

Figure 18.6 Laminate modulus-fiber direction (lb/in.^{2} = 6.9 Pa; 1 in. = 2.54 cm).

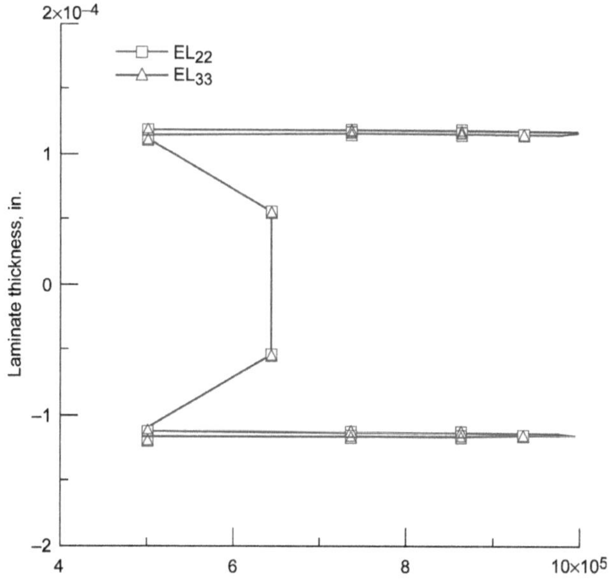

Figure 18.7 Laminate transverse moduli (psi) (1 psi = 6.9 Pa; 1 in. = 2.54 cm).

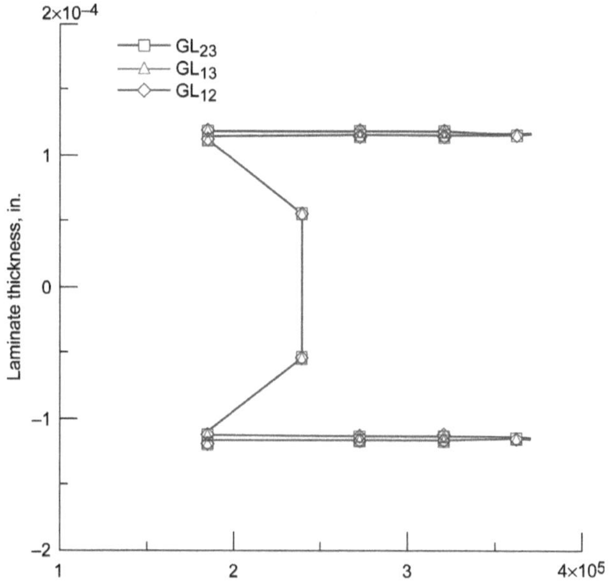

Figure 18.8 Laminate shear moduli (psi) (1 psi = 6.9 Pa; 1 in. = 2.54 cm).

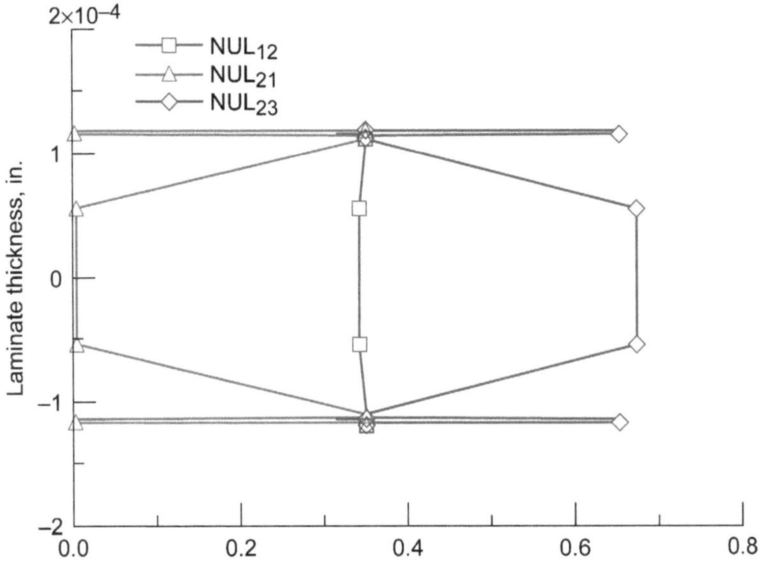

Figure 18.9 Laminate Poisson's ratios (ratios).

The laminate fiber–dominated strengths ($S_{\ell 11T}$ and $S_{\ell 11C}$) are shown in Fig. 18.10. This figure shows the same trend for the spikes in the two end plies ($S_{\ell 11T}$ and $S_{\ell 11C}$) that were described above. The spike effects are not as pronounced in Fig. 18.11, which shows the matrix-dominated strengths ($S_{\ell 22T}$, $S_{\ell 22C}$, $S_{\ell 125}$, $S_{\ell 235}$). It is worth commenting that the results described thus far are independent of loading.

The strains through the laminate thickness are shown in Fig. 18.12. In this figure, the two major assumptions for bending deformation through the thickness are clearly shown. The bending strain due to bending moment is linearly varying, while the Poisson strain (green) is also linearly varying, but with opposite slope, as it should be. On the other hand, the axial strain (blue line) is "o," as it should be, since there is the in-plane axial loading.

The corresponding stresses are shown in Fig. 18.13. Note the two spikes indicate that this laminate behaves more like a sandwich where the core stresses are negligible (Fig. 18.14) compared to the stresses in the two end plies where the stresses are of opposite signs. Note also that the transverse ply stress and the in-plane shear stress

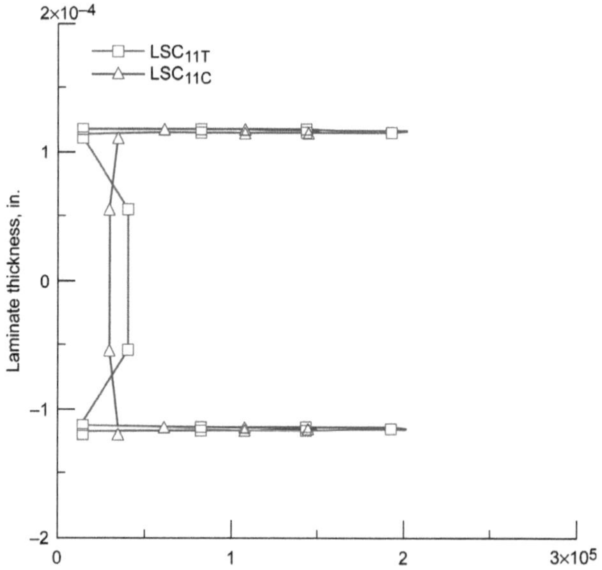

Figure 18.10 Laminate longitudinal strengths (psi) (1 psi = 6.9 Pa; 1 in. = 2.54 cm).

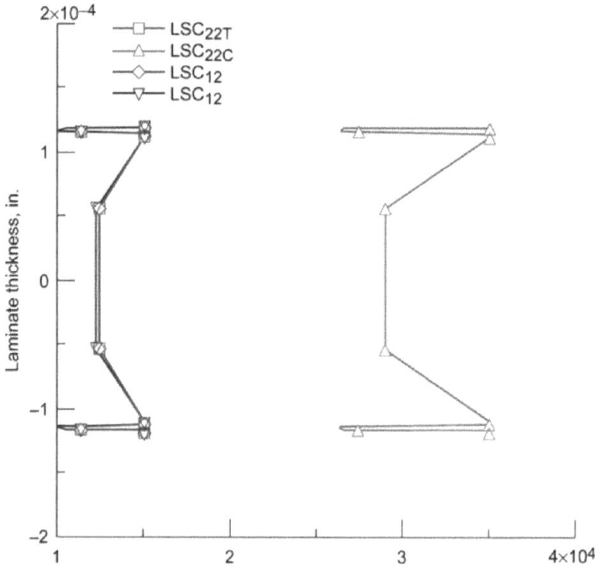

Figure 18.11 Laminate matrix dominated strengths (psi) (psi = 6.9 Pa; 1 in. = 2.54 cm).

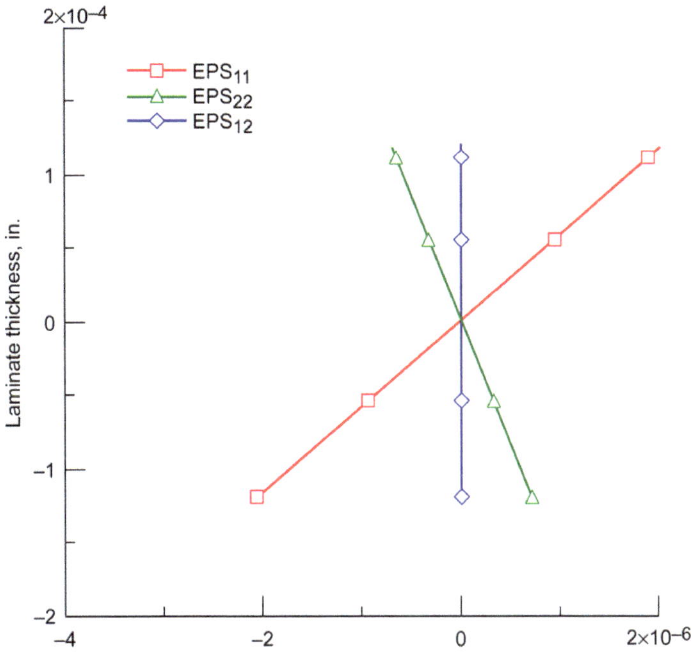

Figure 18.12 Laminate bending strains (in./in.) (in./in. = cm/cm.)

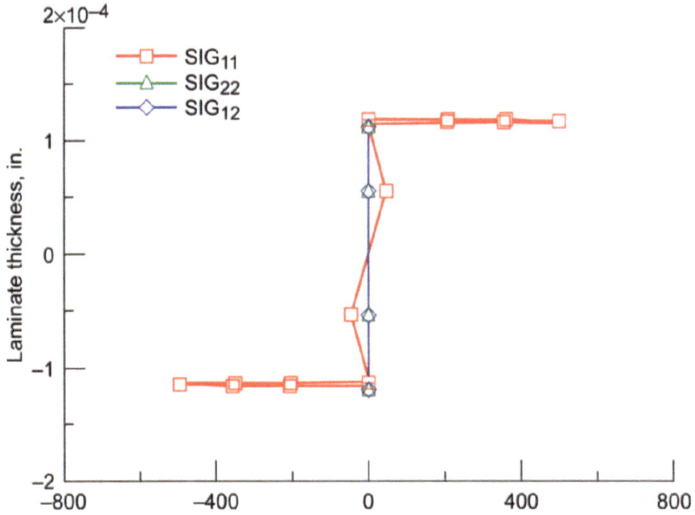

Figure 18.13 Laminated bending stresses (psi) (psi = 6.9 Pa; 1 in. = 2.54 cm).

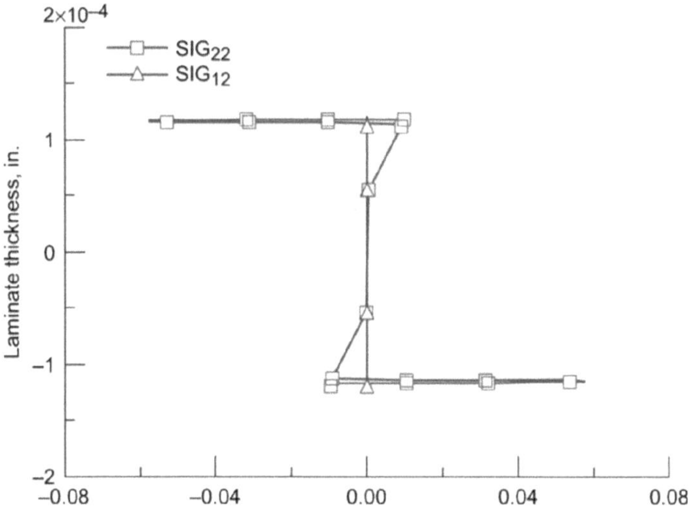

Figure 18.14 Laminated magnified matrix dominated bending stresses (psi) (psi = 6.9 Pa; 1 in. = 2.54 cm).

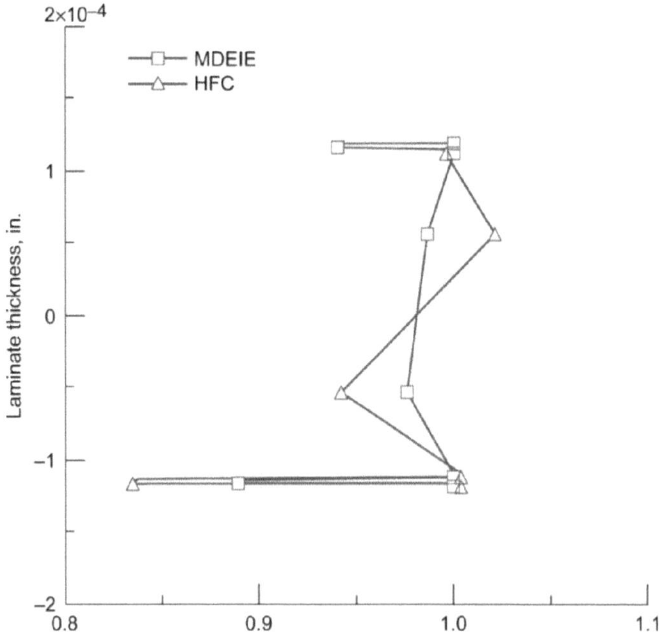

Figure 18.15 Combine stress failure criteria (ratios).

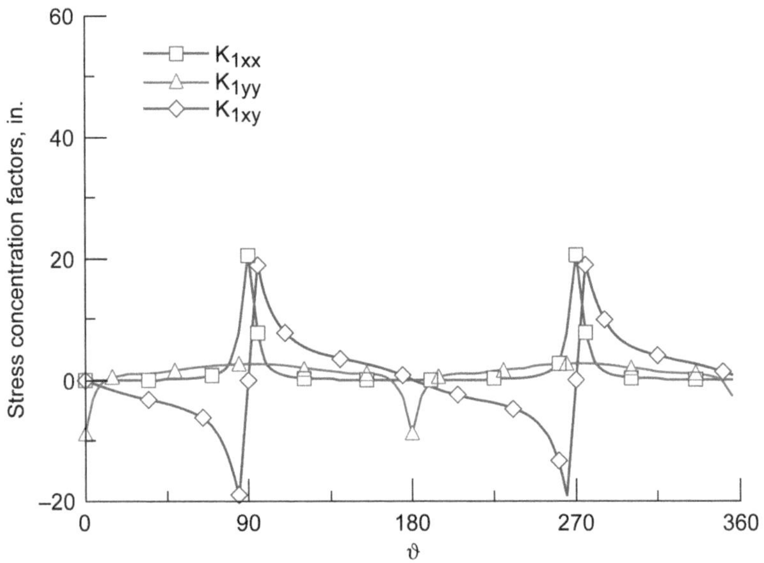

Figure 18.16 Stress concentration factors (ratios).

are near “o,” as they should be, since there are no loads in these two directions. The laminate combined stress failure criteria (modified distortion energy [3, 4] are plotted in Fig. 18.15. The results in this plot show no fracture at the laminate scale.

This local stress concentration factors are shown in Fig. 18.16 around a small hole in the laminate [5]. The plot in Fig. 18.16 illustrates that the stress concentration factors reach maximum at 90° and 270° from the load direction for $+\ \sigma_{xx}$ and $\sigma_{xy}(K_{xy})$. The $\sigma_{yy}(K_{yy})$ stress reaches maximum for 0°, 180°, and back to 360°. Note that maximum for K_{xx} and K_{xy} are equal in magnitude with K_{xg} and have a sharp discontinuity at 90° and 270°, which is that K_{xy} jumps from negative value to equal positive within a few degrees (about 4°). Even if the stresses in Figs. 18.13 and 18.14 are multiplied by 20 (maximum stress concentration factor from Fig. 18.16), the laminate stresses still do not cause any fracture compared with the corresponding strengths in Figs. 18.10 and 18.11.

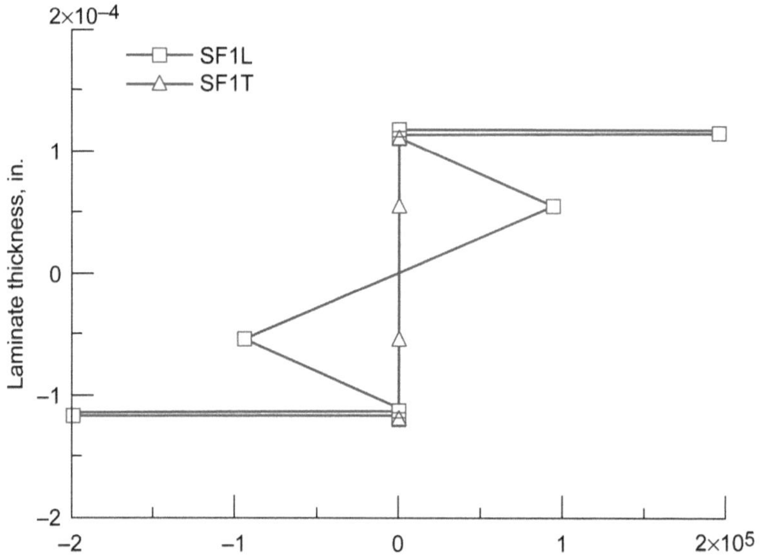

Figure 18.17 Nanofiber dominate stresses (psi) (psi = 6.9 Pa; 1 in. = 2.54 cm).

18.1.2 *Nanostresses and Nanofracture*

The stresses in a fiber in a nanoelement σ_{F1L} and σ_{F1T} are plotted in Fig. 18.17. The ranges for these stresses are −0.138 GPa (−200,000 psi) to 0.138 GPa (200,000 psi). On multiplying these stresses with 20 (stress concentration factor) the stresses becomes 27.6 GPa (4,000,000 psi), which is 5 times the fiber tensile strength of 0.552 GPa (800,000 psi) and 6.7 times the compressive strength in Table 18.1. Therefore, the nanofibers will start breaking all across the bottom and top nanoplies.

Figure 18.18 plots the corresponding stresses σ_{M1L} and σ_{M2T} in a nanomatrix element. These stresses multiplied by the stress intensity factor will not cause any damage in the nanoscale matrix element. The plot for the stresses σ_{M2BL} and σ_{M2BT} is shown in Fig. 18.19. As can be seen in this plot, σ_{M2BL} and σ_{M2BT}, when multiplied by the stress concentration factor (20) from Fig. 18.16, is about (0.552 GPa [800,000]) psi and will definitely crack the matrix. Even without the multiplication of the stress concentration factor, the nanomatrix is the nanoelement of Fig. 18.19 will fracture.

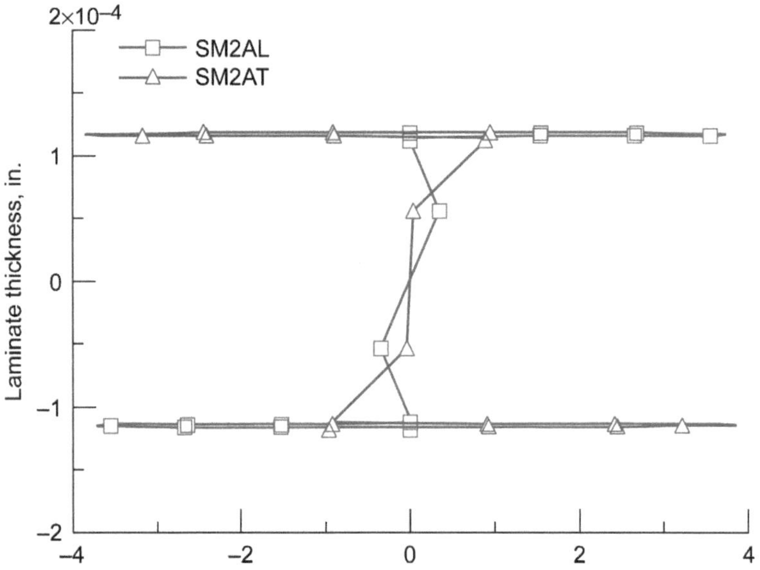

Figure 18.18 Nanomatrix dominated stresses (all matrix region) (psi = 6.9 Pa; 1 in. = 2.54 cm).

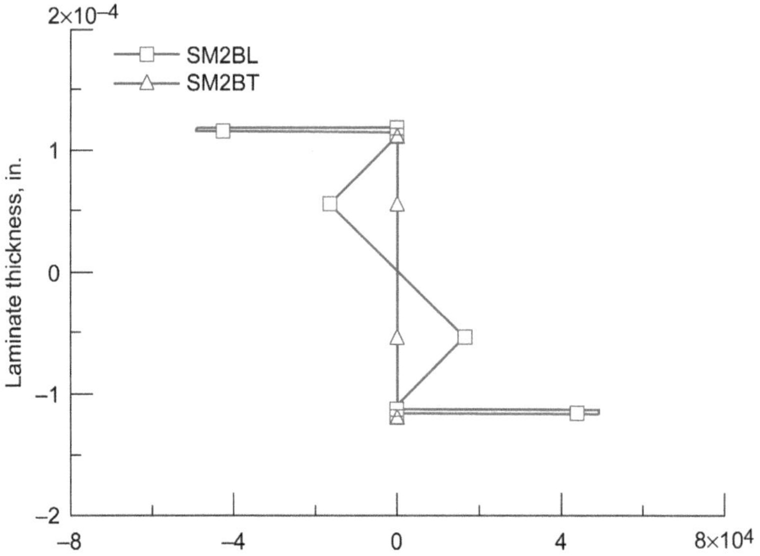

Figure 18.19 Nanomatrix dominated bending stress (matrix between fibers) (psi) (psi = 6.9 Pa; 1 in. = 2.54 cm).

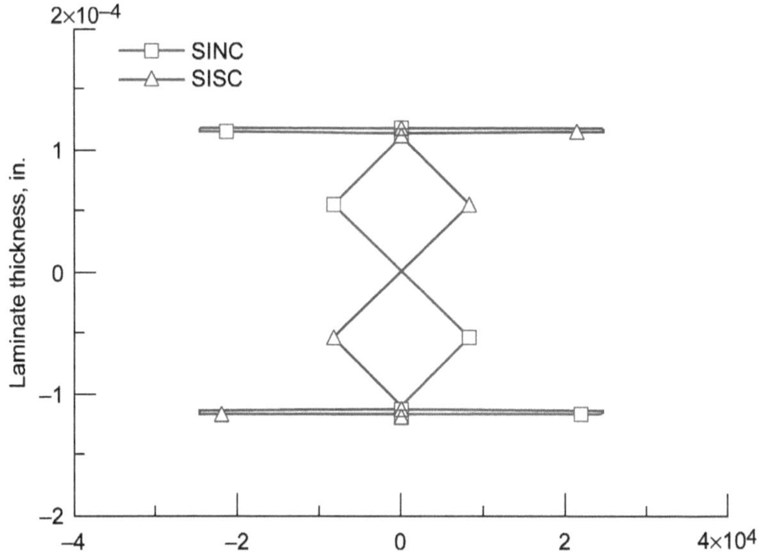

Figure 18.20 Nanomatrix normal and shearing stresses at 45° from the fiber axis (psi) (psi = 6.9 Pa; 1 in. = 2.54 cm).

The normal and shear nanostresses are plotted in Fig. 18.20. As can be seen in this nanoelement, both the normal and the shearing stresses will fracture the matrix in this nanoelement without the effects of stress concentration factors.

The important conclusions from the above discussion are that (i) the matrix will crack at nanoelements even without stress concentration effects and (ii) the fibers will fracture at the nanoscale in the presence of stress concentration.

18.2 Concluding Remarks

The concluding remarks from an investigation on the stresses in a unidirectional nanolaminate (composite) are as follows:

1. The composite mechanics incorporated in a composite mechanics computer code based on constituent materials and a nanoscale element is sufficient for stress analysis of a nanocomposite laminate in bending.

2. The matrix-dominated stresses will fracture at the nanoelement even without stress concentration or residual stresses.
3. The fibers in a nanocomposite nanoelement will fracture when the stress concentration is accounted for.
4. The nanolaminate geometric and material properties are not affected by loading conditions in a linear mechanics nanolaminate theory.
5. The bending stresses in a nanolaminate behave the same as a sandwich where the top and bottom nanoplies exhibit a substantial jump stress from the core material.
6. It is important to evaluate the stresses in a nanoelement even when the nanolaminate composite does not show any stress reaching its limits compared to corresponding strengths.

Appendix

Stress Concentration around a Small Hole in a Composite Plate

$$\frac{\sigma_{c\theta\theta}}{\sigma_{cxx}} = \frac{E_{c\theta\theta}}{E_{cxx}} \left\{ R_o \left[(R_o + R_{DI}) \ \sin^2\phi - \cos^2\phi \right] \cos^2\theta \right. + \left[(1 + R_{DI}) \cos^2\phi - R_o \ \sin^2\phi \right] \sin^2\theta \left. - \frac{R_{DI}}{4} (1 + R_o + R_{DI}) \sin \ 2\phi \ \sin \ 2\theta \right\} \quad \text{(A.1)}$$

$$R_o = \sqrt{E_{cxx}/E_{cyy}} \quad \text{(A.2)}$$

$$R_{DI} = \left[2 \left(\frac{E_{cxx}}{E_{cyy}} - \nu_{cxy} \right) + \frac{E_{cxx}}{G_{cxy}} \right]^{1/2} \quad \text{(A.3)}$$

$$\frac{\sigma_{c\theta\theta}}{\sigma_{cxy}} = \frac{E_{o\theta\theta}}{2E_{cxx}} (1 + R_o + R_{DI}) \{ -R_{DI} \cos \ 2\phi \ \sin \ 2\theta + [(1 + R_o) \cos \ 2\theta \ R_o - 1] \sin 2\phi \} \quad \text{(A.4)}$$

Note: There are no strengths in the above equations, only stiffnesses.

Notation

1. $E_{c\theta\theta}$ = The modulus along the θ which varies as the angle changes from $0°$ to $360°$.
2. γ is the plate structural axis angle.
3. R_o and R_{DI} are defined in the above equations.
4. θ is the angle though which the stress concentration γ is the angle that the principal plate axes are located relative to the loading axes.

References

1. Chamis, C. C., and Manderscheid, J. M. Composites nanomechanics: a mechanics properties prediction, NASA TM. 2007-214673, February 2007.
2. Handler, L. M., and Chamis, C. C. ICAN/JAUA computer code. Webpage ICAN/JAUA (Glenn, R. C.)
3. Chamis, C. C. Failure criteria for filamentary composites. NASA TN D–5367, August 1969.
4. Hoffman, O. The brittle strength of orthotropic materials. *J. Compos. Mater.* **1**(2), April 1967, 200–206.
5. Lekhnitskii, S. C. *Anisotropic Plates*. New York, Gordon & Breach, 1968.

Chapter 19

Probabilistic Simulation for Nanocomposite Fracture

Christos C. Chamis*
Former Senior Aerospace Scientist, NASA Glenn Research Center, Cleveland, OH 44135, USA
christos.c.chamis@aol.com

A unique probabilistic theory is described to predict the uniaxial strengths and fracture properties of nanocomposites. The simulation is based on composite micromechanics with progressive substructuring down to a nanoscale slice of a nanofiber where all the governing equations are formulated. These equations have been programmed in a computer code. That computer code is used to simulate uniaxial strengths and fracture of a nanofiber laminate. The results are presented graphically and discussed with respect to their practical significance. These results show smooth distributions from low probability to high.

19.1 Introduction

The research in the nanoscale technology has exploded in the recent past. An indication of this explosion is that the Society of

*Retired.

Characterization of Nanocomposites: Technology and Industrial Applications
Edited by Frank Abdi and Mohit Garg

ISBN 978-981-4669-02-3 (Hardcover), 978-1-315-36489-6 (eBook)
www.panstanford.com

Aerospace Material and Processing Engineers (SAMPE) Conference is devoting four sessions of about six papers each for the last five years. These papers cover practically all current research activities. Most of the research is devoted to processing because of the difficulties involved in making a useful material [1]. A few investigators have been fortunate to make some testing samples, which they subsequently tested to obtain limited data [2]. A few other investigators researched the characterization of fatigue [3] and creep [4]. A couple of papers explored the construction of nanocomposites for rocket ablative material [5] and for carbon nanotubes for adaptive structures [6]. One paper ventured to describe a computer simulation of macroscopic properties of carbon nanotubes polymer composites [7]. However, there are no results of what special macroscopic properties are included. Srivastava and Wei [7] show one stress–strain curve and citation of several references. One recent article [8] describes multiscale modeling and simulation of nanostructural materials from atomistic to micromechanics. This article does not include information on nanocomposites, but it mentions that mechanistic models will be needed in the end. It is becoming abundantly clear that no holistic approach has been used to investigate the mechanistic prediction of uniaxial strength and fracture.

In this chapter, a unique mechanistic method is described to probabilistically simulate five uniaxial strengths and fracture of a nanofiber uniaxial composite. The mechanistic deterministic simulation of all uniaxial properties is described in a previous paper [9].

19.2 Fundamentals

The fiber alignment with uniform dispersion is not met in nanocomposites. It is assumed here that the fibers are aligned only for predicting "point" through-the-thickness properties. The fussiness can be simulated by estimating the angle of single fibers through the thickness. Therefore, it is assumed that an aligned unidirectional typical section of a nanocomposite is as illustrated schematically in Fig. 19.1 on the left. A nanoply is schematically shown in the

Table 19.1 T300 graphite nanofiber (Pyrograf II) properties (Conversion factors: 110 nm = 2.756×10^{-6} in.; psi = 6.89 Pa; lb/in.^{3} = 1146 kg/cm^3; in./in./°F = (2/5); cm/cm/°F; Btu = 1055 J)

Description	Symbol	Value	Units
Number of fibers per end	Nf	1.0	number
Filament equivalent diameter	df	2.756×10^{-6}	in.
Weight density	Rhof	0.064	lb/in.**3
Normal moduli (11)	Ef11	1.0×10^9	psi
Normal moduli (22)	Ef22	7.0×10^7	psi
Poisson's ratio (12)	Nuf12	0.2	Nondimensional
Poisson's ratio (23)	Nuf23	0.25	Nondimensional
Shear moduli (12)	Gf12	5.0×10^7	psi
Shear moduli (23)	Gf23	3.5×10^7	psi
Thermal expansion coefficient (11)	A1faf11	-5.5×10^{-7}	in./in./°F
Thermal expansion coefficient (22)	Alfaf22	5.6×10^{-6}	in./in./°F
Heat conductivity (11)	Kf11	444.0	Btu/h/ft^2/°F/in.
Heat conductivity (22)	Kf22	4.0	Btu/h/ft^2/°F/in.
Heat capacity	Cf	0.22	Btu/lb/°F
Dielectric strength (11)	KeF11	0.0	V/in.
Dielectric strength (22)	Kef22	0.0	V/in.
Dielectric constant (11)	Gamma11	0.0	in./V
Dielectric constant (22)	Gamma22	0.0	in./V
Capacitance	Cef	0.0	V
Resistivity	Ref	0.0	Ω–in.
Tensile strength	SfT	8.0×10^5	psi
Compressive strength	SiC	6.0×10^5	psi
Shear strength	SfS	4.0×10^5	psi
Normal damping capacity (11)	psi11f	0.38	%Energy
Normal damping capacity (22)	psi22f	6.3	%Energy
Shear damping capacity (12)	psi12f	3.34	%Energy
Shear damping capacity (23)	psi23f	6.3	%Energy
Melting temperature	TMf	6000.0	°F

same figure on the right. The input includes the constituent material properties (Tables 19.1 and 19.2), the fabrication parameters, and the environmental and loading conditions.

The strength prediction is expedited by the following geometric diagrams: An exploded view of nanoscale isolation of a typical part is shown in Fig. 19.2 with nanoscale dimensions. A single nanofiber schematic with substructuring is shown in Fig. 19.3a, and a typical subslice is shown in Fig. 19.3b.

Table 19.2 Intermediate modulus high-strength matrix (epoxy) (Conversion factors: 110 nm = 2.756×10^{-6} in.; psi = 6.89 Pa; lb/in^3 = 1146 kg/cm^3; in./in./°F = (2/5); cm/cm/°C; Btu = 1055 J)

Description	Symbol	Value	Units
Weight density	Rhom	0.044	lb/in.**3
Normal modulus	Em	500000.0	psi
Poisson's ratio	Num	0.35	Nondimensional
Thermal expansion coefficient	Alfa m	3.6×10^{-5}	in./in./°F
Heat conductivity	Km	0.008681	Btu/h/ft^2/°F/in.
Heat capacity	Cm	0.25	Btu/lb/°F
Dielectric strength	Kem	0.0	V/in.
Dielectric constant	Gammam	0.0	in./V
Capacitance	Cem	0.0	V
Resistivity	Rem	0.0	Ω–in.
Moisture expansion coefficient	Betam	0.0033	in./in./%moisture
Diffusivity	Dm	2.16×10^{-7}	in.**2/h
Saturation	Mm	0.0	%moisture
Tensile strength	SmT	15000.0	psi
Compressive strength	SmC	35000.0	psi
Shear strength	SmS	13000.0	psi
Allowable tensile strain	eps mT	0.02	in./in.
Allowable compression strain	eps mC	0.05	in./in.
Allowable shear strain	eps mS	0.035	in./in.
Allowable torsional strain	eps mTOR	0.035	in./in.
Normal damping capacity	psiNM	6.6	%energy
Shear damping capacity	psiSm	6.9	%energy
Void heat conductivity	Kv	0.0012	Btu/h/in./°F
Glass transition temperature	Tgdr	420.0	°F
Melting temperature	TMm	0.0	°F

A nanosubply with its corresponding stresses is shown in Fig. 19.4. The nanomechanics predictive equations are derived by using Fig. 19.4. The equations used are all programmed in ICAN/JAVA [10].

Before describing the results obtained, it is instructive to describe the interphase and how it is modeled. The schematics in Fig. 19.5 show a vertical section (upper figure part) with unit thickness of the nanocomposite and a single fiber in it. As can be seen in the slice (lower figure), the fiber interphase is represented

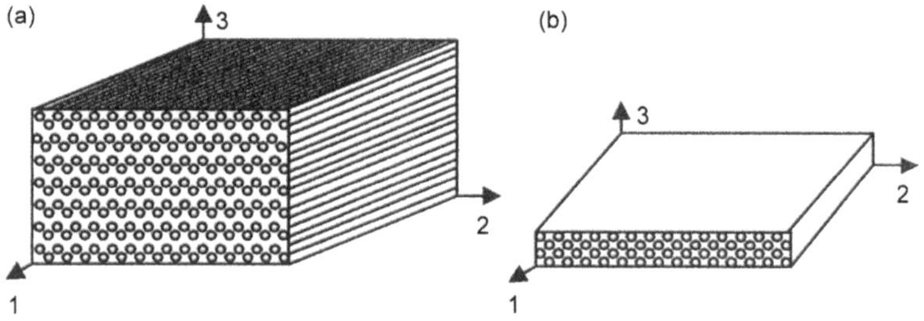

Figure 19.1 Unidirectional nanocomposite typical section. (a) Nanocomposite. (b) Nanoply.

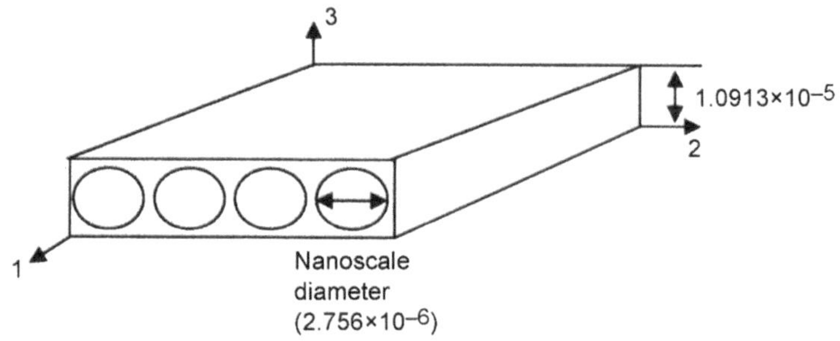

Figure 19.2 Nanoscale isolation of a typical part (units are in in.).

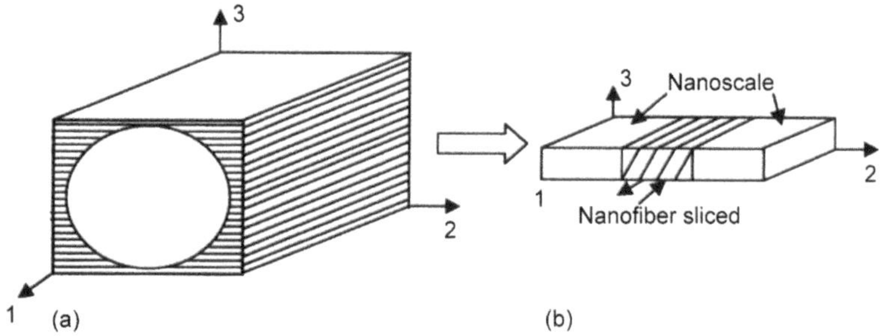

Figure 19.3 Nanofiber substructuring. (a) Several slices through the thickness. (b) Nanofiber sliced.

by a series of progressively larger volume voids, starting with the smallest near the matrix interface and ending with the largest in the fiber interface. It can be seen that the stress in the matrix will be magnified because of the voids. This magnification is shown

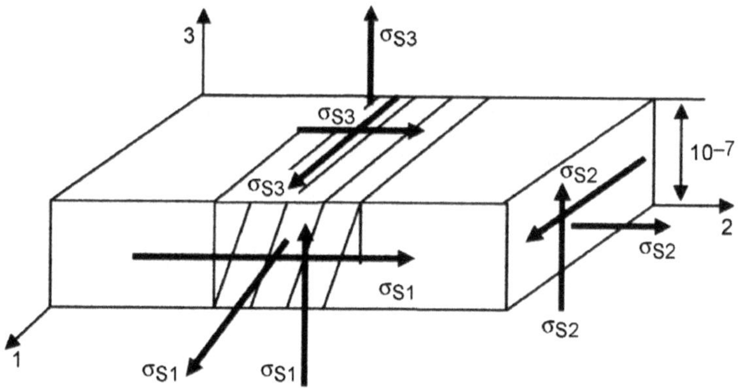

Figure 19.4 Nanostresses on a nanosubply (units are in in.).

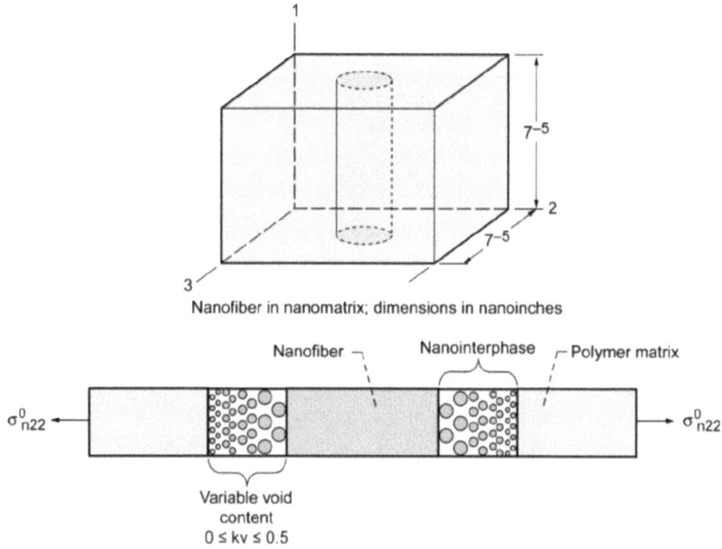

Figure 19.5 Vertical section of a composite nanocell through nanofiber center.

in Fig. 19.6 for a specific nanocomposite with 0.05 fiber volume ratio and with void volume ratio varying from 0.05 to 0.4. The interesting point to note in the lower part of Fig. 19.5 is that the matrix is continuous even though it is filled with progressively larger nanovoids; otherwise the stresses will not be continuous in the

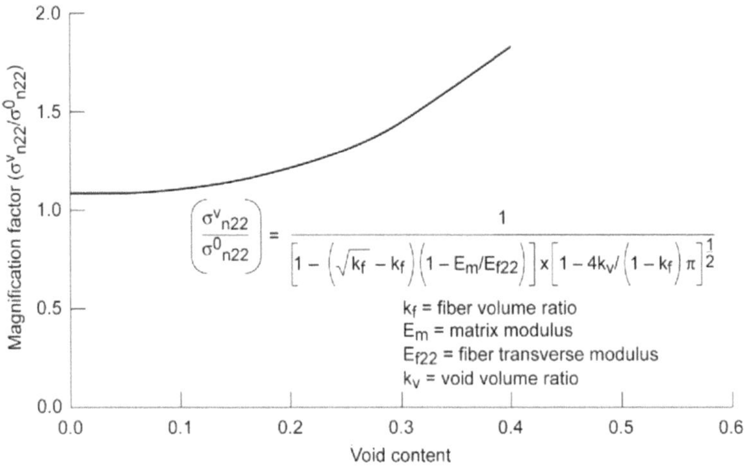

Figure 19.6 Nanocomposite magnification factor.

matrix. It is instructive to elaborate a bit further with the geometry of Fig. 19.5 (lower part). Filling up a conventional ply that is 0.005 in. thick and 1 in. wide will require about 1×10^6 nanofibers, a very large number indeed. The magnification factor of the voids effect in the interphase is shown in Fig. 19.6. As can be seen in Fig. 19.6, the magnification factor increases from a value of about 1.1 to a maximum of about 2. Therefore, the maximum void effect will be nearest to the fiber interface. The author is not aware of any experimental data that shows these effects. It is important to note that this is the only mechanics approach to the interphase description.

19.3 Results and Discussion

In this section the probabilistic results are presented and discussed starting with the large voids in the interphase. The probabilistic void magnification factor is shown graphically in Fig. 19.7. It can be seen in this figure that the larger the void content the greater the deviation. The leftmost figure is closest to the matrix interphase fiber interface, while the rightmost curve is closest to the interphase interface. The respective scatter is about 0.1 for

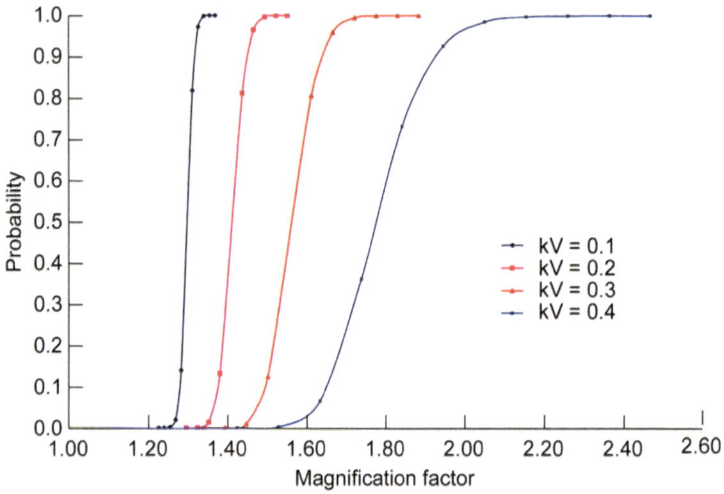

Figure 19.7 Probabilistic magnification factor of voids in the interphase.

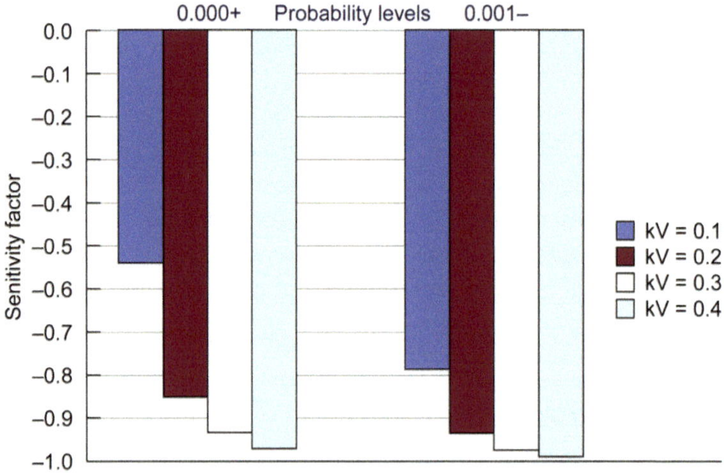

Figure 19.8 Void sensitivities on the interphase magnification factor.

the curve closest to the matrix to about 1 for the curve closest to the fiber. The corresponding sensitivities are shown in Fig. 19.8. It can be seen in this figure that the void sensitivities on the magnification factor is large. The probabilistic void effects on the

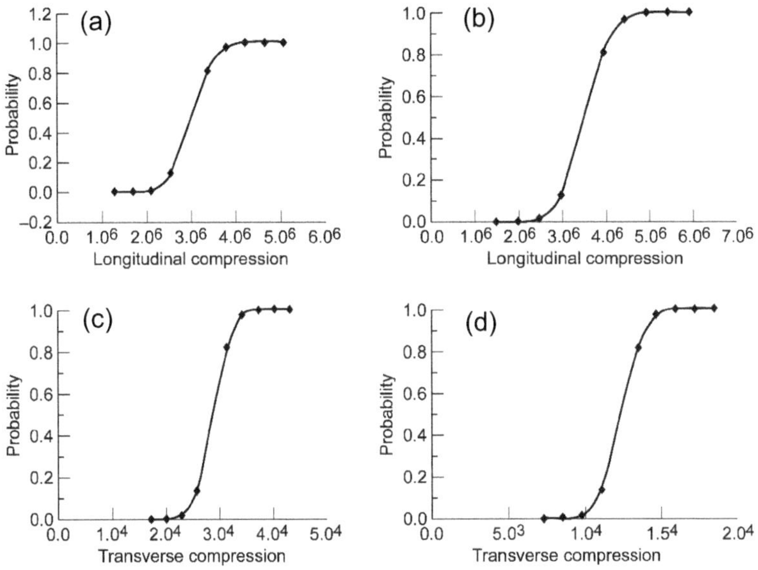

Figure 19.9 Probabilistically plotted nanouniaxial strengths.

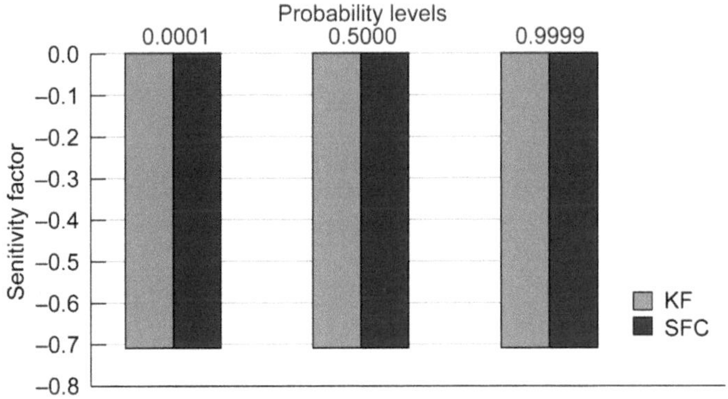

Figure 19.10 Probabilistic sensitivities for nanolongitudinal uniaxial strength.

uniaxial strengths are plotted in Fig. 19.9. Figure 19.9a shows the spread in the longitudinal tensile strength; Fig. 19.9b, in the longitudinal compressive strength; Fig. 19.9c, in the transverse tensile strength; and Fig. 19.9d in the transverse compressive

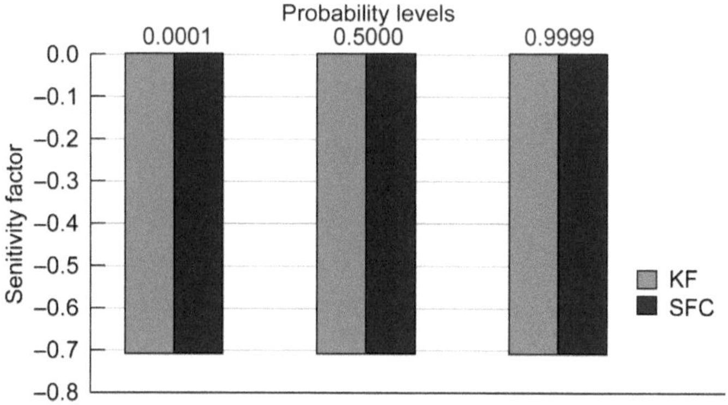

Figure 19.11 Probabilistic sensitivities for the nanocompressive uniaxial strength.

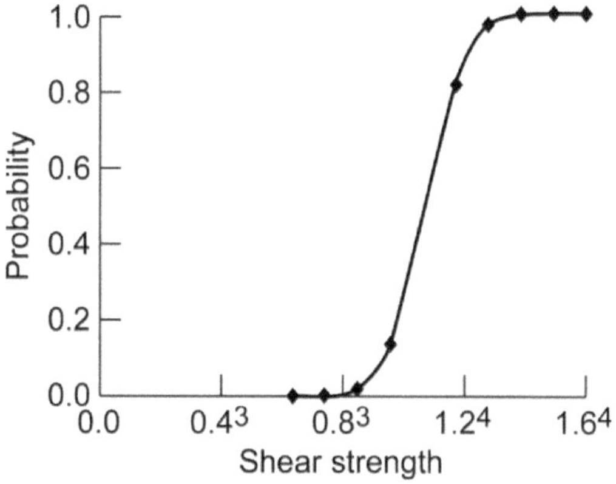

Figure 19.12 Probabilistically plotted intralaminar uniaxial shear strength.

strength. It can be seen in Fig. 19.9 that the distribution for the two longitudinal strengths is relatively large. It is from 150 to 650 ksi, for tensile strength and with a distribution of about 500 ksi, and for the compressive strength is from 140 to 500 ksi or a distribution of about 360 ksi. The corresponding probabilistic sensitivities are plotted in Fig. 19.10 for tensile and Fig. 19.11 for compressive. It can be seen in these two figures that there is no difference in the sensitivities for the three probabilities. The probabilistic intralaminar shear strength is plotted in Fig. 19.12. The distribution

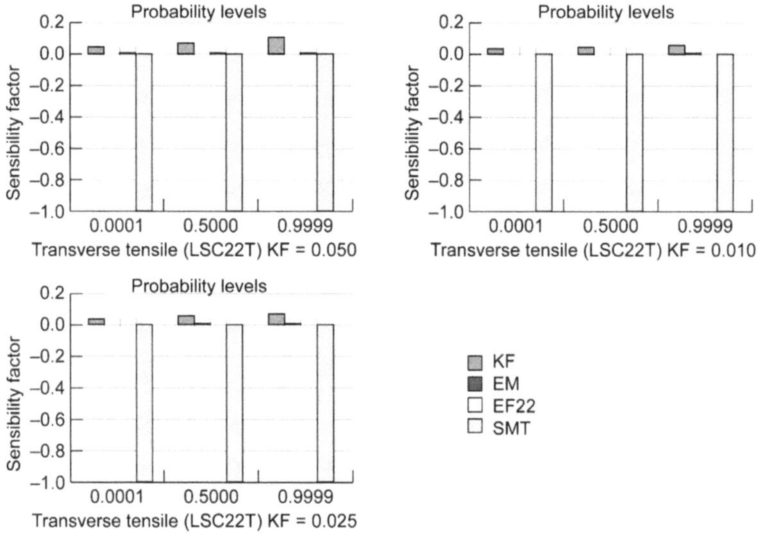

Figure 19.13 Probabilistic sensitivities for nanouniaxial transverse tensile strength for three different probabilities.

in this strength is from about 6000 to ~16,000 psi or ~10 ksi spread. It is a relatively wide distribution from the lowest probability to the highest. The corresponding probability sensitivities are plotted in Fig. 19.13 for uniaxial nanotransverse tensile strength. Note that these probabilistics are for 0.0001, 0.50, and 0.9999. They are about the same and may be easily interchangeable as well as for three fiber volume ratios.

The respective sensitivities for the other nanouniaxial strengths are the same and are not shown. The fracture for uniaxial nanofiber composites are the same as their respective uniaxial nanofiber uniaxial strengths.

19.4 Concluding Remarks

The salient remarks from an investigation to characterize an aligned monofiber nanolaminate are as follows:

1. The characterization for the nanolaminate (composite) was based on a series of progressive substructuring down a sliced single-nano diameter fiber.
2. The theoretical development and all the equations are included in a computer code called ICAN/JAVA.
3. The uniaxial strength and fracture includes two fabrication parameters, 5-nanouniaxial strengths/fracture.
4. The nanolaminate investigated consists of single nanofiber laminate with 0.05 fiber volume ratio.
5. The effects of the interphase are especially important and are represented by progressively large amounts of nanovoids from the matrix interface to the fiber interphase.
6. The probabilistic evaluation characterizes the effects of uncertainties in all participating variables.
7. The voids uncertainties indicate as the void volume ratio increases the distribution increases as well.
8. The voids contribute significantly to matrix-dominated strengths/fracture—that is, longitudinal compression, transverse tension and compression, and all intralaminar/interlaminar properties.

References

1. M. Jose, J. Tyner, E. Nyairo, D. Dean. Synthesis and processing of aligned carbon nanotube based fibers, SAMPE, 49th Int. Symp. Exhib. Conf. Proc., 2004.
2. G. Ayalasomayajuala, A. Garg, S. Kapila, K. Chandrashekhara, V. Flanigan. Fabrication and evaluation of rice hull derived nano silica composites, SAMPE, 49th Int. Symp. Exhib. Conf. Proc., 2004.
3. T. Karaki, J. P. Killgore, J. C. Seferis. Characterization of fatigue behavior of polynanomeric matrix composites, SAMPE, 49th Int. Symp. Exhib. Conf. Proc., 2004.
4. A. Ranade, N. A. D'Souza, K. Nayak, B. Gnade, D. Fairbrother. Correlation between creep-recovery, crystallization and dispersion of linear low density polyethylene nanocomposite films, SAMPE, 48th Int. Symp. Exhib. Conf. Proc., **48**(1), 2164–2176, 2003.

5. J. H. Koo, H. Stretz, J. Weispfenning, Z. P. Luo, W. Wootan. Nanocomposite rocket ablative materials: subscale ablation test, SAMPE, 49th Int. Symp. Exhib. Conf. Proc., 2004.
6. S. Muhle, H. P. Monner, P. Wiersch. Carbon-nanotubes for adaptive structures, SAMPE, *48th Int. Symp. Exhib.* **48**(1), 1181–1190, 2003.
7. D. Srivastava, C. Wei. Computer simulations of macroscopic properties of carbon-nanotube polymer composites, SAMPE, *48th Int. Symp. Exhib.* **48**(1), 2153–2163, 2003.
8. T. S. Gates et al. Computational materials: multi-scale modeling and simulation of nanostructural materials, *Compos. Sci. Technol.* **65**(15/16), 2416–2434, 2005.
9. C. C. Chamis. Composite nanomechanics properties prediction, NASA/TM—2007-214673.
10. L. M. Handler, C. C. Chamis. ICAN/JAVA computer code.

Chapter 20

Material Characterization and Microstructural Assessment: Fatigue Curve S-N Development Using Fracture Mechanics

Hamid Saghizadeh
Boeing Technical Fellow - TF Boeing Commercial Airplanes, 2060 East Imperial Highway, El Segundo, CA 90245, USA
hamid.saghizadeh@boeing.com

20.1 The Evolution of Structural Design: Aerospace Structures

With regard to the evolution of structural design, let us first summarize where we were and where we are heading (Fig. 20.1).

Our first approach was based on trial and error—based on successful heritage design and lessons learned. Then, the 19th century saw the advent of the quantitative design approach, based on elasticity and the stress–strain concept and high margin of safety Then, Inglis arrived at estimating stresses at the edge of a hole or crack and the existence of singularity, which was then resolved by

Characterization of Nanocomposites: Technology and Industrial Applications
Edited by Frank Abdi and Mohit Garg

ISBN 978-981-4669-02-3 (Hardcover), 978-1-315-36489-6 (eBook)
www.panstanford.com

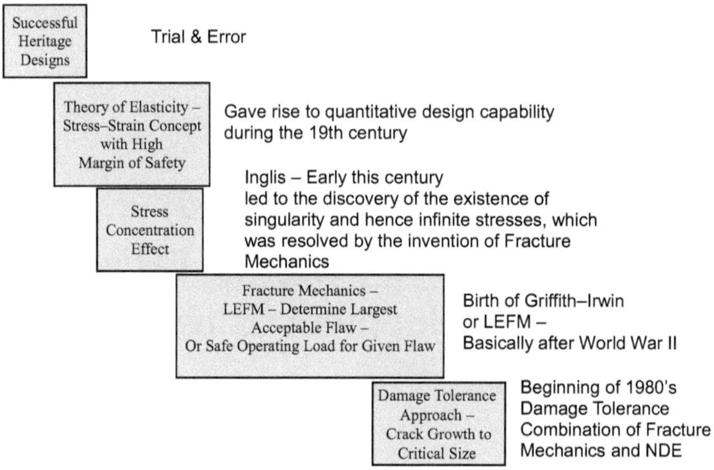

Figure 20.1 Structural design process: failure mechanisms.

the invention of fracture mechanics. Then, after World War I, came linear elastic fracture mechnics (LEFM), based on Griffith and Irwin approaches. Then, by the beginning of the 1980s, we started using more of the concept of damage tolerance (DT) and non-destructive examination (NDE).

What is it that we are doing today and where are we heading?

The microstructural behavior of materials, which include molecular dynamics (MD), molecular manufacturing (MM), nanotechnology, and virtual testing, and the use of computational techniques and simulation to reduce costly and time-consuming tests are the subject of our work (Fig. 20.2).

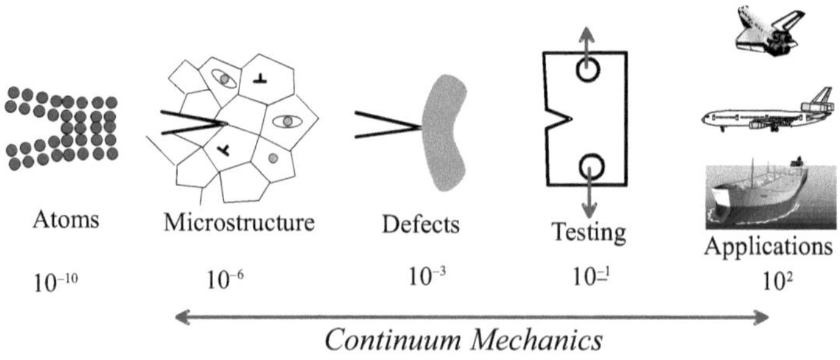

Figure 20.2 Continuum mechanics: failure mechanisms.

20.2 Engineering Structures: Microstructural Assessment

Unique energy dissipation mechanisms exist within solids. These mechanisms are essentially linear to critical and are not available for the analysis of propagation in inelastic solids such as glassy polymer systems (polymer composite materials and adhesively bonded joints).

It is important to make a clear distinction between the classification of fracture problems based on global deformation mode and those based on local fracture mechanisms. In addition, we will discuss the recent research in the micromechanics of fracture processes that has made the link between fracture mechanics parameters and local irreversible damage mechanisms clearer. It has also increased our awareness toward the development of microstructural models for fracture to relate local and global fracture behavior of materials.

As we progress, more sophisticated models for material behavior need to be incorporated into fracture mechanics analyses and modeling. For example, time-dependent nonlinear material behavior such as viscoplasticity and viscoelasticity are critical factors when creep failure is predominant, and similarly, there are other related parametric characterizations when addressing composite materials as well.

20.3 Fundamental Behavior of Solids

All solids appear to deform by two fundamental modes, dilatation (change in size) and distortion (change in shape) (Fig. 20.3). The expressions for these deformations come from continuum mechanics. The differences between solids are manifest at the microstructural (including molecular) level.

20.4 Dependence on Materials Science

Applied mechanics and materials science need to be involved in advanced design. Reducing weight, mitigation of testing and improved performance require that the material be designed to help meet these goals.

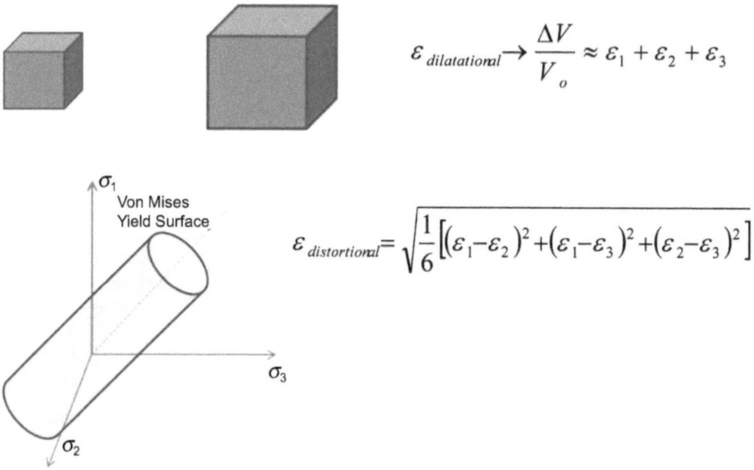

Figure 20.3 Fundamental deformation mechanisms for all solids: dilatational and distortional.

20.5 The Nature of Fracture: Macroscopic vs. Microscopic

It is important to make a clear distinction between classification of fracture problems on the basis of global deformation mode and those based on local fracture mechanisms. Transition from square to slant fracture (transition in the stress field at the tip of the crack from plane strain to plane stress).

20.5.1 *Macroscopic Scale*

Fracture can be described in terms of the macroscopic or microscopic details of the fracture surface.

Figure 20.5 shows the schematic of fracture surface appearance on a macroscopic scale ranging from 100% square to 100% slant. The slant fracture may form in either symmetrical or asymmetrical form.

On a macroscopic scale, fracture surfaces can be described by the percentage of square (brittle, cleavage) or slant (ductile, oblique) fracture.

In fracture tests, on high-strength steels, and alloys of titanium or aluminum, there is a transition from square to slant fracture as

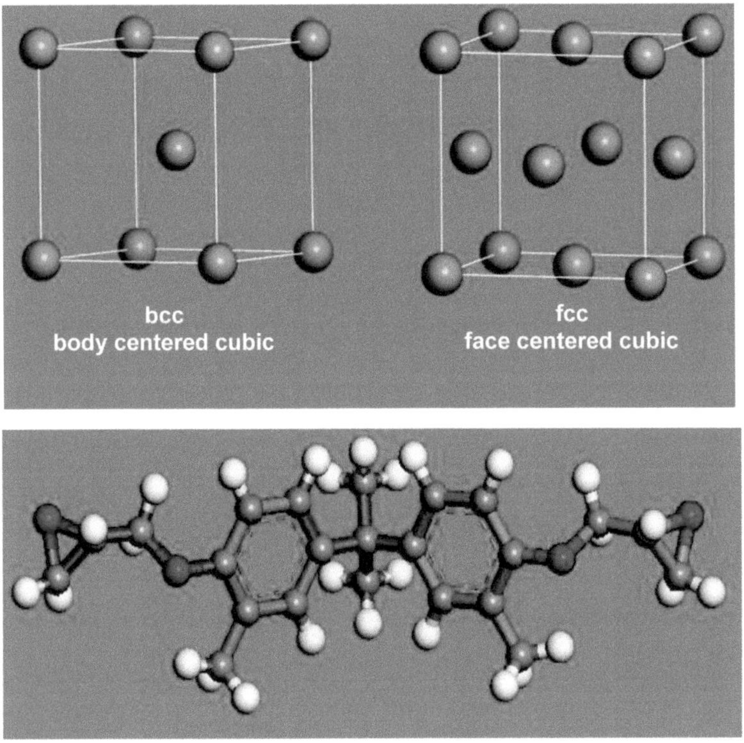

Figure 20.4 Cubic crystal system.

the plate thickness is decreased. This is related to a transition in the stress field at the tip of the crack from plane strain to plane stress as the plate thickness is decreased.

20.5.2 *Microscopic Scale*

Local (microscopic) fracture mechanisms: fracture mechanisms in a zone very near to the tip of the crack.

20.6 Composites Contain Multiple Materials

Inherently, all materials at the micro level can be considered as composites.

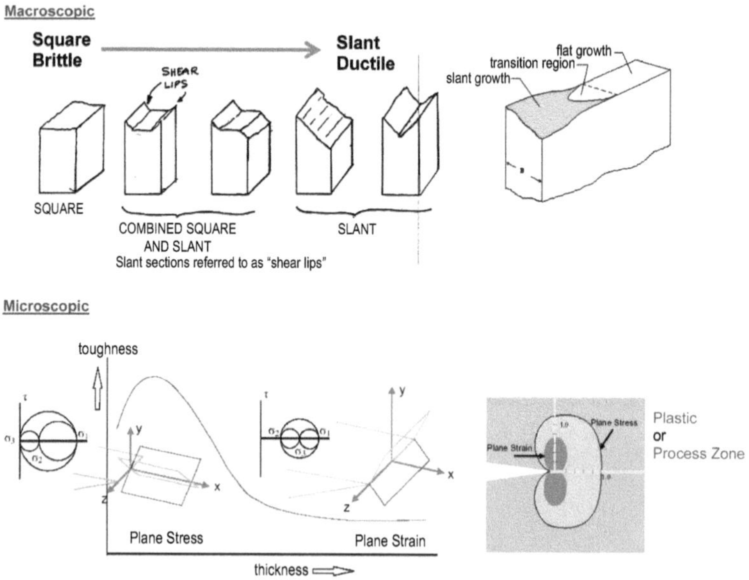

Figure 20.5 Schematic of fracture surface appearance on a macroscopic scale.

Composites, by definition, contain two or more different materials. For carbon fiber composites, those materials are mainly a polymeric matrix and a fiber reinforcement. For an efficient analysis methodology, the deformation of the composite system must be considered. Micromechanical enhancement allows one to consider the deformation of the resin and the fiber separately.

20.7 Thermosetting Polymers Are Glassy Polymers

Polymer deformation is not rate dependent, but the stress state is (always deformation state dependent). Yield stress is, therefore, no longer a single value. However, yield strain is. This means when we need to describe the critical point for a polymer, we need to do it as a strain measure (Fig. 20.6).

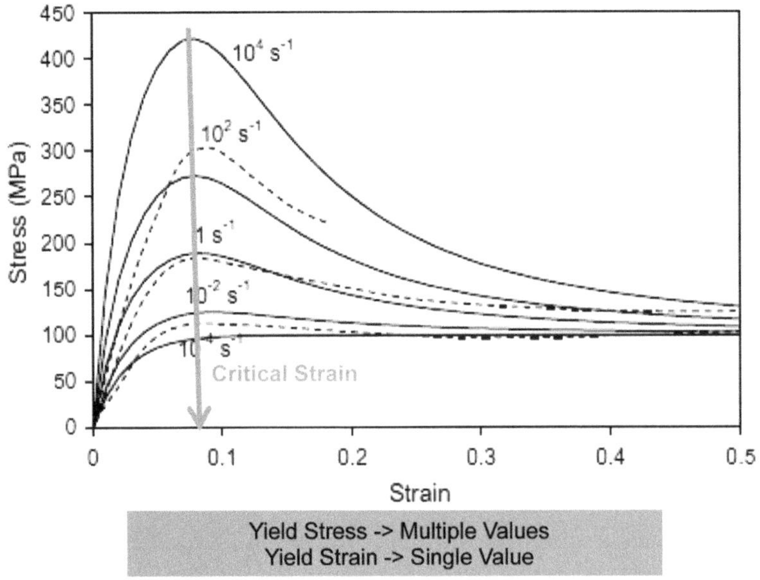

Figure 20.6 Thermosetting polymers: deformation rate.

20.8 Deformation in a Polymer

Two types of deformation occur in polymers, as described earlier.

Dilatation is controlled by the cohesive forces that hold the polymer together. This means that non-bonded interactions in the polymers are the primary controlling interactions at the strain levels of interest.

Distortion is controlled by molecular rearrangements. This means that bonded interactions control the behavior.

By understanding that the critical behavior of polymers (and therefore thermosetting polymers) is either dilatational or distortional, we can design simple experiments and atom-level simulation to calculate these properties.

With the additional understanding of how the arrangement of the atoms affects distortion and dilatational behavior, we are better able to select and design polymers for optimal composite performance; e.g., distortional resin.

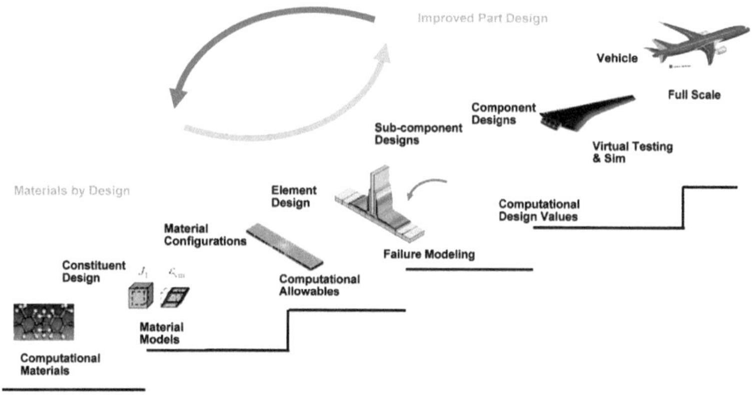

Figure 20.7 Material to part: design process.

The understanding of how the components of a material behaves at the various levels allows for hierarchical multiscale modeling. This enables targeted design and analysis at all levels from materials to parts (Fig. 20.7).

20.9 Multiscale Modeling and Simulation

An example of making a tube and polymer through the Accelrys Material Studio (Fig. 20.8):

> Make an amorphous cell (1), use sketch to make a bond from 1 to 2 (2), functionalize (3), minimize (4).
>
> Apply the coarse grain technique to link nano, meso, and macro scales (5) and verify with test data (6).

An Illustration of the mapping of the atomistic polymer to a coarse-grained bead-spring model is shown (Fig. 20.8). One bead represents one chemical repeat unit, and a spring spans two carbon–carbon bonds on the backbone. The angle between three CG beads results from two successive atomistic torsions.

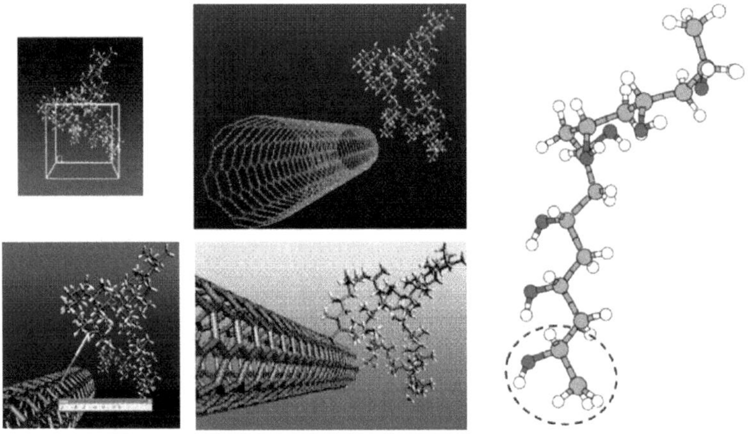

Figure 20.8 Mapping of the atomistic polymer to a coarse-grained bead-spring model.

20.10 Multiscale Modeling (Linking Nano, Micro, and Macro)

Multiscale modeling allows us to predict material properties or system behavior at each region of length and time scales (Fig. 20.10). Based on the knowledge of the atomistic structure and properties of elementary processes, through different levels of information and modeling, the distinguished levels include

- Level of quantum mechanical models (information about electrons is included)
- Level of molecular dynamics models (information about individual atoms is included)
- Mesoscale or nano level (information about groups of atoms and molecules is included)
- Level of continuum models
- Level of device models

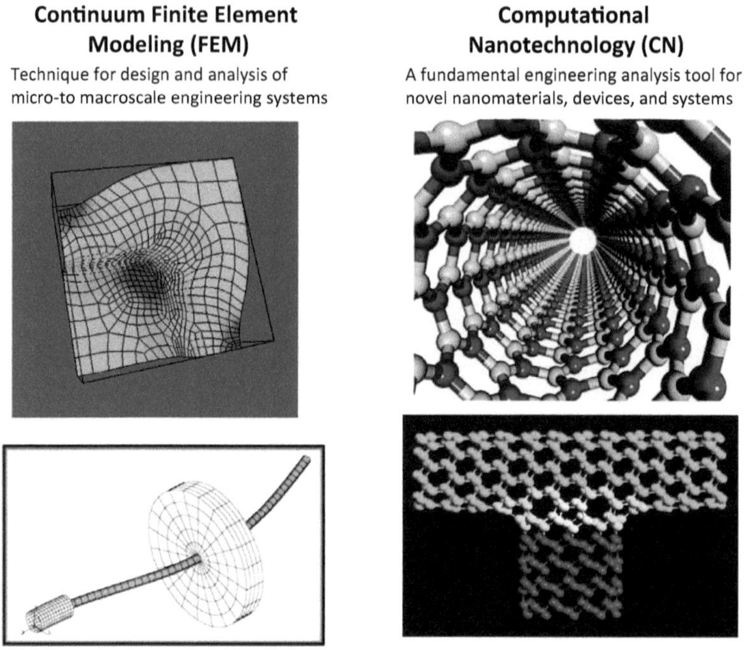

Figure 20.9 Computational methods.

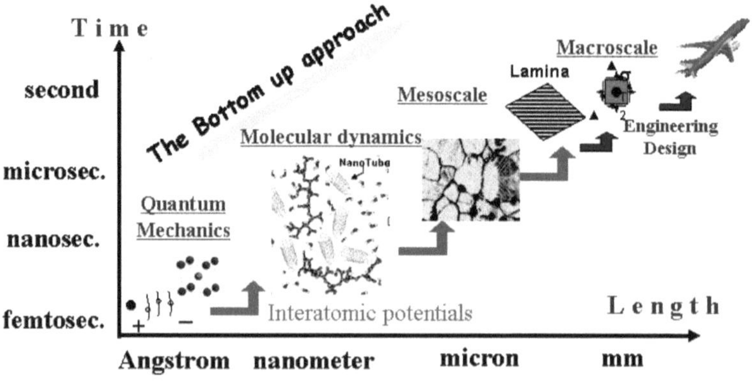

Figure 20.10 Linking nano, micro, and macro: bottom-up approach.

20.11 What Is the Future of Aerospace Materials?

- Advanced composite/nanocomposite materials
- What is the significance of CNTs in the aerospace applications?
- Why nanoparticle aerospace polymer interfaces matter?
- **The bottom-up approach: nanoscale to macroscale**
- How to predict material allowables and design values
- How to eliminate (minimize) costly and time-consuming tests

20.12 The Bottom-Up Approach: Nanoscale to Macroscale

(1) To enhance the mechanical properties of aerospace materials by reinforcing nanomaterials in the matrix: the interface region (Figs. 20.11 and 20.12)

(2) To understand material properties at each region of length and time scales

(3) To develop a workable formalism to bridge length and time scales and relate properties at the nanoscale to the properties of macroscale structures

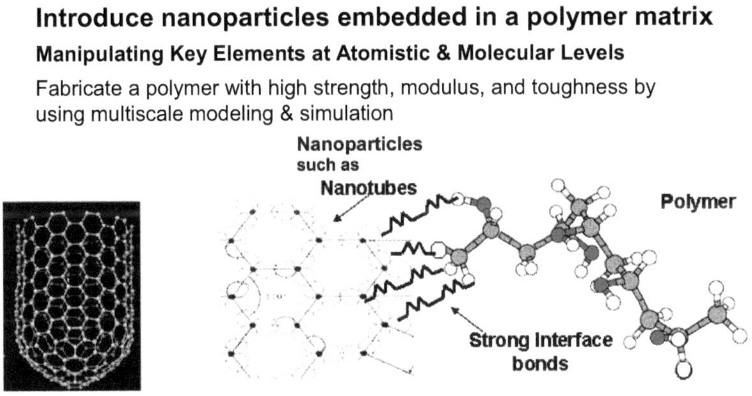

Figure 20.11 Developing light-weight material with high strength, modulus, and toughness.

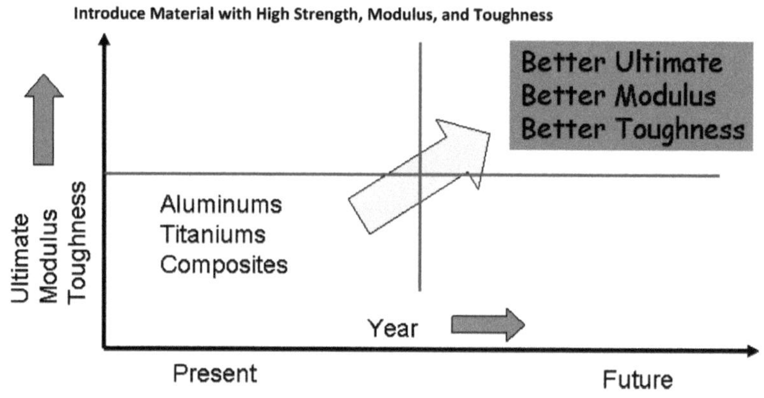

Figure 20.12 Enhancing the mechanical properties of aerospace polymers by reinforcing nanomaterials in the matrix.

(4) To have models that predict the properties and effects of nanomaterials integrated in larger structures (specifically at the interface)

Introduce Material with High Strength, Modulus, and Toughness

Fabricate a polymer with high strength, modulus, and toughness

- This can be done by
  - ◇ manipulating key elements at the atomistic and molecular levels
  - ◇ using a multiscale modeling and simulation tool (e.g., Accelrys)
- Introduce nanoparticles embedded in a polymer matrix (Fig. 20.13)
  - ◇ Nanoparticles have good mechanical properties
    - ∗ Carbon nanotubes possess exceptionally superior mechanical properties like high elastic modulus and tensile strength
    - ∗ They seem to be the ideal reinforcements for polymer composites, especially for enhancement of mechanical properties

Table 20.1 Material properties: comparison of metals to CNTs

Material	Aluminum	Steel	Glass fibers	Carbon fibers	Silicon carbide fibers	Carbon Nanotubes
Young's modulus, *E* (GPa)	71	215	72	300	450	**1000**
Tensile strength, Ftu (GPa)	0.65	0.9	3	5	10	**200**

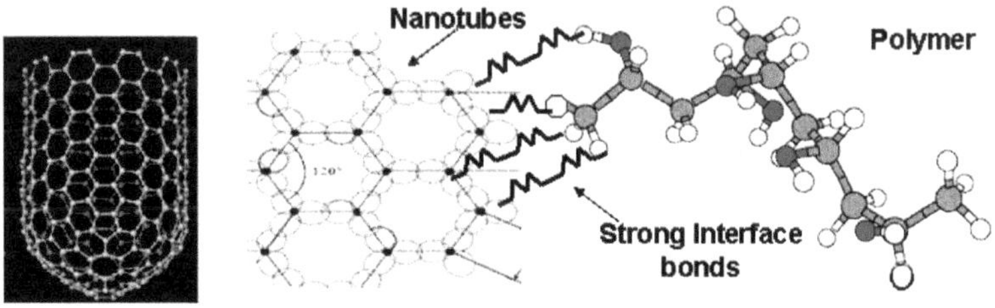

Figure 20.13 Nanotube–polymer interface.

20.13 Nanotube–Polymer Interface

The superiority of the mechanical properties of nanotubes alone does not ensure mechanically superior composites

- The composite properties are strongly influenced by the mechanics that govern the nanotube–polymer interface (strong bonds)

The structural strength characteristics of nanocomposite materials greatly depend on the mechanical load transfer from the matrix to the nanotube and the strength of the interface bonds.

20.14 Dispersion of Nanotubes (Improper Dispersion Can Initiate Defects)

Carbon nanotubes can be aligned during processing by mechanical stretching and/or application of electric or magnetic field (exploiting the electrical and magnetic properties of carbon nanotube) (Fig. 20.14).

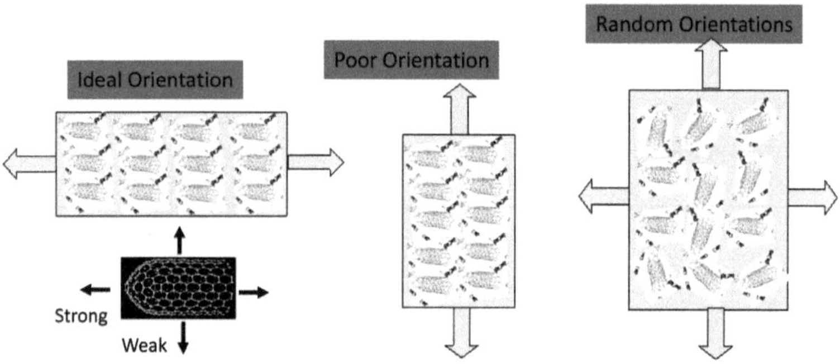

Figure 20.14 Dispersion of nanotubes.

20.15 To Understand Material Properties at Each Region of Length and Time Scales

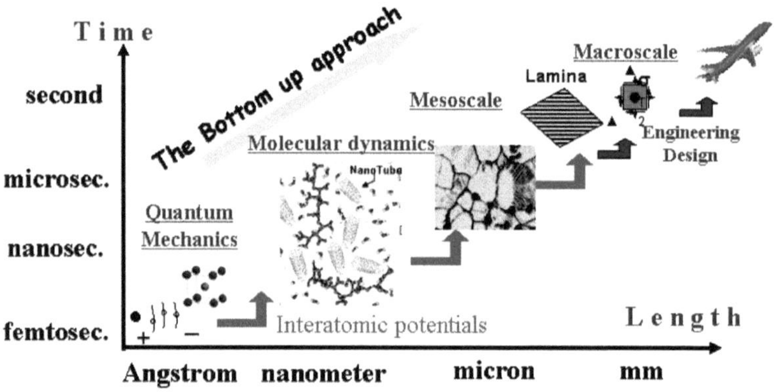

Figure 20.15 Multiscale modeling (linking nano-micro-macro).

20.16 Chemistry of the Interface (Functionalize)

- Further research needs to be conducted to study the chemistry of the interface and hence functionalize the polymer with appropriate molecules

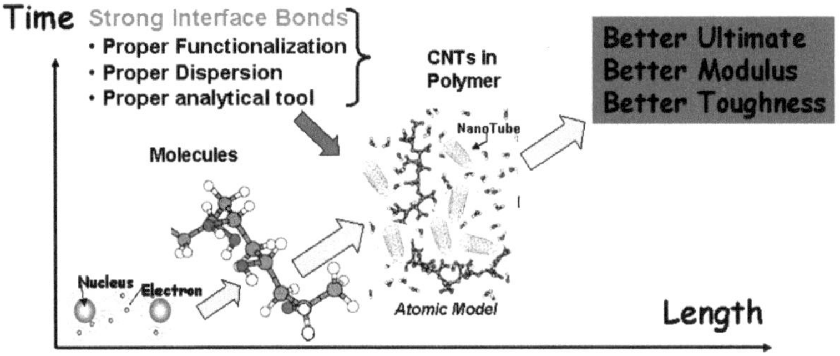

Figure 20.16 Manipulating key parameters at atomistic and molecular levels.

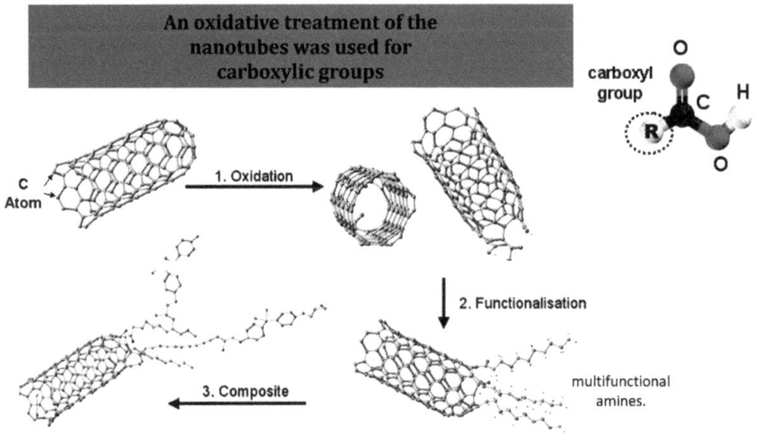

Figure 20.17 Functionalization of CNTs with polymers (enhancement of mechanical properties).

- A tailored functionalization can lead to the formation of covalent bonds between CNTs and the polymeric matrix, resulting in a strengthened interface (Fig. 20.17)
- Proper functionalization improves the interfacial stress transfer and positively affect the dispersibility of the nanofiller

20.17 Strong Interface Bonds between Nanoparticles and Resin: Failure Initiation

Interface Region: Failure Initiation

- Correlate nanointerface properties with hierarchical assembly conditions
- Predict mechanical properties using the bottom-up approach
- Poor interface bonds will cause crack initiation and failure (Fig. 20.18)

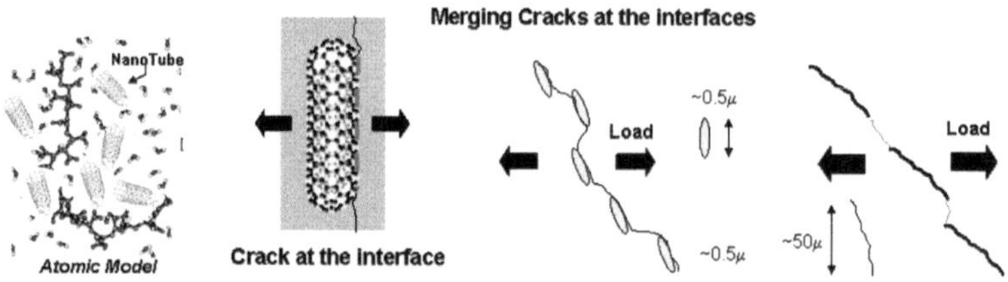

Figure 20.18 Interface region: failure initiation.

20.18 Bridging the Length and Time Scales (From Nano to Macro: The Coarse Grain Technique)

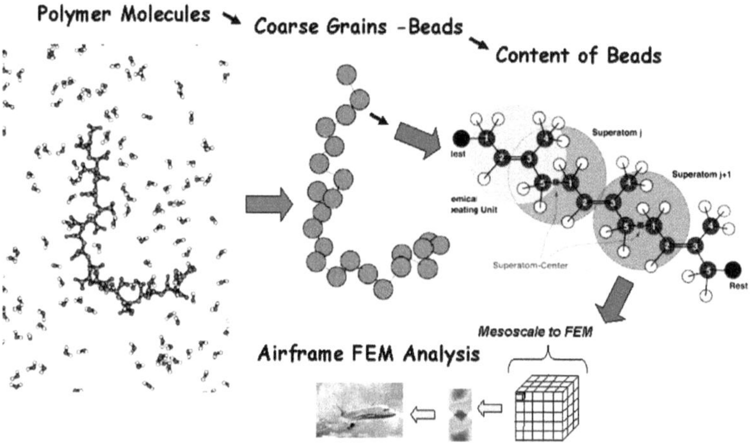

Figure 20.19 Coarse grain technique.

20.19 Summary and Conclusion

- **Using multiscale modeling and simulation as well as analytical closedform solutions, we can**
  (1) develop ultralight, strong, and durable materials tailored to our needs for optimum usage
  (2) predict material allowables and design values
  (3) eliminate (minimize) costly and time-consuming tests
- **Nanometals (CNT/NGP/CU) application**

  To replace Cu wires
- **Nanowelding application**

  Technique to substantially improve bonding metal to composite matrix

20.20 Final Objective

Develop an analytical tool that can predict fatigue and fracture allowables without conducting costly and time-consuming ASTM tests.

- **Approach**
  - ◇ Develop concepts that can define **crack initiation for high-cycle fatigue**
  - ◇ Apply energy release rate concept
    - ∗ Use material fracture toughness data
    - ∗ Use fatigue crack growth rate data
  - ◇ Develop material **fracture toughness**
  - ◇ Apply the extended Griffith theory
    - ∗ Use the full stress–strain curve
  - ◇ Develop material **fatigue crack growth rate data**
    - ∗ Use the extended Griffith theory
    - ∗ Use prediction on the Paris and threshold regions
- **Strategic Alignment**
  - ◇ Bring structures into conceptual design to lower design risks and add credibility
  - ◇ Explore rapid parametric analysis and design

- **Opportunities**
  - ◇ Leverage global-local processes developed elsewhere
  - ◇ Leverage past integrated design process developments
  - ◇ Leverage D&DT developments
- **Material is condensed energy**
- **A void in a body of material causes reduction in its overall energy**
- **A void in a body of material is basically proportional to the energy released during the generation of that void and creation of any new surfaces**
- **The difference between a cracked and uncracked body of material can be observed as simply the difference in their energy levels**
- **A. A. Griffith (1893–1963)**
  - ◇ Equilibrium state of a solid body deformed by specified forces or surface displacements is such that the *potential energy of the system is minimum*
  - ◇ Fracture can only occur if the solid can pass from the unbroken to broken condition by a process involving *a continuous decrease in potential energy*
  - ◇ The potential energy of the system would normally be taken as the sum of the potential energy of the external load and the elastic strain energy stored in the structure
  - ◇ In fracture calculations we also need to consider the energy to create fresh surface: material surface energy
- **Potential energy of the system**

Change in energy between the uncracked and the cracked system is of interest.

The decrease in the potential energy of the applied loads when a crack is introduced or extends under constant load is merely twice the increase in the strain energy.

$$\Delta U = U_{\text{Uncracked}} - U_{\text{Cracked}} = (\pi a^2\sigma^2/E) - 4a\eta \qquad \text{Plane stress}$$

$$= (\pi a^2\sigma^2(1-\nu^2)/E) - 4a\eta \quad \text{Plane strain}$$

U = Potential energy, a = 1/2 crack length, E = Young's modulus, η = Material surface energy σ = Stress level, ν = Poisson's ratio
Relate S/N data as a factor of the energy ratio.

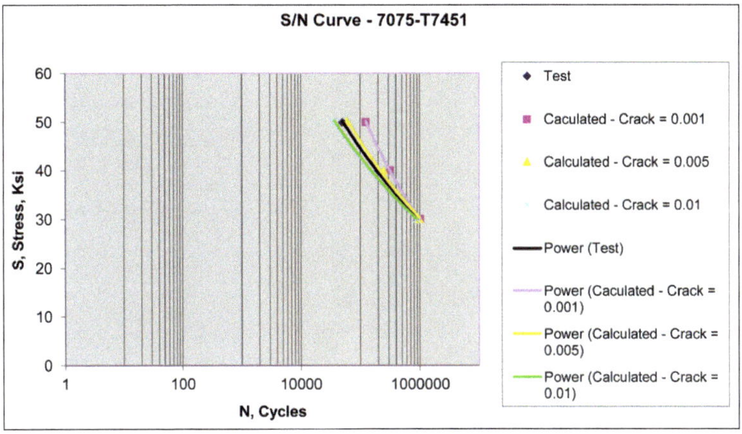

Figure 20.20 S/N Curve for Cracked and Uncracked Specimen - 7075-T7451, $K_t = 1$, $R = 0$.

$$\alpha = (S/N\text{ data})_{\text{Uncracked}}/(S/N\text{ data})_{\text{Cracked}}$$

Assume $\alpha \equiv U^*{}_1/U^*{}_2$, where

$$U^*{}_1 = 2\pi a^2\sigma^2(1-\nu^2)/E \; U*_2 = (\pi a^2\sigma^2(1-\nu^2)/E) + 4a\eta$$

Employ

Paris law $\quad \mathrm{d}a/\mathrm{d}N = C(\Delta K)^m$

$$\Delta K = f(\lambda)\Delta\sigma\sqrt{\pi a},\; f(\lambda) = 1$$

Assume crack growth to be exponential type

$$a = a_i e^{kN}$$

$$\alpha = \frac{\Omega(2\pi a^2\sigma^2(1-\nu^2)/E)}{((\pi a^2\sigma^2(1-\nu^2)/E) + 4a\eta)}$$

where

Ω and Θ are defined as $\Omega = \sigma_f/(\sigma\, e^{\Theta})$, & $\Theta = p(1.50\,\sigma\sqrt{a})/\sigma_f$

$\sigma_f = (\sigma_y + \sigma_u)/2 =$ Flow Stress

$\eta =$ surface energy

$\sigma_y =$ Yield Strength

$\sigma_u =$ Ultimate Strength

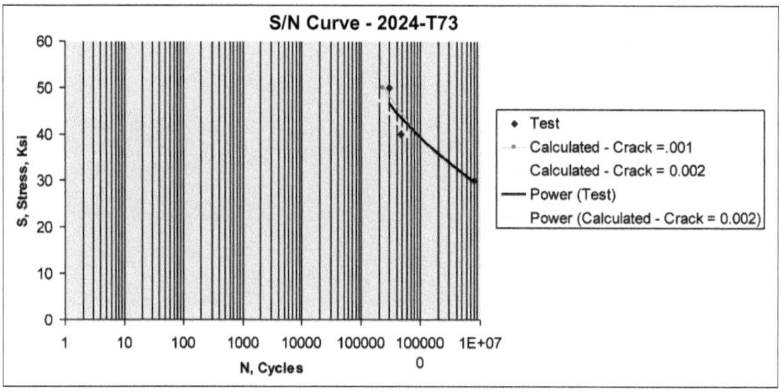

Figure 20.21 S/N Curve for Cracked and Uncracked Specimen - 2024-T73, $K_t = 1$, $R = 0$.

σ = External Stress – Tensile

E = Young's modulus

$a = \frac{1}{2}$ crack length

ν = Poisson's ratio

p = Plasticity effect

20.21 Final Summary and Conclusion

- Using the basic concept of fracture mechanics and energy method, an innovative methodology is developed to arrive at a Stress Shift Factor to establish S/N Curve using Fracture Mechanics (LEFM) and material properties
- This development is very significant and extremely useful for life assessment of aircraft and space structures. With the implementation of the proposed analytical method, considerable cost saving can be made by eliminating the laboratory testing at the expense of analysis
- Upon further improvement, a software package will be developed which generates the S-N curve for metallic material. The completed software will be used throughout the Boeing Company "Go common/Go lean."

- Continuation of this work will be related to studying the fatigue failure of composite material subjected to cyclic load environment
- Using Multi-Scale Modeling and Simulation as well as Analytical Closed Form Solutions, we are able to:
  - ◇ Develop ultra light, strong and durable materials tailored to our needs for optimum usage
  - ◇ Predict material allowables and design values
  - ◇ Eliminate (minimize) costly and time consuming tests

Virtual Testing: Fatigue S-N Curve, Strain and Surface Energy Concept

Using the basic concept of fracture mechanics and energy method, an innovative methodology is developed to arrive at a Stress Shift Factor to establish S/N Curve using Fracture Mechanics (LEFM) and material properties

This development is very significant and extremely useful for life assessment of aircraft and space structures. With the implementation of the proposed analytical method, considerable cost saving can be made by eliminating the laboratory testing at the expense of analysis

Development of α as a function of material properties as well as crack size and geometry is a very complex and challenging task. It involves understanding of material properties at the crystalline and fine levels

At small cracks, S/N data for cracked and uncracked specimen tend to merge together. α is relatively small for small cracks

As the crack size increases α's accuracy decreases

$$\alpha = \Omega(2\pi a^2\sigma^2(1-\nu^2)/E)/((\pi a^2\sigma^2(1-\nu^2)/E)+4a\eta)$$

where

Ω and Θ are defined as, $\Omega = \sigma_f/(\sigma e^{\Theta})$, & $\Theta = p(1.50\,\sigma\sqrt{a})/\sigma_f$

$\sigma_f = (\sigma_y + \sigma_u)/2 =$ Flow stress

$\eta =$ Surface energy

$\sigma_y =$ Yield strength

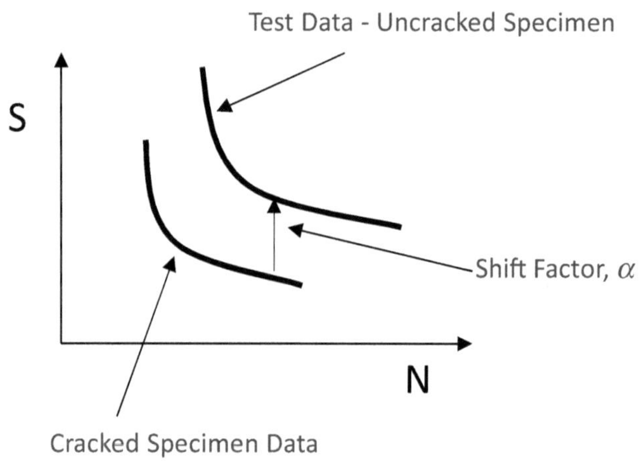

Figure 20.22 S/N Curve and α.

σ_u = Ultimate strength
σ = External stress (tensile)
E = Young's modulus
$a = \frac{1}{2}$ Crack length
ν = Poisson's ratio
p = Plasticity effect

20.22 Overall and Concluding Summary

(1) In principle, all properties of all materials and phenomena are describable by quantum mechanics.
(2) The fundamental deformation mechanisms in solids are dilatational and distortion, which are governed by the microstructure of the material and environmental factors, including load and geometry.
(3) It is not all about material properties. It is more about material behavior.
(4) Application of both applied mechanics and materials science is needed in advanced design to properly understand and assess the critical failure mechanisms as part of the design process.

(5) As we progress from global to micro-mechanism behavior, we will identify and implement the related unique parameters for an optimum design.
(6) Ultralight and durable materials with high strength and stiffness are desirable for aerospace and aircraft parts. This can be accomplished through the multiscale modeling and simulation technique.
(7) Through multiscale modeling and simulation technique, it is possible to predict material allowables (including DaDT) without conducting significant tests. With this technique, Boeing can significantly reduce the cost and time of testing.
(8) The fatigue S-N virtual testing proposed is very significant and extremely useful for life assessment of aircraft and space structures. With the implementation of the proposed analytical method, considerable cost can be saved by eliminating the laboratory testing at the expense of analysis.
(9) The implementation of the techniques proposed including advanced composites can save significant cost and schedule improvements on programs.

Further Reading

1. W. K. Liu, E. G. Karpov, S. Zhang, and H. S. Park (2004). An introduction to computational nanomechanics and materials. *Comput. Methods Appl. Mech. Eng.* **193**, 1529–1578.
2. A. T. Sears (2008). Carbon nanotube mechanics: continuum model development from molecular mechanics virtual experiments. Dissertation submitted to the faculty of the Virginia Polytechnic Institute and State University for the Degree of Doctor of Philosophy in Department of Engineering Mechanics.
3. T. L. Anderson (2005). *Fracture Mechanics: Fundamentals and Applications*, 3rd Ed. (CRC Press, Taylor & Francis Group, Boca Raton, FL).
4. C. Y. Li, and T. W. Chou (2003). A multiscale modeling of the interfacial load transfer in carbon nanotube/polymer composites. *J. Nanosci. Nanotechnol.* **3**, 423–430.
5. J. L. Fattebert, and F. Gygi (2004). Theme article – realistic modeling of nanostructures using density functional theory. *Computer Phys. Commun.* **162**, 24.

6. I. Finnie (1982). UC Berkeley, Course Material.
7. R. Ritchie (1981). UC Berkeley, Course Material.
8. G. Kresse, and J. Furthmüller (1996). Efficiency of ab-initio total energy calculations for metals and semiconductors using a plane-wave basis set. *Comput. Mater. Sci.* **6**, 15–50.
9. K. Mylvaganam, and L.C. Zhang (2007). Ballistic resistance capacity of carbon nanotubes. *Nanotechnology* **18**, 475701.
10. J. W. Dally, and W. F. Riley (1991). *Experimental Stress Analysis,* 3rd Ed. (McGraw Hill).
11. M. F. Kanninen, and C. H. Popelar (1985). *Advanced Fracture Mechanics,* Oxford Engineering Science Series 15 (Oxford University Press, NY).
12. A. Saxena (1998). *Nonlinear Fracture Mechanics for Engineers* (CRC Press, Boca Raton, FL).
13. R. Vaia, and E. P. Giannelis (2001). Polymer nanocomposites: status and opportunities. *MRS Bull.* **26**(5), 394–401.
14. B. Farahmand, G. Bockrath, and J. Glassco (1997). *Fatigue and Fracture Mechanics of High Risk Parts: Application of LEFM & FMDM Theory* (Chapman & Hall, New York).
15. A. Rosankis, K. Ravi-Chandra, and Y. Rajapakse (1988). Analytical, *Numerical and Experimental Aspects of Three Dimensional Fracture Process* (ASME AMD-Vol. 91, New York).
16. R. W. Hertzberg (1996). *Deformation* and *Fracture Mechanics of Engineering Materials,* 4th Ed. (John Wiley and Sons).
17. A. T. Zehnder (2013). Griffith theory of fracture. In *Encyclopedia of Tribology* Q. J. Wang, and Y.-W. Chung, eds. (Springer, US), pp. 1570–1573.
18. Q. Yan, J. Wu, G. Zhou, W. Duan, and B. L. Gu (2005). Ab initio study of transport properties of multiwalled carbon nanotubes. *Phys. Rev. B* **72**, 155425.
19. R. P. Feynman (1960). There's plenty of room at the bottom. *Eng. Sci.* **23**, 22–36.
20. Q. Yao, and J. Qu (2002). Interfacial versus cohesive failure on polymer-metal interfaces in electronic packaging–effects of interface roughness. *J. Electron. Packag.* **124**, 127–134.
21. R. Keunings (2004). Micro-macro methods for the multiscale simulation of viscoelastic flow using molecular models of kinetic theory. In *Rheology Reviews 2004*, D. M. Binding, and K. Walters, eds. (British Society of Rheology), pp. 67–98.

Index